Tensors

AND THEIR APPLICATIONS

Tensors

AND THEIR APPLICATIONS

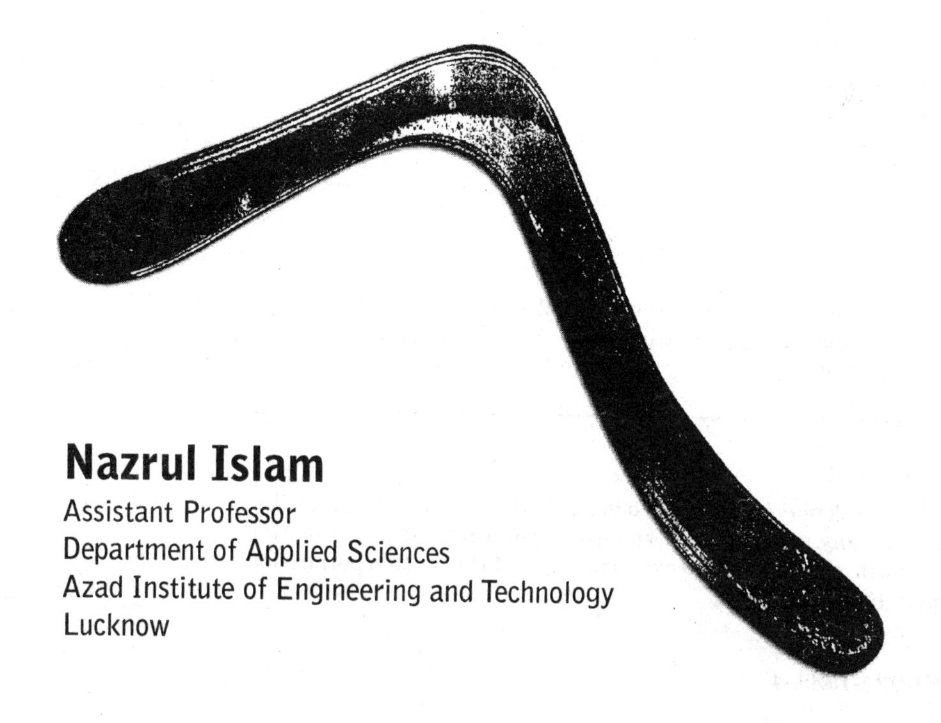

Nazrul Islam
Assistant Professor
Department of Applied Sciences
Azad Institute of Engineering and Technology
Lucknow

NEW AGE INTERNATIONAL (P) LIMITED, PUBLISHERS
(formerly Wiley Eastern Limited)
New Delhi • Bangalore • Chennai • Cochin • Guwahati • Hyderabad
Jalandhar • Kolkata • Lucknow • Mumbai • Ranchi

Published by New Age International (P) Ltd., Publishers
First Edition : 2006

ISBN : 81-224-1838-4

Rs. 145.00

C-06-01-572

Printed in India at Sandeep Printers, Delhi.
Typeset at Akriti, New Delhi.

PUBLISHING FOR ONE WORLD

NEW AGE INTERNATIONAL (P) LIMITED, PUBLISHERS
(formerly Wiley Eastern Limited)
4835/24, Ansari Road, Daryaganj, New Delhi - 110002
Visit us at **www.newagepublishers.com**

To

My parents

FOREWORD

It gives me great pleasure to write the foreword to Dr. Nazrul Islam's book entitled "Tensors and Their Applications. I know the author as a research scholar who has worked with me for several years. This book is a humble step of efforts made by him to prove him to be a dedicated and striving teacher who has worked relentlessly in this field.

This book fills the gap as methodology has been explained in a simple manner to enable students to understand easily. This book will prove to be a complete book for the students in this field.

Ram Nivas
Professor,
Department of Mathematics and Astronomy,
Lucknow University,
Lucknow

PREFACE

'Tensors' were introduced by Professor Gregorio Ricci of University of Padua (Italy) in 1887 primarily as extension of vectors. A quantity having magnitude only is called Scalar and a quantity with magnitude and direction both, called Vector. But certain quantities are associated with two or more directions, such a quantity is called Tensor. The stress at a point of an elastic solid is an example of a Tensor which depends on two directions one normal to the area and other that of the force on it.

Tensors have their applications to Riemannian Geometry, Mechanics, Elasticity, Theory of Relativity, Electromagnetic Theory and many other disciplines of Science and Engineering.

This book has been presented in such a clear and easy way that the students will have no difficulty in understanding it. The definitions, proofs of theorems, notes have been given in details.

The subject is taught at graduate/postgraduate level in almost all universities.

In the end, I wish to thank the publisher and the printer for their full co-operation in bringing out the book in the present nice form.

Suggestions for further improvement of the book will be gratefully acknowledged.

Dr. Nazrul Islam

CONTENTS

CHAPTER – 1

PRELIMINARIES

1.1 *n*-DIMENSIONAL SPACE

In three dimensional rectangular space, the coordinates of a point are (x, y, z). It is convenient to write (x^1, x^2, x^3) for (x, y, z). The coordinates of a point in four dimensional space are given by (x^1, x^2, x^3, x^4). In general, the coordinates of a point in n-dimensional space are given by $(x^1, x^2, x^3, \ldots, x^n)$ such n-dimensional space is denoted by V_n.

1.2 SUPERSCRIPT AND SUBSCRIPT

In the symbol A^{ij}_{kl}, the indices i, j written in the upper position are called *superscripts* and k, l written in the lower position are called *subscripts*.

1.3 THE EINSTEIN'S SUMMATION CONVENTION

Consider the sum of the series $S = a_1x^1 + a_2x^2 + \ldots + a_nx^n = \sum_{i=1}^{n} a_ix^i$. By using summation convention, drop the sigma sign and write convention as

$$\sum_{i=1}^{n} a_ix^i = a_ix^i$$

This convention is called *Einstein's Summation Convention* and stated as

"If a suffix occurs twice in a term, once in the lower position and once in the upper position then that suffix implies sum over defined range."

If the range is not given, then assume that the range is from 1 to n.

1.4 DUMMY INDEX

Any index which is repeated in a given term is called a *dummy index* or *dummy suffix*. This is also called *Umbral* or *Dextral Index*.

e.g. Consider the expression $a_i x^i$ where i is dummy index; then

$$a_i x^i = a_1x^1 + a_2x^2 + \cdots + a_nx^n$$

and $$a_j x^j = a_1 x^1 + a_2 x^2 + \cdots + a_n x^n$$

These two equations prove that

$$a_i x^i = a_j\, x^j$$

So, any dummy index can be replaced by any other index ranging the same numbers.

1.5 FREE INDEX

Any index occurring only once in a given term is called a *Free Index.*

e.g. Consider the expression $a_i^j x^i$ where j is free index.

1.6 KRÖNECKER DELTA

The symbol δ^i_j, called Krönecker Delta (a German mathematician Leopold Krönecker, 1823-91 A.D.) is defined by

$$\delta^i_j = \begin{cases} 1 & \text{if } i = j \\ 0 & \text{if } i \neq j \end{cases}$$

Similarly δ_{ij} and δ^{ij} are defined as

$$\delta^{ij} = \begin{cases} 1 & \text{if } i = j \\ 0 & \text{if } i \neq j \end{cases}$$

and $$\delta_{ij} = \begin{cases} 1 & \text{if } i = j \\ 0 & \text{if } i \neq j \end{cases}$$

Properties

1. If $x^1, x^2, \ldots x^n$ are independent coordinates, then

$$\frac{\partial x^i}{\partial x^j} = 0 \quad \text{if} \quad i \neq j$$

$$\frac{\partial x^i}{\partial x^j} = 1 \quad \text{if} \quad i = j$$

This implies that

$$\frac{\partial x^i}{\partial x^j} = \delta^i_j$$

It is also written as $\dfrac{\partial x^i}{\partial x^k}\dfrac{\partial x^k}{\partial x^j} = \delta^i_j$.

2. $\delta^i_i = \delta^1_1 + \delta^2_2 + \delta^3_3 + \cdots + \delta^n_n$ (by summation convention)

$$\delta^i_i = 1 + 1 + 1 + \cdots + 1$$

$$\delta^i_i = n$$

3. $a^{ij}\delta^j_k = a^{ik}$

Since $$a^{3j}\delta^j_2 = a^{31}\delta^1_2 + a^{32}\delta^2_2 + a^{33}\delta^3_2 + \cdots + a^{3n}\delta^n_2 \quad \text{(as } j \text{ is dummy index)}$$

$$= a^{32} \quad (\text{as } \delta_2^1 = \delta_2^3 = \cdots = \delta_2^n = 0 \text{ and } \delta_2^2 = 1)$$

In general,

$$a^{ij}\delta_k^j = a^{i1}\delta_k^1 + a^{i2}\delta_k^2 + a^{i3}\delta_k^3 + \cdots + a^{ik}\delta_k^k + \cdots + a^{in}\delta_k^n$$

$$a^{ij}\delta_k^j = a^{ik} \quad (\text{as } \delta_k^1 = \delta_k^2 = \cdots = \delta_k^n = 0 \text{ and } \delta_k^k = 1)$$

4.
$$\delta_j^i \delta_k^j = \delta_k^i$$

$$\delta_j^i \delta_k^j = \delta_1^i \delta_k^1 + \delta_2^i \delta_k^2 + \delta_3^i \delta_k^3 + \cdots + \delta_i^i \delta_k^i + \cdots + \delta_n^i \delta_k^n$$

$$= \delta_k^i \quad (\text{as } \delta_1^i = \delta_2^i = \delta_3^i = \cdots = \delta_n^i = 0 \text{ and } \delta_i^i = 1)$$

EXAMPLE 1

Write $\dfrac{d\phi}{dt} = \dfrac{\partial\phi}{\partial x^1}\dfrac{dx^1}{dt} + \dfrac{\partial\phi}{\partial x^2}\dfrac{dx^2}{dt} + \cdots + \dfrac{\partial\phi}{\partial x^n}\dfrac{dx^n}{dt}$ using summation convention.

Solution

$$\frac{d\phi}{dt} = \frac{\partial\phi}{\partial x^1}\frac{dx^1}{dt} + \frac{\partial\phi}{\partial x^2}\frac{dx^2}{dt} + \cdots + \frac{\partial\phi}{\partial x^n}\frac{dx^n}{dt}$$

$$\frac{d\phi}{dt} = \frac{\partial\phi}{\partial x^i}\frac{dx^i}{dt}$$

EXAMPLE 2

Expand: (*i*) $a_{ij}\, x^i x^j$; (*ii*) $g_{lm}\, g_{mp}$

Solution

(*i*)
$$a_{ij}x^i x^j = a_{1j}x^1 x^j + a_{2j}x^2 x^j + \cdots + a_{nj}x^n x^j$$
$$= a_{11}x^1 x^1 + a_{22}x^2 x^2 + \cdots + a_{nn}x^n x^n$$
$$a_{ij}x^i x^j = a_{11}(x^1)^2 + a_{22}(x^2)^2 + \cdots + a_{nn}(x^n)^2$$

(as i and j are dummy indices)

(*ii*) $g_{lm}g_{mp} = g_{l1}g_{1p} + g_{l2}g_{2p} + \cdots + g_{ln}g_{np}$, as m is dummy index.

EXAMPLE 3

If a_{ij} are constant and $a_{ij} = a_{ji}$, calculate:

(*i*) $\dfrac{\partial}{\partial x_k}(a_{ij}x_i x_j)$ (*ii*) $\dfrac{\partial}{\partial x_k \partial x_l}(a_{ij}x_i x_j)$

Solution

(*i*)
$$\frac{\partial}{\partial x_k}(a_{ij}x_i x_j) = a_{ij}\frac{\partial}{\partial x_k}(x_i x_j)$$

$$= a_{ij}x_i \frac{\partial x_j}{\partial x_k} + a_{ij}x_j \frac{\partial x_i}{\partial x_k}$$

$$= a_{ij}x_i\delta_{jk} + a_{ij}x_j\delta_{ik}, \qquad \text{as} \quad \frac{\partial x_j}{\partial x_k} = \delta_{jk}$$

$$= (a_{ij}\delta_{jk})x_i + (a_{ij}\delta_{ik})x_j$$

$$= a_{ik}x_i + a_{kj}x_j \qquad \text{as } a_{ij}\,\delta_{jk} = a_{ik}$$

$$= a_{ik}x_i + a_{ki}x_i \qquad \text{as } j \text{ is dummy index}$$

$$\frac{\partial(a_{ij}x_ix_j)}{\partial_{x_k}} = 2a_{ik}x_i \qquad \text{as given } a_{ik} = a_{ki}$$

(*ii*)
$$\frac{\partial(a_{ij}x_ix_j)}{\partial x_k} = 2a_{ik}x_i$$

Differentiating it w.r.t. x_l:

$$\frac{\partial^2(a_{ij}x_ix_j)}{\partial x_k \partial x_l} = 2a_{ik}\frac{\partial x_i}{\partial x_l}$$

$$= 2a_{ik}\delta_l^i$$

$$\frac{\partial^2(a_{ij}x_ix_j)}{\partial x_k \partial x_l} = 2a_{lk} \qquad \text{as } a_{ik}\delta_l^i = a_{lk}.$$

EXAMPLE 4

If
$$a_{ij}x^ix^j = 0$$
where a_{ij} are constant then show that
$$a_{ij} + a_{ji} = 0$$

Solution

Given

$$a_{ij}x^ix^j = 0$$

$$\Rightarrow \qquad a_{lm}x^lx^m = 0 \qquad \text{since } i \text{ and } j \text{ are dummy indices}$$

Differentiating it w.r.t. x^i partially,

$$\frac{\partial}{\partial x_i}(a_{lm}x^lx^m) = 0$$

$$a_{lm}\frac{\partial}{\partial x_i}(x^lx^m) = 0$$

$$a_{lm}\frac{\partial x^l}{\partial x_i}x^m + a_{lm}\frac{\partial x^m}{\partial x_i}x^l = 0$$

Since
$$\frac{\partial x^l}{\partial x_i} = \delta_i^l \quad \text{and} \quad \frac{\partial x^m}{\partial x_i} = \delta_i^m$$

$$a_{lm}\delta_i^l x^m + a_{lm}\delta_i^m x^l = 0$$

$$a_{im}x^m + a_{li}x^l = 0$$

as $a_{lm}\delta_i^l = a_{im}$ and $a_{lm}\delta_i^m = a_{li}$.

Differentiating it w.r.t. x_j partially

$$a_{im}\frac{\partial x^m}{\partial x_j} + a_{li}\frac{\partial x^l}{\partial x_j} = 0$$

$$a_{im}\delta_j^m + a_{li}\delta_j^l = 0$$

$$a_{ij} + a_{ji} = 0$$

Proved.

EXERCISES

1. Write the following using the summation convention.
   (*i*) $(x^1)^2 + (x^2)^2 + (x^3)^2 + \cdots + (x^n)^2$
   (*ii*) $ds^2 = g_{11}(dx^1)^2 + g_{22}(dx^2)^2 + \cdots + g_{nn}(dx^n)^2$
   (*iii*) $a_1x^1x^3 + a_2x^2x^3 + \cdots + a_nx^nx^3$
2. Expand the following:
   (*i*) $a_{ij}x^j$ (*ii*) $\frac{\partial}{\partial x^i}(\sqrt{g}a^i)$ (*iii*) $A_i^k B^i$
3. Evaluate:
   (*i*) $x^j\delta_j^i$ (*ii*) $\delta_j^i\delta_k^j\delta_l^k$ (*iii*) $\delta_j^i\delta_i^j$
4. Express $b^{ij}y_iy_j$ in the terms of x variables where $y_i = c_{ij}x_j$ and $b^{ij}c_{ik} = \delta_k^i$.

ANSWERS

1. (*i*) x^ix^j (*ii*) $ds^2 = g_{ij}\,dx^idx^j$ (*iii*) $a_ix^ix^3$.
2. (*i*) $a_{i1}x^1 + a_{i2}x^2 + a_{i3}x^3 + \cdots + a_{in}x^n$
   (*ii*) $\frac{\partial}{\partial x^1}(\sqrt{g}a^1) + \frac{\partial}{\partial x^2}(\sqrt{g}a^2) + \cdots + \frac{\partial}{\partial x^n}(\sqrt{g}a^n)$
   (*iii*) $A_1^kB^1 + A_2^kB^2 + \ldots + A_n^kB^n$
3. (*i*) x^i (*ii*) δ_l^i (*iii*) n
4. $C_{ij}x_ix_j$

CHAPTER – 2

TENSOR ALGEBRA

2.1 INTRODUCTION

A scalar (density, pressure, temperature, etc.) is a quantity whose specification (in any coordinate system) requires just one number. On the other hand, a vector (displacement, acceleration, force, etc.) is a quantity whose specification requires three numbers, namely its components with respect to some basis. Scalers and vectors are both special cases of a more general object called a tensor of order n whose specification in any coordinate system requires 3^n numbers, called the components of tensor. In fact, scalars are tensors of order zero with $3^\circ = 1$ component. Vectors are tensors of order one with $3^1 = 3$ components.

2.2 TRANSFORMATION OF COORDINATES

In three dimensional rectangular space, the coordinates of a point are (x, y, z) where x, y, z are real numbers. It is convenient to write (x^1, x^2, x^3) for (x, y, z) or simply x^i where $i = 1, 2, 3$. Similarly in n- dimensional space, the coordinate of a point are n-independent variables $(x^1, x^2, ..., x^n)$ in X-coordinate system. Let $(\bar{x}^1, \bar{x}^2, ..., \bar{x}^n)$ be coordinate of the same point in Y-coordinate system.

Let $\bar{x}^1, \bar{x}^2, \ldots, \bar{x}^n$ be independent single valued function of $x^1, x^2, ..., x^n$, so that,

$$\bar{x}^1 = \bar{x}^1(x^1, x^2, ..., x^n)$$

$$\bar{x}^2 = \bar{x}^2(x^1, x^2, ..., x^n)$$

$$\bar{x}^3 = \bar{x}^3(x^1, x^2, ..., x^n)$$

$$\vdots \qquad \vdots$$

$$\bar{x}^n = \bar{x}^n(x^1, x^2, ..., x^n)$$

or

$$\bar{x}^i = \bar{x}^i(x^1, x^2, ..., x^n); \quad i = 1, 2, ..., n \qquad ...(1)$$

Solving these equations and expressing x^i as functions of $\bar{x}^1, \bar{x}^2, ..., \bar{x}^n$, so that

$$x^i = x^i(\bar{x}^1, \bar{x}^2, ..., \bar{x}^n); \qquad i = 1, 2, ..., n$$

The equations (1) and (2) are said to be a transformation of the coordinates from one coordinate system to another

2.3 COVARIANT AND CONTRAVARIANT VECTORS (TENSOR OF RANK ONE)

Let $(x^1, x^2, ..., x^n)$ or x^i be coordinates of a point in X-coordinate system and $(\bar{x}^1, \bar{x}^2, ..., \bar{x}^n)$ or $\bar{x}^i$ be coordinates of the same point in the Y-coordinate system.

Let A^i, $i = 1, 2, ..., n$ (or $A^1, A^2, ..., A^n$) be n functions of coordinates $x^1, x^2, ..., x^n$ in X-coordinate system. If the quantities A^i are transformed to $\bar{A}^i$ in Y-coordinate system then according to the law of transformation

$$\bar{A}^i = \frac{\partial \bar{x}^i}{\partial x^j} A^j \quad \text{or} \quad A^j = \frac{\partial x^j}{\partial \bar{x}^i} \bar{A}^i$$

Then A^i are called components of contravariant vector.

Let A_i, $i = 1, 2, ..., n$ (or $A_1, A_2, ..., A_n$) be n functions of the coordinates $x^1, x^2, ..., x^n$ *in* X-coordinate system. If the quantities A_i are transformed to $\bar{A}_i$ in Y-coordinate system then according to the law of transformation

$$\bar{A}_i = \frac{\partial x^j}{\partial \bar{x}^i} A_j \quad \text{or} \quad A_j = \frac{\partial \bar{x}^i}{\partial x^j} \bar{A}_i$$

Then A^i are called components of covariant vector.

The contravariant (or covariant) vector is also called a contravariant (or covariant) tensor of rank one.

Note: A superscript is always used to indicate contravariant component and a subscript is always used to indicate covariant component.

EXAMPLE 1

If x^i be the coordinate of a point in n-dimensional space show that dx^i are component of a contravariant vector.

Solution

Let $x^1, x^2, ..., x^n$ or x^i are coordinates in X-coordinate system and $\bar{x}^1, \bar{x}^2, ..., \bar{x}^n$ or $\bar{x}^i$ are coordinates in Y-coordinate system.

If

$$\bar{x}^i = \bar{x}^i(x^1, x^2, ..., x^n)$$

$$d\bar{x}^i = \frac{\partial \bar{x}^i}{\partial x^1} dx^1 + \frac{\partial \bar{x}^i}{\partial x^2} dx^2 + \cdots + \frac{\partial \bar{x}^i}{\partial x^n} dx^n$$

$$d\bar{x}^i = \frac{\partial \bar{x}^i}{\partial x^j} dx^j$$

It is law of transformation of contravariant vector. So, dx^i are components of a contravariant vector.

EXAMPLE 2

Show that $\frac{\partial \phi}{\partial x^i}$ is a covariant vector where ϕ is a scalar function.

Solution

Let $x^1, x^2, ..., x^n$ or x^i are coordinates in X-coordinate system and $\bar{x}^1, \bar{x}^2, ..., \bar{x}^n$ or $\bar{x}^i$ are coordinates in Y-coordinate system.

Consider $\phi(\bar{x}^1, \bar{x}^2, ..., \bar{x}^n) = \phi(x^1, x^2, ..., x^n)$

$$\partial \phi = \frac{\partial \phi}{\partial x^1} \partial x^1 + \frac{\partial \phi}{\partial x^2} \partial x^2 + \cdots + \frac{\partial \phi}{\partial x^n} \partial x^n$$

$$\frac{\partial \phi}{\partial \bar{x}^i} = \frac{\partial \phi}{\partial x^1} \frac{\partial x^1}{\partial \bar{x}^i} + \frac{\partial \phi}{\partial x^2} \frac{\partial x^2}{\partial \bar{x}^i} + \cdots + \frac{\partial \phi}{\partial x^n} \frac{\partial x^n}{\partial \bar{x}^i}$$

$$\frac{\partial \phi}{\partial \bar{x}^i} = \frac{\partial \phi}{\partial x^j} \frac{\partial x^j}{\partial \bar{x}^i}$$

or

$$\frac{\partial \phi}{\partial \bar{x}^i} = \frac{\partial x^j}{\partial \bar{x}^i} \frac{\partial \phi}{\partial x^j}$$

It is law of transformation of component of covariant vector. So, $\frac{\partial \phi}{\partial x^i}$ is component of covariant vector.

EXAMPLE 3

Show that the velocity of fluid at any point is a component of contravariant vector

or

Show that the component of tangent vector on the curve in n-dimensional space are component of contravariant vector.

Solution

Let $\frac{dx^1}{dt}, \frac{dx^2}{dt}, ..., \frac{dx^n}{dt}$ be the component of the tangent vector of the point $(x^1, x^2, ..., x^n)$ *i.e.*, $\frac{dx^i}{dt}$ be the component of the tangent vector in X-coordinate system. Let the component of tangent vector of the point $(\bar{x}^1, \bar{x}^2, ..., \bar{x}^n)$ in Y-coordinate system are $\frac{d\bar{x}^i}{dt}$. Then $\bar{x}^1, \bar{x}^2, ..., \bar{x}^n$ or $\bar{x}^i$ being a function of $x^1, x^2, ..., x^n$ which is a function of t. So,

$$\frac{d\bar{x}^i}{dt} = \frac{\partial\bar{x}^i}{dt^1}\frac{dx^1}{dt} + \frac{\partial\bar{x}^i}{dx^2}\frac{dx^2}{dt} + \cdots + \frac{\partial\bar{x}^i}{dx^n}\frac{dx^n}{dt}$$

$$\frac{d\bar{x}^i}{dt} = \frac{\partial\bar{x}^i}{dx^j}\frac{dx^j}{dt}$$

It is law of transformation of component of contravariant vector. So, $\frac{dx^i}{dt}$ is component of contravariant vector.

i.e. the component of tangent vector on the curve in n-dimensional space are component of contravariant vector.

2.4 CONTRAVARIANT TENSOR OF RANK TWO

Let A^{ij} $(i, j = 1, 2, ..., n)$ be n^2 functions of coordinates $x^1, x^2, ..., x^n$ in X-coordinate system. If the quantities A^{ij} are transformed to $\bar{A}^{ij}$ in Y-coordinate system having coordinates $\bar{x}^1, \bar{x}^2, ..., \bar{x}^n$. Then according to the law of transformation

$$\bar{A}^{ij} = \frac{\partial\bar{x}^i}{\partial x^k}\frac{\partial\bar{x}^j}{\partial x^l}A^{kl}$$

Then A^{ij} are called components of Contravariant Tensor of rank two.

2.5 COVARIANT TENSOR OF RANK TWO

Let A_{ij} $(i, j = 1, 2, ..., n)$ be n^2 functions of coordinates $x^1, x^2, ..., x^n$ in X-coordinate system. If the quantities A_{ij} are transformed to A_{ij} in Y-coordinate system having coordinates $\bar{x}^1, \bar{x}^2, ..., \bar{x}^n$, then according to the law of transformation,

$$\bar{A}_{ij} = \frac{\partial x^k}{\partial\bar{x}^i}\frac{\partial x^l}{\partial\bar{x}^j}A_{kl}$$

Then A_{ij} called components of covariant tensor of rank two.

2.6 MIXED TENSOR OF RANK TWO

Let A^i_j $(i, j = 1, 2, ..., n)$ be n^2 functions of coordinates $x^1, x^2, ..., x^n$ in X-coordinate system. If the quantities A^i_j are transformed to $\bar{A}^i_j$ in Y-coordinate system having coordinates $\bar{x}^1, \bar{x}^2, ..., \bar{x}^n$, then according to the law of transformation

$$\bar{A}^i_j = \frac{\partial\bar{x}^i}{\partial x^k}\frac{\partial x^l}{\partial\bar{x}}A^k_l$$

Then A^i_j are called components of mixed tensor of rank two.

Note: (*i*) The rank of the tensor is defined as the total number of indices per component.

(*ii*) Instead of saying that "A^{ij} are the components of a tensor of rank two" we shall often say "A^{ij} is a tensor of rank two."

THEOREM 2.1 *To show that the Krönecker delta is a mixed tensor of rank two.*

Solution

Let X and Y be two coordinate systems. Let the component of Kronecker delta in X-coordinate system δ^i_j and component of Krönecker delta in Y-coordinate be $\bar{\delta}^i_j$, then according to the law of transformation

$$\bar{\delta}^i_j = \frac{\partial \bar{x}^i}{\partial \bar{x}^j} = \frac{\partial \bar{x}^i}{\partial x^k}\frac{\partial x^l}{\partial \bar{x}^j}\frac{\partial x^k}{\partial x^l}$$

$$\bar{\delta}^i_j = \frac{\partial \bar{x}^i}{\partial x^k}\frac{\partial x^l}{\partial \bar{x}^j}\delta^k_l$$

This shows that Krönecker δ^i_j is mixed tensor of rank two.

EXAMPLE 4

If A_i is a covariant tensor, then prove that $\dfrac{\partial A_i}{\partial x^j}$ do not form a tensor.

Solution

Let X and Y be two coordinate systems. As given A_i is a covariant tensor. Then

$$\bar{A}_i = \frac{\partial x^k}{\partial \bar{x}^i}A_k$$

Differentiating it w.r.t. $\bar{x}^j$

$$\frac{\partial \bar{A}_i}{\partial \bar{x}^j} = \frac{\partial}{\partial \bar{x}^j}\left(\frac{\partial x^k}{\partial \bar{x}^i}A_k\right)$$

$$\frac{\partial \bar{A}_i}{\partial \bar{x}^j} = \frac{\partial x^k}{\partial \bar{x}^i}\frac{\partial A_k}{\partial \bar{x}^j} + A_k\frac{\partial^2 x^k}{\partial \bar{x}^i \partial \bar{x}^j} \quad \text{...(1)}$$

It is not any law of transformation of tensor due to presence of second term. So, $\dfrac{\partial A_i}{\partial x^j}$ is not a tensor.

THEOREM 2.2 *To show that* δ^i_j *is an invariant i.e., it has same components in every coordinate system.*

Proof: Since δ^i_j is a mixed tensor of rank two, then

$$\bar{\delta}^i_j = \frac{\partial \bar{x}^i}{\partial x^k}\frac{\partial x^l}{\partial \bar{x}^j}\delta^k_l$$

$$= \frac{\partial \bar{x}^i}{\partial x^k}\left(\frac{\partial x^l}{\partial \bar{x}^j}\delta_l^k\right)$$

$$= \frac{\partial \bar{x}^i}{\partial x^k}\frac{\partial x^k}{\partial \bar{x}^j}, \text{ as } \frac{\partial x^l}{\partial \bar{x}^j}\delta_l^k = \frac{\partial x^k}{\partial \bar{x}^j}$$

$$\bar{\delta}_j^i = \frac{\partial \bar{x}^i}{\partial \bar{x}^j} = \delta_j^i \text{ , as } \frac{\partial \bar{x}^i}{\partial \bar{x}^j} = \delta_j^i$$

So, δ_j^i is an invariant.

THEOREM 2.3 *Prove that the transformation of a contravariant vector is transitive.*

or

Prove that the transformation of a contravariant vector form a group.

Proof: Let A^i be a contravariant vector in a coordinate system $x^i (i = 1,2,...,n)$. Let the coordinates x^i be transformed to the coordinate system $\bar{x}^i$ and $\bar{x}^i$ be transformed to $\bar{\bar{x}}^i$.

When coordinate x^i be transformed to $\bar{x}^i$, the law of transformation of a contravariant vector is

$$\bar{A}^p = \frac{\partial \bar{x}^p}{\partial x^q}A^q \qquad \text{... (1)}$$

When coordinate $\bar{x}^i$ be transformed to $\bar{\bar{x}}^i$, the law of transformation of contravariant vector is

$$\bar{\bar{A}}^i = \frac{\partial \bar{\bar{x}}^i}{\partial \bar{x}^p}\bar{A}^p$$

$$\bar{\bar{A}}^i = \frac{\partial \bar{\bar{x}}^i}{\partial \bar{x}^i}\frac{\partial \bar{x}^p}{\partial x^q}A^q \text{ from (1)}$$

$$\bar{\bar{A}}^i = \frac{\partial \bar{\bar{x}}^i}{\partial x^q}A^q$$

This shows that if we make direct transformation from x^i to $\bar{\bar{x}}^i$, we get same law of transformation. This property is called that transformation of contravariant vectors is transitive or form a group.

THEOREM 2.4 *Prove that the transformation of a covariant vector is transitive.*

or

Prove that the transformation of a covariant vector form a group.

Proof: Let A_i be a covariant vector in a coordinate system $x^i (i = 1, 2, ..., n)$. Let the coordinates x^i be transformed to the coordinate system $\bar{x}^i$ and $\bar{x}^i$ be transformed to $\bar{\bar{x}}^i$.

When coordinate x^i be transformed to $\bar{x}^i$, the law of transformation of a covariant vector is

$$\bar{A}_p = \frac{\partial x^q}{\partial \bar{x}^p}A_q \qquad \text{... (1)}$$

When coordinate $\bar{x}^i$ be transformed to $\bar{\bar{x}}^i$, the law of transformation of a covariant vector is

$$\bar{\bar{A}}_i = \frac{\partial \bar{x}^p}{\partial \bar{\bar{x}}^i}\bar{A}_p$$

$$\overline{\overline{A}}_i = \frac{\partial \overline{x}^p}{\partial \overline{\overline{x}}^i} \frac{\partial x^q}{\partial \overline{x}^p} A_q$$

$$\overline{\overline{A}}_i = \frac{\partial x^q}{\partial \overline{\overline{x}}^i} A_q$$

This shows that if we make direct transformation from x^i to $\overline{\overline{x}}^i$, we get same law of transformation. This property is called that transformation of covariant vectors is transitive or form a group.

THEOREM 2.5 *Prove that the transformations of tensors form a group*

or

Prove that the equations of transformation a tensor (Mixed tensor) posses the group property.

Proof: Let A^i_j be a mixed tensor of rank two in a coordinate system $x^i (i = 1, 2, ..., n)$. Let the coordinates x^i be transformed to the coordinate system $\overline{x}^i$ and $\overline{x}^i$ be transformed to $\overline{\overline{x}}^i$.

When coordinate x^i be transformed to $\overline{x}^i$, the transformation of a mixed tensor of rank two is

$$\overline{A}^p_q = \frac{\partial \overline{x}^p}{\partial x^r} \frac{\partial x^s}{\partial \overline{x}^q} A^r_s \qquad \text{... (1)}$$

When coordinate $\overline{x}^i$ be transformed to $\overline{\overline{x}}^i$, the law of transformation of a mixed tensor of rank two is

$$\overline{\overline{A}}^i_j = \frac{\partial \overline{\overline{x}}^i}{\partial \overline{x}^p} \frac{\partial \overline{x}^q}{\partial \overline{\overline{x}}^j} \overline{A}^p_q$$

$$= \frac{\partial \overline{\overline{x}}^i}{\partial \overline{x}^p} \frac{\partial \overline{x}^q}{\partial \overline{\overline{x}}^j} \frac{\partial \overline{x}^p}{\partial x^r} \frac{\partial x^s}{\partial \overline{x}^q} A^r_s \text{ from (1)}$$

$$\overline{\overline{A}}^i_j = \frac{\partial \overline{\overline{x}}^i}{\partial x^r} \frac{\partial x^s}{\partial \overline{\overline{x}}^j} A^r_s$$

This shows that if we make direct transformation from x^i to $\overline{\overline{x}}^i$, we get same law of transformation. This property is called that transformation of tensors form a group.

THEOREM 2.6 *There is no distinction between contravariant and covariant vectors when we restrict ourselves to rectangular Cartesian transformation of coordinates.*

Proof: Let $P(x, y)$ be a point with respect to the rectangular Cartesian axes X and Y. Let $(\overline{x}, \overline{y})$ be the coordinate of the same point P in another rectangular cartesian axes $\overline{X}$ and $\overline{Y}$, Let (l_1, m_1) and (l_2, m_2) be the direction cosines of the axes $\overline{X}$, $\overline{Y}$ respectively. Then the transformation relations are given by

$$\left.\begin{aligned} \overline{x} &= l_1 x + m_1 y \\ \overline{y} &= l_2 x + m_2 y \end{aligned}\right\} \qquad \text{...(1)}$$

and solving these equations, we have

$$\left.\begin{aligned} x &= l_1 \overline{x} + l_2 \overline{y} \\ y &= m_1 \overline{x} + m_2 \overline{y} \end{aligned}\right\} \qquad \text{...(2)}$$

put $x = x^1,\ y = x^2,\ \overline{x} = \overline{x}^1,\ \overline{y} = \overline{x}^2$

Consider the contravariant transformation

$$\overline{A}^i = \frac{\partial \overline{x}^i}{\partial x^j} A^j; \quad j = 1,2$$

$$\overline{A}^i = \frac{\partial \overline{x}^i}{\partial x^1} A^1 + \frac{\partial \overline{x}^i}{\partial x^2} A^2$$

for $i = 1,2$.

$$\overline{A}^1 = \frac{\partial \overline{x}^1}{\partial x^1} A^1 + \frac{\partial \overline{x}^1}{\partial x^2} A^2$$

$$\overline{A}^2 = \frac{\partial \overline{x}^2}{\partial x^1} A^1 + \frac{\partial \overline{x}^2}{\partial x^2} A^2$$

From (1) $\frac{\partial \overline{x}}{\partial x} = l_1$, but $x = x^1$, $y = x^2$, $\overline{x} = \overline{x}^1$, $\overline{y} = \overline{x}^2$

Then

$$\frac{\partial \overline{x}}{\partial x} = \frac{\partial \overline{x}^1}{\partial x^1} = l_1.$$

Similarly,

$$\left.\begin{aligned} &\frac{\partial \overline{x}}{\partial y} = m_1 = \frac{\partial \overline{x}^1}{\partial x^2}; \\ &\frac{\partial \overline{y}}{\partial x} = l_2 = \frac{\partial \overline{x}^2}{\partial x^1}; \quad \frac{\partial \overline{y}}{\partial y} = m_2 = \frac{\partial \overline{x}^2}{\partial x^2} \end{aligned}\right\} \qquad \text{...(3)}$$

So, we have

$$\left.\begin{aligned} \overline{A}^1 &= l_1 A^1 + m_1 A^2 \\ \overline{A}^2 &= l_2 A^1 + m_2 A^2 \end{aligned}\right\} \qquad \text{..(4)}$$

Consider the covariant transformation

$$\overline{A}_i = \frac{\partial x^j}{\partial \overline{x}^i} A_j; \quad j = 1,2$$

$$\overline{A}_i = \frac{\partial x^1}{\partial \overline{x}^i} A_1 + \frac{\partial x^2}{\partial \overline{x}^1} A_2$$

for $i = 1, 2$.

$$\overline{A}_1 = \frac{\partial x^1}{\partial \overline{x}^1} A_1 + \frac{\partial x^2}{\partial \overline{x}^1} A_2$$

$$\overline{A}_2 = \frac{\partial \overline{x}^1}{\partial \overline{x}^2} A_1 + \frac{\partial x^2}{\partial \overline{x}^2} A_2$$

From (3)

$$\left.\begin{aligned} \overline{A}_1 &= l_1 A_1 + m_1 A_2 \\ \overline{A}_2 &= l_2 A_1 + m_2 A_2 \end{aligned}\right\} \qquad \text{...(5)}$$

So, from (4) and (5), we have

$$\overline{A}^1 = \overline{A}_1 \text{ and } \overline{A}^2 = \overline{A}_2$$

Hence the theorem is proved.

2.7 TENSORS OF HIGHER ORDER

(a) Contravariant tensor of rank r

Let $A^{i_1 i_2 \ldots i_r}$ be n^r function of coordinates $x^1, x^2, \ldots, x^n$ in X-coordinates system. If the quantities $A^{i_1 i_2 \ldots i_r}$ are transformed to $\overline{A}^{i_1 i_2 \ldots i_r}$ in Y-coordinate system having coordinates $\overline{x}^1, \overline{x}^2, \ldots, \overline{x}^n$. Then according to the law of transformation

$$\overline{A}^{i_1 i_2 \ldots i_r} = \frac{\partial \overline{x}^{i_1}}{\partial x^{p_1}} \frac{\partial \overline{x}^{i_2}}{\partial x^{p_2}} \cdots \frac{\partial \overline{x}^{i_r}}{\partial x^{p_r}} A^{p_1 p_2 \ldots p_r}$$

Then $A^{i_1 i_2 \ldots i_r}$ are called components of contravariant tensor of rank r.

(b) Covariant tensor of rank s

Let $A_{j_1 j_2 \ldots j_s}$ be n^s functions of coordinates $x^1, x^2, \ldots, x^n$ in X-coordinate system. If the quantities $A_{j_1 j_2 \ldots j_s}$ are transformed to $\overline{A}_{j_1 j_2 \ldots j_s}$ in Y- coordinate system having coordinates $\overline{x}^1, \overline{x}^2, \ldots, \overline{x}^n$. Then according to the law of transformation

$$\overline{A}_{j_1 j_2 \ldots j_s} = \frac{\partial x^{q_1}}{\partial \overline{x}^{j_1}} \frac{\partial x^{q_2}}{\partial \overline{x}^{j_2}} \cdots \frac{\partial x^{q_s}}{\partial \overline{x}^{j_s}} A_{q_1, q_2, \ldots, q_s}$$

Then $A_{j_1 j_2 \ldots j_s}$ are called the components of covariant tensor of rank s.

(c) Mixed tensor of rank r + s

Let $A^{i_1 i_2 \ldots i_r}_{j_1 j_2 \ldots j_s}$ be n^{r+s} functions of coordinates $x^1, x^2, \ldots, x^n$ in X-coordinate system. If the quantities $A^{i_1 i_2 \ldots i_r}_{j_1 j_2 \ldots j_s}$ are transformed to $\overline{A}^{i_1 i_2 \ldots i_r}_{j_1 j_2 \ldots j_s}$ in Y-coordinate system having coordinates $\overline{x}^1, \overline{x}^2, \ldots, \overline{x}^n$. Then according to the law of transformation

$$\overline{A}^{i_1 i_2 \ldots i_r}_{j_1 j_2 \ldots j_s} = \frac{\partial \overline{x}^{i_1}}{\partial x^{p_1}} \frac{\partial \overline{x}^{i_2}}{\partial x^{p_2}} \frac{\partial \overline{x}^{i_r}}{\partial x^{p_r}} \frac{\partial x^{q_1}}{\partial \overline{x}^{j_1}} \frac{\partial x^{q_2}}{\partial \overline{x}^{j_2}} \cdots \frac{\partial x^{q_s}}{\partial \overline{x}^{j_s}} A^{p_1 p_2 \ldots p_r}_{q_1 q_2 \ldots q_s}$$

Then $A^{i_1 i_2 \ldots i_r}_{j_1 j_2 \ldots j_s}$ are called component of mixed tensor of rank $r+s$.

A tensor of type $A^{j_1 j_2 \ldots j_s}_{i_1 i_2 \ldots i_r}$ is known as tensor of type (r, s), In (r,s), the first component r indicates the rank of contravariant tensor and the second component s indicates the rank of covariant tensor.

Thus the tensors A_{ij} and A^{ij} are type (0, 2) and (2, 0) respectively while tensor A^i_j is type (1, 1).

EXAMPLE

A^{ijk}_{lm} is a mixed tensor of type (3, 2) in which contravariant tensor of rank three and covariant tensor of rank two. Then according to the law of transformation

$$\overline{A}^{ijk}_{lm} = \frac{\partial \overline{x}^i}{\partial x^\alpha}\frac{\partial \overline{x}^j}{\partial x^\beta}\frac{\partial \overline{x}^k}{\partial x^\gamma}\frac{\partial x^a}{\partial \overline{x}^l}\frac{\partial x^b}{\partial \overline{x}^m} A^{\alpha\beta\gamma}_{ab}$$

2.8 SCALAR OR INVARIANT

A function $\phi\ (x^1, x^2, ..., x^n)$ is called Scalar or an invariant if its original value does not change upon transformation of coördinates from $x^1, x^2, ..., x^n$ to $\overline{x}^1, \overline{x}^2, ..., \overline{x}^n$. *i.e.*

$$\phi(x^1, x^2,..., x^n) = \overline{\phi}(\overline{x}^1, \overline{x}^2,..., \overline{x}^n)$$

Scalar is also called tensor of rank zero.

For example, $A^i B_i$ is scalar.

2.9 ADDITION AND SUBTRACTION OF TENSORS

THEOREM 2.7 *The sum (or difference) of two tensors which have same number of covariant and the same contravariant indices is again a tensor of the same rank and type as the given tensors.*

Proof: Consider two tensors $A^{i_1i_2...i_r}_{j_1j_2...j_s}$ and $B^{i_1i_2...i_r}_{j_1j_2...j_s}$ of the same rank and type (*i.e.*, covariant tensor of rank s and contravariant tensor of rank r.). Then according to the law of transformation

$$\overline{A}^{i_1i_2...i_r}_{j_1j_2...j_s} = \frac{\partial \overline{x}^{i_1}}{\partial x^{p_1}}\frac{\partial \overline{x}^{i_2}}{\partial x^{p_2}}\cdots\frac{\partial \overline{x}^{i_r}}{\partial x^{p_r}}\frac{\partial x^{q_1}}{\partial \overline{x}^{j_1}}\frac{\partial x^{q_2}}{\partial \overline{x}^{j_2}}\cdots\frac{\partial x^{q_s}}{\partial \overline{x}^{j_s}}\cdots A^{p_1p_2...p_r}_{q_1q_2...q_s}$$

and

$$\overline{B}^{i_1i_2...i_r}_{j_1j_2...j_s} = \frac{\partial \overline{x}^{i_1}}{\partial x^{p_1}}\frac{\partial \overline{x}^{i_2}}{\partial x^{p_2}}\cdots\frac{\partial \overline{x}^{i_r}}{\partial x^{p_r}}\frac{\partial x^{q_1}}{\partial \overline{x}^{j_1}}\frac{\partial x^{q_2}}{\partial \overline{x}^{j_2}}\cdots\frac{\partial x^{q_s}}{\partial \overline{x}^{j_s}}\cdots B^{p_1p_2...p_r}_{q_1q_2...q_s}$$

Then

$$\overline{A}^{i_1i_2...i_r}_{j_1j_2...j_s} \pm \overline{B}^{i_1i_2...i_r}_{j_1j_2...j_s} = \frac{\partial \overline{x}^{i_1}}{\partial x^{p_1}}\frac{\partial \overline{x}^{i_2}}{\partial x^{p_2}}\cdots\frac{\partial \overline{x}^{i_r}}{\partial x^{p_r}}\frac{\partial x^{q_1}}{\partial \overline{x}^{j_1}}\frac{\partial x^{q_2}}{\partial \overline{x}^{j_2}}\cdots\frac{\partial x^{q_s}}{\partial \overline{x}^{j_s}}\left(A^{p_1p_2...p_r}_{q_1q_2...q_s} \pm B^{p_1p_2...p_r}_{q_1q_2...q_s}\right)$$

If

$$\overline{A}^{i_1i_2...i_r}_{j_1j_2...j_s} \pm \overline{B}^{i_1i_2...i_r}_{j_1j_2...j_s} = \overline{C}^{i_1i_2...i_r}_{j_1j_2...j_s}$$

and

$$A^{p_1p_2...p_r}_{q_1q_2...q_s} \pm B^{p_1p_2...p_r}_{q_1q_2...q_s} = C^{p_1p_2...p_r}_{q_1q_2...q_s}$$

So,

$$\overline{C}^{i_1i_2...i_r}_{j_1j_2...j_s} = \frac{\partial \overline{x}^{i_1}}{\partial x^{p_1}}\frac{\partial \overline{x}^{i_2}}{\partial x^{p_2}}\cdots\frac{\partial \overline{x}^{i_r}}{\partial x^{p_r}}\frac{\partial x^{q_1}}{\partial \overline{x}^{j_1}}\frac{\partial x^{q_2}}{\partial \overline{x}^{j_2}}\cdots\frac{\partial x^{q_s}}{\partial \overline{x}^{j_s}} C^{p_1,p_2,...,p_r}_{q_1,q_2,...,q_s}$$

This is law of transformation of a mixed tensor of rank $r+s$. So, $\overline{C}^{i_1,i_2,...,i_r}_{j_1,j_2,...,j_s}$ is a mixed tensor of rank $r+s$ or of type (r, s).

EXAMPLE 5

If A_k^{ij} and B_n^{lm} are tensors then their sum and difference are tensors of the same rank and type.

Solution

As given A_k^{ij} and B_k^{ij} are tensors. Then according to the law of transformation

$$\overline{A}_k^{ij} = \frac{\partial \overline{x}^i}{\partial x^p}\frac{\partial \overline{x}^j}{\partial x^q}\frac{\partial x^r}{\partial \overline{x}^k} A_r^{pq}$$

and

$$\overline{B}_k^{ij} = \frac{\partial \overline{x}^i}{\partial x^p}\frac{\partial \overline{x}^j}{\partial x^q}\frac{\partial x^r}{\partial \overline{x}^k} B_r^{pq}$$

then

$$\overline{A}_k^{ij} \pm \overline{B}_k^{ij} = \frac{\partial \overline{x}^i}{\partial x^p}\frac{\partial \overline{x}^j}{\partial x^q}\frac{\partial x^r}{\partial \overline{x}^k}\left(A_r^{pq} \pm B_r^{pq}\right)$$

If

$$\overline{A}_k^{ij} \pm \overline{B}_k^{ij} = \overline{C}_k^{ij} \text{ and } A_r^{pq} \pm B_r^{pq} = C_r^{pq}$$

So,

$$\overline{C}_k^{ij} = \frac{\partial \overline{x}^i}{\partial x^p}\frac{\partial \overline{x}^j}{\partial x^q}\frac{\partial x^r}{\partial \overline{x}^k} C_r^{pq}$$

The shows that C_k^{ij} is a tensor of same rank and type as A_k^{ij} and B_k^{ij}.

2.10 MULTIPLICATION OF TENSORS (OUTER PRODUCT OF TENSOR)

THEOREM 2.8 *The multiplication of two tensors is a tensor whose rank is the sum of the ranks of two tensors.*

Proof: Consider two tensors $A_{j_1j_2...j_s}^{i_1i_2...i_r}$ (which is covariant tensor of rank s and contravariant tensor of rank r) and $B_{l_1l_2...l_n}^{k_1k_2...k_m}$ (which is covariant tensor of rank m and contravariant tensor of rank n). Then according to the law of transformation.

$$\overline{A}_{j_1j_2...j_s}^{i_1i_2...i_r} = \frac{\partial \overline{x}^{i_1}}{\partial x^{p_1}}\frac{\partial \overline{x}^{i_2}}{\partial x^{p_2}}\cdots\frac{\partial \overline{x}^{i_r}}{\partial x^{p_r}}\frac{\partial x^{q_1}}{\partial \overline{x}^{j_1}}\frac{\partial x^{q_2}}{\partial \overline{x}^{j_2}}\cdots\frac{\partial x^{q_r}}{\partial \overline{x}^{j_s}} A_{q_1q_2...q_s}^{p_1p_2...p_r}$$

and

$$\overline{B}_{l_1l_2...l_n}^{k_1k_2...k_m} = \frac{\partial \overline{x}^{k_1}}{\partial x^{\alpha_1}}\frac{\partial \overline{x}^{k_2}}{\partial x^{\alpha_2}}\cdots\frac{\partial \overline{x}^{k_m}}{\partial x^{\alpha_m}}\frac{\partial x^{\beta_1}}{\partial \overline{x}^{l_1}}\frac{\partial x^{\beta_2}}{\partial \overline{x}^{l_2}}\cdots\frac{\partial x^{\beta_n}}{\partial \overline{x}^{l_n}} B_{\beta_1\beta_2...\beta_n}^{\alpha_1\alpha_2...\alpha_m}$$

Then their product is

$$\overline{A}_{j_1j_2...j_s}^{i_1i_2...i_r}\ \overline{B}_{l_1l_2...l_n}^{k_1k_2...k_m} = \frac{\partial \overline{x}^{i_1}}{\partial x^{p_1}}\cdots\frac{\partial \overline{x}^{i_r}}{\partial x^{p_r}}\frac{\partial x^{q_1}}{\partial \overline{x}^{j_1}}\cdots\frac{\partial x^{q_s}}{\partial \overline{x}^{j_s}}\frac{\partial \overline{x}^{k_1}}{\partial x^{\alpha_1}}\cdots\frac{\partial \overline{x}^{k_m}}{\partial x^{\alpha_m}}\cdot\frac{\partial x^{\beta_1}}{\partial \overline{x}^{l_1}}\cdots\frac{\partial x^{\beta_n}}{\partial \overline{x}^{l_m}}$$

$$A_{q_1q_2...q_s}^{p_1p_2...p_r}\ B_{\beta_1\beta_2...\beta_n}^{\alpha_1\alpha_2...\alpha_m}$$

If

$$\overline{C}^{i_1 i_2 \ldots i_r k_1 k_2 \ldots k_m}_{j_1 j_2 \ldots j_s l_1 l_2 \ldots l_n} = \overline{A}^{i_1 i_2 \ldots i_r}_{j_1 j_2 \ldots j_s} \overline{B}^{k_1 k_2 \ldots k_n}_{l_1 l_2 \ldots l_n}$$

and

$$C^{p_1 p_2 \ldots p_r \alpha_1 \alpha_2 \ldots \alpha_m}_{q_1 q_2 \ldots q_s \beta_1 \beta_2 \ldots \beta_n} = A^{p_1 p_2 \ldots, p_r}_{q_1 q_2 \ldots, q_s} B^{\alpha_1 \alpha_2 \ldots \alpha_m}_{\beta_1 \beta_2 \ldots \beta_n}$$

So,

$$\overline{C}^{i_1 i_2 \ldots i_r k_1 k_2 \ldots k_m}_{j_1 j_2 \ldots j_s l_1 l_2 \ldots l_n} = \frac{\partial \overline{x}^{i_1}}{\partial x^{p_1}} \cdots \frac{\partial \overline{x}^{i_r}}{\partial x^{p_r}} \cdot \frac{\partial \overline{x}^{q_1}}{\partial \overline{x}^{j_1}} \cdots \frac{\partial x^{q_s}}{\partial \overline{x}^{j_s}} \frac{\partial \overline{x}^{k_1}}{\partial x^{\alpha_1}}$$

$$\cdots \frac{\partial \overline{x}^{k_m}}{\partial x^{\alpha_m}} \cdot \frac{\partial x^{\beta_1}}{\partial \overline{x}^{l_1}} \cdots \frac{\partial x^{\beta_n}}{\partial \overline{x}^{l_n}} \; C^{p_1 p_2 \ldots p_r \alpha_1 \alpha_2 \ldots \alpha_h}_{q_1 q_2 \ldots q_s \beta_1 \beta_2 \ldots \beta_n}$$

This is law of transformation of a mixed tensor of rank $r+m+s+n.$. So, $\overline{C}^{i_1 i_2 \ldots i_r k_1 k_2 \ldots k_m}_{j_1 j_2 \ldots j_s l_1 l_2 \ldots l_n}$ is a mixed tensor of rank $r+m+s+n$. or of type $(r+m, s+n)$. Such product is called outer product or open proudct of two tensors.

THEOREM 2.9 *If A^i and B_j are the components of a contravariant and covariant tensors of rank one then prove that $A^i B_j$ are components of a mixed tensor of rank two.*

Proof: As A^i is contravariant tensor of rank one and B_j is covariant tensor of rank one. Then according to the law of transformation

$$\overline{A}^i = \frac{\partial \overline{x}^i}{\partial x^k} A^k \qquad \ldots(1)$$

and

$$\overline{B}_j = \frac{\partial x^l}{\partial \overline{x}^j} B_l \qquad \ldots(2)$$

Multiply (1) and (2), we get

$$\overline{A}^i \overline{B}_j = \frac{\partial \overline{x}^i}{\partial x^k} \frac{\partial x^l}{\partial \overline{x}^j} A^k B_l$$

This is law of transformation of tensor of rank two. So, $A^i B_j$ are mixed tensor of rank two. Such product is called outer product of two tensors.

EXAMPLE 6

Show that the product of two tensors A^i_j and B^{kl}_m is a tensor of rank five.

Solution

As A^i_j and B^{kl}_m are tensors. Then by law of transformation

$$\overline{A}^i_j = \frac{\partial \overline{x}^i}{\partial x^p} \frac{\partial x^q}{\partial \overline{x}^j} A^p_q \quad \text{and} \quad \overline{B}^{kl}_m = \frac{\partial \overline{x}^k}{\partial x^r} \frac{\partial \overline{x}^l}{\partial x^s} \frac{\partial x^t}{\partial \overline{x}^m} B^{rs}_t$$

Multiplying these, we get

$$\overline{A}^i_j \overline{B}^{kl}_m = \frac{\partial \overline{x}^i}{\partial x^p}\frac{\partial x^q}{\partial \overline{x}^j}\frac{\partial \overline{x}^k}{\partial x^r}\frac{\partial \overline{x}^l}{\partial x^s}\frac{\partial x^t}{\partial \overline{x}^m} A^p_q B^{rs}_t$$

This is law of transformation of tensor of rank five. So, $A^i_j B^{kl}_m$ is a tensor of rank five.

2.11 CONTRACTION OF A TENSOR

The process of getting a tensor of lower order (reduced by 2) by putting a covariant index equal to a contravariant index and performing the summation indicated is known as *Contraction.*

In other words, if in a tensor we put one contravariant and one covariant indices equal, the process is called contraction of a tensor.

For example, consider a mixed tensor A^{ijk}_{lm} of order five. Then by law of transformation,

$$\overline{A}^{ijk}_{lm} = \frac{\partial \overline{x}^i}{\partial x^p}\frac{\partial \overline{x}^j}{\partial x^q}\frac{\partial \overline{x}^k}{\partial x^r}\frac{\partial x^s}{\partial \overline{x}^l}\frac{\partial x^t}{\partial \overline{x}^m} A^{pqr}_{st}$$

Put the covariant index l = contravariant index i, so that

$$\overline{A}^{ijk}_{im} = \frac{\partial \overline{x}^i}{\partial x^p}\frac{\partial \overline{x}^j}{\partial x^q}\frac{\partial \overline{x}^k}{\partial x^r}\frac{\partial x^s}{\partial \overline{x}^i}\frac{\partial x^t}{\partial \overline{x}^m} A^{pqr}_{st}$$

$$= \frac{\partial \overline{x}^j}{\partial x^q}\frac{\partial \overline{x}^k}{\partial x^r}\frac{\partial x^s}{\partial x^p}\frac{\partial x^t}{\partial \overline{x}^m} A^{pqr}_{st}$$

$$= \frac{\partial \overline{x}^j}{\partial x^q}\frac{\partial \overline{x}^k}{\partial x^r}\frac{\partial x^t}{\partial \overline{x}^m} \delta^s_p A^{pqr}_{st} \qquad \text{Since } \frac{\partial x^s}{\partial x^p} = \delta^s_p$$

$$\overline{A}^{ijk}_{im} = \frac{\partial \overline{x}^j}{\partial x^q}\frac{\partial \overline{x}^k}{\partial x^r}\frac{\partial x^t}{\partial \overline{x}^m} A^{pqr}_{pt}$$

This is law of transformation of tensor of rank 3. So, A^{ijk}_{im} is a tensor of rank 3 and type (1, 2) while A^{ijk}_{lm} is a tensor of rank 5 and type (2, 3). It means that contraction reduces rank of tensor by two.

2.12 INNER PRODUCT OF TWO TENSORS

Consider the tensors A^{ij}_k and B^l_{mn} if we first form their outer product $A^{ij}_k B^l_{mn}$ and contract this by putting $l = k$ then the result is $A^{ij}_k B^k_{mn}$ which is also a tensor, called the inner product of the given tensors.

Hence the inner product of two tensors is obtained by first taking outer product and then contracting it.

EXAMPLE 7

If A^i and B_i are the components of a contravariant and covariant tensors of rank are respectively then prove that A^iB_i is scalar or invariant.

Solution

As A^i and B_i are the components of a contravariant and covariant tensor of rank one respectively, then according to the law of the transformation

$$\overline{A}^i = \frac{\partial \overline{x}^i}{\partial x^p} A^p \text{ and } \overline{B}_i = \frac{\partial x^q}{\partial \overline{x}^i} B_q$$

Multiplying these, we get

$$\overline{A}^i \overline{B}^i = \frac{\partial \overline{x}^i}{\partial x^p} \frac{\partial x^q}{\partial \overline{x}^i} A^p B_q$$

$$= \frac{\partial x^q}{\partial x^p} A^p B_q, \text{ since } \frac{\partial x^q}{\partial x^p} = \delta_p^q$$

$$= \delta_p^q A^p B_q$$

$$\overline{A}^i \overline{B}_i = A^p B_p$$

This shows that $A^i B_i$ is scalar or Invariant.

EXAMPLE 8

If A_j^i is mixed tensor of rank 2 and B_m^{kl} is mixed tensor of rank 3. Prove that $A_j^i B_m^{jl}$ is a mixed tensor of rank 3.

Solution

As A_j^i is mixed tensor of rank 2 and B_m^{kl} is mixed tensor of rank 3. Then by law of transformation

$$\overline{A}_j^i = \frac{\partial \overline{x}^i}{\partial x^p} \frac{\partial x^q}{\partial \overline{x}^j} A_q^p \text{ and } \overline{B}_m^{kl} = \frac{\partial \overline{x}^k}{\partial x^r} \frac{\partial \overline{x}^l}{\partial x^s} \frac{\partial x^t}{\partial \overline{x}^m} B_t^{rs} \quad ...(1)$$

Put $k = j$ then

$$\overline{B}_m^{jl} = \frac{\partial \overline{x}^j}{\partial x^r} \frac{\partial \overline{x}^l}{\partial x^s} \frac{\partial x^t}{\partial \overline{x}^m} B_t^{rs} \quad ...(2)$$

Multiplying (1) & (2) we get

$$\overline{A}_j^i \overline{B}_m^{jl} = \frac{\partial \overline{x}^i}{\partial x^p} \frac{\partial x^q}{\partial \overline{x}^j} \frac{\partial \overline{x}^j}{\partial x^r} \frac{\partial \overline{x}^l}{\partial x^s} \frac{\partial x^t}{\partial \overline{x}^m} A_q^p B_t^{rs}$$

$$= \frac{\partial \overline{x}^i}{\partial x^p} \frac{\partial \overline{x}^l}{\partial x^s} \frac{\partial x^t}{\partial \overline{x}^m} \delta_r^q A_q^p B_t^{rs} \quad \text{since } \frac{\partial x^q}{\partial \overline{x}^j} \frac{\partial \overline{x}^j}{\partial x^r} = \frac{\partial x^q}{\partial x^r} = \delta_r^q$$

$$\overline{A}_j^i \overline{B}_m^{jl} = \frac{\partial \overline{x}^i}{\partial x^p} \frac{\partial \overline{x}^l}{\partial x^s} \frac{\partial x^t}{\partial \overline{x}^m} A_q^p B_t^{qs} \quad \text{since } \delta_r^q B_t^{rs} = B_t^{qs}$$

This is the law of transformation of a mixed tensor of rank three. Hence $A_j^i B_m^{jl}$ is a mixed tensor of rank three.

2.13 SYMMETRIC TENSORS

A tensor is said to be symmetric with respect to two contravariant (or two covariant) indices if its components remain unchanged on an interchange of the two indices.

EXAMPLE

(1) The tensor A^{ij} is symmetric if $A^{ij} = A^{ji}$

(2) The tensor A^{ijk}_{lm} is symmetric if $A^{ijk}_{lm} = A^{jik}_{lm}$

THEOREM 2.10 *A symmetric tensor of rank two has only* $\frac{1}{2}n(n+1)$ *different components in n-dimensional space.*

Proof: Let A^{ij} be a symmetric tensor of rank two. So that $A^{ij} = A^{ji}$.

$$\text{The component of } A^{ij} \text{ are } \begin{bmatrix} A^{11} & A^{12} & A^{13} & \cdots & A^{1n} \\ A^{21} & A^{22} & A^{23} & \cdots & A^{2n} \\ A^{31} & A^{32} & A^{33} & \cdots & A^{3n} \\ \vdots & \vdots & \vdots & \cdots & \vdots \\ A^{n1} & A^{n2} & A^{n3} & \cdots & A^{nn} \end{bmatrix}$$

i.e., A^{ij} will have n^2 components. Out of these n^2 components, n components $A^{11}, A^{22}, A^{33}, ..., A^{nn}$ are different. Thus remaining components are $(n^2 - n)$. In which $A^{12} = A^{21}$, $A^{23} = A^{32}$ etc. due to symmetry.

So, the remaining different components are $\frac{1}{2}(n^2 - n)$. Hence the total number of different components

$$= n + \frac{1}{2}(n^2 - n) = \frac{1}{2}n(n+1)$$

2.14 SKEW-SYMMETRIC TENSOR

A tensor is said to be skew-symmetric with respect to two contravariant (or two covariant) indices if its components change sign on interchange of the two indices.

EXAMPLE

(*i*) The tensor A^{ij} is Skew-symmetric of $A^{ij} = -A^{ji}$

(*ii*) The tensor A^{ijk}_{lm} is Skew-symmetric if $A^{ijk}_{lm} = -A^{jik}_{lm}$

THEOREM 2.11 *A Skew symmetric tensor of second order has only* $\frac{1}{2}n(n-1)$ *different non-zero components.*

Proof: Let A^{ij} be a skew-symmetric tensor of order two. Then $A^{ij} = -A^{ji}$.

The components of A^{ij} are $\begin{bmatrix} 0 & A^{12} & A^{13} & \cdots & A^{1n} \\ A^{21} & 0 & A^{23} & \cdots & A^{2n} \\ A^{31} & A^{32} & 0 & \cdots & A^{3n} \\ \vdots & \vdots & \vdots & \cdots & \vdots \\ A^{n1} & A^{n2} & A^{n3} & \cdots & 0 \end{bmatrix}$

$$\left[\text{Since } A^{ii} = -A^{ii} \Rightarrow 2A^{ii} = 0 \Rightarrow A^{ii} = 0 \Rightarrow A^{11} = A^{22} = \cdots A^{nn} = 0\right]$$

i.e., A^{ij} will have n^2 components. Out of these n^2 components, n components $A^{11}, A^{22}, A^{33}, ..., A^{nn}$ are zero. Omitting there, then the remaining components are $n^2 - n$. In which $A^{12} = -A^{21}$, $A^{13} = -A^{31}$ etc. Ignoring the sign. Their remaining the different components are $\frac{1}{2}(n^2 - n)$.

Hence the total number of different non-zero components $= \frac{1}{2}n(n-1)$

Note: Skew-symmetric tensor is also called anti-symmetric tensor.

THEOREM 2.12 *A covariant or contravariant tensor of rank two say A_{ij} can always be written as the sum of a symmetric and skew-symmetric tensor.*

Proof: Consider a covariant tensor A_{ij}. We can write A_{ij} as

$$A_{ij} = \frac{1}{2}(A_{ij} + A_{ji}) + \frac{1}{2}(A_{ij} - A_{ji})$$

$$A_{ij} = S_{ij} + T_{ij}$$

where $$S_{ij} = \frac{1}{2}(A_{ij} + A_{ji}) \text{ and } T_{ij} = \frac{1}{2}(A_{ij} - A_{ji})$$

Now,

$$S_{ji} = \frac{1}{2}(A_{ji} + A_{ij})$$

$$S_{ji} = S_{ij}$$

So, S_{ij} is symmetric tensor.

and

$$T_{ij} = \frac{1}{2}(A_{ij} + A_{ji})$$

$$T_{ji} = \frac{1}{2}(A_{ji} - A_{ij})$$

$$= -\frac{1}{2}(A_{ij} - A_{ji})$$

$$T_{ji} = -T_{ij}$$

or $$T_{ij} = -T_{ji}$$

So, T_{ij} is Skew-symmetric Tensor.

EXAMPLE 9

If $\phi = a_{jk} A^j A^k$. Show that we can always write $\phi = b_{jk} A^j A^k$ where b_{jk} is symmetric.

Solution

As given

$$\phi = a_{jk} A^j A^k \quad \text{...(1)}$$

Interchange the indices i and j

$$\phi = a_{kj} A^k A^j \quad \text{...(2)}$$

Adding (1) and (2),

$$2\phi = (a_{jk} + a_{kj}) A^j A^k$$

$$\phi = \frac{1}{2}(a_{jk} + a_{kj}) A^j A^k$$

$$\phi = b_{jk} A^j A^k$$

where $b_{jk} = \frac{1}{2}(a_{jk} + a_{kj})$

To show that b_{jk} is symmetric.

Since

$$b_{jk} = \frac{1}{2}(a_{jk} + a_{kj})$$

$$b_{kj} = \frac{1}{2}(a_{kj} + a_{jk})$$

$$= \frac{1}{2}(a_{jk} + a_{kj})$$

$$b_{kj} = b_{jk}$$

So, b_{jk} is Symmetric.

EXAMPLE 10

If T_i be the component of a covariant vector show that $\left(\frac{\partial T_i}{\partial x^j} - \frac{\partial T_j}{\partial x^i}\right)$ are component of a Skew-symmetric covariant tensor of rank two.

Solution

As T_i is covariant vector. Then by the law of transformation

$$\overline{T}_i = \frac{\partial x^k}{\partial \overline{x}^i} T_k$$

Differentiating it w.r.t. to $\bar{x}^j$ partially,

$$\frac{\partial \bar{T}_i}{\partial \bar{x}^j} = \frac{\partial}{\partial \bar{x}^j}\left(\frac{\partial x^k}{\partial \bar{x}^i} T_k\right)$$

$$= \frac{\partial^2 x^k}{\partial \bar{x}^j \partial \bar{x}^i} T_k + \frac{\partial x^k}{\partial \bar{x}^i} \frac{\partial T_k}{\partial \bar{x}^j}$$

$$\frac{\partial \bar{T}_j}{\partial \bar{x}^j} = \frac{\partial^2 x^k}{\partial \bar{x}^j \partial \bar{x}^i} T_k + \frac{\partial x^k}{\partial \bar{x}^i} \frac{\partial x^l}{\partial \bar{x}^j} \frac{\partial T_k}{\partial \bar{x}^l} \qquad ...(1)$$

Similarly,

$$\frac{\partial \bar{T}_j}{\partial \bar{x}^i} = \frac{\partial^2 x^k}{\partial \bar{x}^i \partial \bar{x}^j} T_k + \frac{\partial x^k}{\partial \bar{x}^j} \frac{\partial x^l}{\partial \bar{x}^i} \frac{\partial T_k}{\partial x^l}$$

Interchanging the dummy indices k & l

$$\frac{\partial \bar{T}_j}{\partial \bar{x}^i} = \frac{\partial^2 x^k}{\partial \bar{x}^i \partial \bar{x}^j} T_k + \frac{\partial x^k}{\partial \bar{x}^i} \frac{\partial x^l}{\partial \bar{x}^j} \frac{\partial T_l}{\partial x^k} \qquad ...(2)$$

Substituting (1) and (2), we get

$$\frac{\partial \bar{T}_i}{\partial \bar{x}^j} - \frac{\partial \bar{T}_j}{\partial \bar{x}^i} = \frac{\partial x^k}{\partial \bar{x}^i} \frac{\partial x^l}{\partial \bar{x}^j}\left(\frac{\partial T_k}{\partial x^l} - \frac{\partial T_l}{\partial x^k}\right)$$

This is law of transformation of covariant tensor of rank two. So, $\frac{\partial T_i}{\partial x^j} - \frac{\partial T_j}{\partial x^i}$ are component of a covariant tensor of rank two.

To show that $\frac{\partial T_i}{\partial x^j} - \frac{\partial T_j}{\partial x^i}$ is Skew-symmetric tensor.

Let

$$A_{ij} = \frac{\partial T_i}{\partial x^j} - \frac{\partial T_j}{\partial x^i}$$

$$A_{ji} = \frac{\partial T_j}{\partial x^i} - \frac{\partial T_i}{\partial x^j}$$

$$= -\left(\frac{\partial T_i}{\partial x^j} - \frac{\partial T_j}{\partial x^i}\right)$$

$$A_{ji} = -A_{ij}$$

or

$$A_{ij} = -A_{ji}$$

So, $A_{ij} = \frac{\partial T_i}{\partial x^j} - \frac{\partial T_j}{\partial x^i}$ is Skew-symmetric.

So, $\frac{\partial T_i}{\partial x^j} - \frac{\partial T_j}{\partial x^i}$ are component of a Skew-symmetric covariant tensor of rank two.

2.15 QUOTIENT LAW

By this law, we can test a given quantity is a tensor or not. Suppose given quantity be A and we do not know that A is a tensor or not. To test A, we take inner product of A with an arbitrary tensor, if this inner product is a tensor then A is also a tensor.

Statement

If the inner product of a set of functions with an atbitrary tensor is a tensor then these set of functions are the components of a tensor.

The proof of this law is given by the following examples.

EXAMPLE 11

Show that the expression $A(i,j,k)$ is a covariant tensor of rank three if $A(i,j,k)B^k$ is covariant tensor of rank two and B^k is contravariant vector

Solution

Let X and Y be two coordinate systems.

As given $A(i, j, k)B^k$ is covariant tensor of rank two then

$$\overline{A}(i,j,k)\overline{B}^k = \frac{\partial x^p}{\partial \overline{x}^i}\frac{\partial x^q}{\partial \overline{x}^j}A(p,q,r)B^r \qquad ...(1)$$

Since B^k is contravariant vector. Then

$$\overline{B}^k = \frac{\partial \overline{x}^k}{\partial x^r}B^r \quad \text{or} \quad B^r = \frac{\partial x^r}{\partial \overline{x}^k}\overline{B}^k.$$

So, from (1)

$$\overline{A}(i,j,k)\overline{B}^k = \frac{\partial x^p}{\partial \overline{x}^i}\frac{\partial x^q}{\partial \overline{x}^j}A(p,q,r)\frac{\partial x^r}{\partial \overline{x}^k}\overline{B}^k$$

$$\overline{A}(i,j,k)\overline{B}^k = \frac{\partial x^p}{\partial \overline{x}^i}\frac{\partial x^q}{\partial \overline{x}^j}\frac{\partial x^r}{\partial \overline{x}^k}A(p,q,r)\overline{B}^k$$

$$\overline{A}(i,j,k) = \frac{\partial x^p}{\partial \overline{x}^i}\frac{\partial x^q}{\partial \overline{x}^i}\frac{\partial x^r}{\partial \overline{x}^k}A(p,q,r)$$

As $\overline{B}^k$ is arbitrary.

So, $A(i,j,k)$ is covariant tensor of rank three.

EXAMPLE 12

If $A(i, j, k)A^iB^jC_k$ is a scalar for arbitrary vectors A^i, B^j, C_k. Show that $A(i, j, k)$ is a tensor of type (1, 2).

Solution

Let X and Y be two coordinate systems. As given $A(i,j,k)A^iB^jC_k$ is scalar. Then

$$\overline{A}(i,j,k)\,\overline{A}^i\overline{B}^j\overline{C}_k = A(p,q,r)A^pB^qC_r \qquad ...(1)$$

Since A^i, B^i and C_k are vectors. Then

$$\overline{A}^i = \frac{\partial \overline{x}^i}{\partial x^p} A^p \quad \text{or} \quad A^p = \frac{\partial x^p}{\partial \overline{x}^i} \overline{A}^i$$

$$\overline{B}^j = \frac{\partial \overline{x}^j}{\partial x^q} B^q \quad \text{or} \quad B^q = \frac{\partial x^q}{\partial \overline{x}^j} \overline{B}^j$$

$$\overline{C}^k = \frac{\partial \overline{x}^k}{\partial x^r} C^r \quad \text{or} \quad C^r = \frac{\partial x^r}{\partial \overline{x}^k} \overline{C}^k$$

So, from (1)

$$\overline{A}(i,j,k)\, \overline{A}^i \overline{B}^j \overline{C}_k = A(p,q,r) \frac{\partial x^p}{\partial \overline{x}^i} \frac{\partial x^q}{\partial \overline{x}^j} \frac{\partial \overline{x}^k}{\partial x^r} \overline{A}^i \overline{B}^j \overline{C}_k$$

As $\overline{A}^i, \overline{B}^j, \overline{C}_k$ are arbitrary.

Then

$$\overline{A}(i,j,k) = \frac{\partial x^p}{\partial \overline{x}^i} \frac{\partial x^q}{\partial \overline{x}^j} \frac{\partial \overline{x}^k}{\partial x^r} A(p,q,r)$$

So, $A(i, j, k)$ is tensor of type (1, 2).

2.16 CONJUGATE (OR RECIPROCAL) SYMMETRIC TENSOR

Consider a covariant symmetric tensor A_{ij} of rank two. Let d denote the determinant $|A_{ij}|$ with the elements A_{ij} *i.e.*, $d = |A_{ij}|$ and $d \neq 0$.

Now, define A^{ij} by

$$A^{ij} = \frac{\text{Cofactor of } A_{ij} \text{ is the determinant } |A_{ij}|}{d}$$

A^{ij} is a contravariant symmetric tensor of rank two which is called conjugate (or Reciprocal) tensor of A_{ij}.

THEOREM 2.13 *If B_{ij} is the cofactor of A_{ij} in the determinant $d = |A_{ij}| \neq 0$ and A^{ij} defined as*

$$A^{ij} = \frac{B_{ij}}{d}$$

Then prove that $A_{ij}A^{kj} = \delta_i^k$.

Proof: From the properties of the determinants, we have two results.

(*i*) $A_{ij}B_{ij} = d$

$$\Rightarrow \qquad A_{ij}\frac{B_{ij}}{d} = 1$$

$$A_{ij}A^{ij} = 1, \quad \text{given} \quad A^{ij} = \frac{B_{ij}}{d}$$

(*ii*) $A_{ij}B_{kj} = 0$

$$A_{ij}\frac{B_{kj}}{d} = 0, \qquad d \neq 0$$

$$A_{ij}A^{kj} = 0 \qquad \text{if } i \neq k$$

from (*i*) & (*ii*)

$$A_{ij}A^{kj} = \begin{cases} 1 \text{ if } i = k \\ 0 \text{ if } i \neq k \end{cases}$$

i.e., $$A_{ij}A^{kj} = \delta_i^k$$

2.17 RELATIVE TENSOR

If the components of a tensor $A^{i_1 i_2 \ldots i_r}_{j_1 j_2 \ldots j_s}$ transform according to the equation

$$\bar{A}^{k_1 k_2 \ldots k_r}_{l_1 l_2 \ldots l_s} = \left|\frac{\partial x}{\partial \bar{x}}\right|^{\omega} A^{i_1 i_2 \ldots i_r}_{j_1 j_2 \ldots j_s} \frac{\partial \bar{x}^{k_1}}{\partial x^{i_1}}\frac{\partial \bar{x}^{k_2}}{\partial x^{i_2}} \cdots \frac{\partial \bar{x}^{k_r}}{\partial x^{i_r}} \cdot \frac{\partial x^{j_1}}{\partial \bar{x}^{l_r}}\frac{\partial x^{j_2}}{\partial \bar{x}^{l_2}} \cdots \frac{\partial x^{j_s}}{\partial \bar{x}^{l_s}}$$

Hence $A^{i_1 i_2 \ldots i_r}_{j_1 j_2 \ldots j_r}$ is called a relative tensor of weight ω, where $\left|\frac{\partial x}{\partial \bar{x}}\right|$ is the Jacobian of transformation. If $\omega = 1$, the relative tensor is called a tensor density. If $w = 0$ then tensor is said to be absolute.

MISCELLANEOUS EXAMPLES

1. Show that there is no distinction between contravariant and covariant vectors when we restrict ourselves to transformation of the type

$$\bar{x}^i = a^i_m x^m + b^i;$$

where *a*'s and *b*'s are constants such that

$$a^i_r a^i_m = \delta^r_m$$

Solution

Given that

$$\bar{x}^i = a^i_m x^m + b^i \qquad \ldots(1)$$

or $$a^i_m x^m = \bar{x}^i - b^i \qquad \ldots(2)$$

Multiplying both sides (2) by a^i_r, we get

$$a^i_r a^i_m x^m = a^i_r \bar{x}^i - b^i a^i_r$$

$$\delta^r_m x^m = a^i_r \bar{x}^i - b^i a^i_r \text{ as given } a^i_r a^i_m = \delta^r_m$$

$$x^r = a^i_r \bar{x}^i - b^i a^i_r \text{ as } \delta^r_m x^m = x^r$$

or $$x^s = a^i_s \bar{x}^i - b^i a^i_s$$

Differentiating Partially it w.r.t. to $\bar{x}^i$

$$\frac{\partial x^s}{\partial \bar{x}^i} = a^i_s \qquad \ldots(3)$$

Now, from (1)

$$\bar{x}^i = a^i_s x^s + b^i$$

$$\frac{\partial \bar{x}^i}{\partial x^s} = a^i_s \qquad ...(4)$$

The contravariant transformation is

$$\bar{A}^i = \frac{\partial \bar{x}^i}{\partial x^s} A^s = a^i_s A^s \qquad ...(5)$$

The covariant transformation is

$$\bar{A}_i = \frac{\partial x^s}{\partial \bar{x}^i} A_s = a^i_s A_s \qquad ...(6)$$

Thus from (5) and (6), it shows that there is no distinction between contravariant and covariant tensor law of transformation

2. If the tensors a_{ij} and g_{ij} are symmetric and u^i, v^i are components of contravariant vectors satisfying the equations

$$(a_{ij} - kg_{ij})u^i = 0, \quad i, j = 1, 2, ..., n$$

$$(a_{ij} - k'g_{ij})v^i = 0, \quad k \neq k'.$$

Prove that $g_{ij}u^i v^j = 0$, $a_{ij}u^i v^j = 0$.

Solution

The equations are

$$(a_{ij} - kg_{ij})u^i = 0 \qquad ...(1)$$

$$(a_{ij} - k'g_{ij})v^i = 0 \qquad ...(2)$$

Multiplying (1) and (2) by u^j and v^j respectively and subtracting, we get

$$a_{ij}u^i v^j - a_{ij}v^i u^j - kg_{ij}u^i v^j + k'g_{ij}u^j v^i = 0$$

Interchanging i and j in the second and fourth terms,

$$a_{ij}u^i v^j - a_{ji}v^j u^i - kg_{ij}u^i v^j + k'g_{ji}u^i v^j = 0$$

As a_{ij} and g_{ij} is symmetric *i.e.*, $a_{ij} = a_{ji}$ & $g_{ij} = g_{ji}$

$$-kg_{ij}v^j u^i + k'g_{ij}u^i v^j = 0$$

$$(k' - k)g_{ij}u^i v^j = 0$$

$$g_{ij}u^i v^j = 0 \text{ since } k \neq k' \Rightarrow k - k' \neq 0$$

Multiplying (1) by v^j, we get

$$a_{ij}v^j u^i - kg_{ij}u^i v^j = 0$$

$$a_{ij}u^i v^j = 0 \text{ as } g_{ij}u^i v^j = 0.$$

Proved.

3. If A_{ij} is a Skew-Symmetric tensor prove that

$$(\delta^i_j \delta^k_l + \delta^i_l \delta^k_j) A_{ik} = 0$$

Solution

Given A_{ij} is a Skew-symmetric tensor then $A_{ij} = -A_{ji}$.

Now,

$$\begin{aligned}(\delta^i_j \delta^k_l + \delta^i_l \delta^k_j) A_{ik} &= \delta^i_j \delta^k_l A_{ik} + \delta^i_l \delta^k_j A_{ik} \\ &= \delta^i_j A_{il} + \delta^i_l A_{ij} \\ &= A_{jl} + A_{lj} \\ (\delta^i_j \delta^k_l + \delta^i_l \delta^k_j) A_{ik} &= 0 \quad \text{as } A_{jl} = -A_{lj}\end{aligned}$$

4. If a_{ij} is symmetric tensor and b_i is a vector and $a_{ij}b_k + a_{jk}b_i + a_{ki}b_j = 0$ then prove that $a_{ij} = 0$ or $b_k = 0$.

Solution

The equation is

$$a_{ij}b_k + a_{jk}b_i + a_{ki}b_j = 0$$

$$\Rightarrow \quad \bar{a}_{ij}\bar{b}_k + \bar{a}_{jk}\bar{b}_i + \bar{a}_{ki}\bar{b}_j = 0$$

By tensor law of transformation, we have

$$a_{pq}\frac{\partial x^p}{\partial \bar{x}^i}\frac{\partial x^q}{\partial \bar{x}^j} b_r \frac{\partial x^r}{\partial \bar{x}^k} + a_{pq}\frac{\partial x^p}{\partial \bar{x}^j}\frac{\partial x^q}{\partial \bar{x}^k} b_r \frac{\partial x^r}{\partial \bar{x}^i} + a_{pq}\frac{\partial x^p}{\partial \bar{x}^k}\frac{\partial x^q}{\partial \bar{x}^i} b_r \frac{\partial x^r}{\partial \bar{x}^j} = 0$$

$$a_{pq}b_r\left[\frac{\partial x^p}{\partial \bar{x}^i}\frac{\partial x^q}{\partial \bar{x}^j}\frac{\partial x^r}{\partial \bar{x}^k} + \frac{\partial x^p}{\partial \bar{x}^j}\frac{\partial x^q}{\partial \bar{x}^k}\frac{\partial x^r}{\partial \bar{x}^i} + \frac{\partial x^p}{\partial \bar{x}^k}\frac{\partial x^q}{\partial \bar{x}^i}\frac{\partial x^r}{\partial \bar{x}^j}\right] = 0$$

$$\Rightarrow \quad a_{pq}b_r = 0 \Rightarrow a_{pq} = 0 \text{ or } b_r = 0$$

$$\Rightarrow a_{ij} = 0 \text{ or } b_k = 0$$

5. If $a_{mn}x^m x^n = b_{mn}x^m x^n$ for arbitrary values of x^r, show that $a_{(mn)} = b_{(mn)}$ *i.e.*, $a_{mn} + a_{nm} = b_{mn} + b_{nm}$

If a_{mn} and b_{mn} are symmetric tensors then further show the $a_{mn} = b_{mn}$.

Solution

Given

$$a_{mn}x^m x^n = b_{mn}x^m x^n$$

$$(a_{mn} - b_{mn})x^m x^n = 0$$

Differentiating w.r.t. x^i partially

$$(a_{in} - b_{in})x^n + (a_{mi} - b_{mi})x^m = 0$$

Differentiating again w.r.t. x^j partially

$$(a_{ij} - b_{ij}) + (a_{ji} - b_{ji}) = 0$$

$$a_{ij} + a_{ji} = b_{ij} + b_{ji}$$

or $$a_{mn} + a_{nm} = b_{mn} + b_{nm} \text{ or } a_{(mn)} = b_{(mn)}$$

Also, since a_{mn} and b_{mn} are symmetric then $a_{mn} = a_{nm}$, $b_{mn} = b_{nm}$.

So,

$$2a_{mn} = 2b_{mn}$$

$$a_{mn} = b_{mn}$$

EXERCISES

1. Write down the law of transformation for the tensors

   (i) A_{ij}

   (ii) B_k^{ij}

   (iii) C_{lm}^{ijk}

2. If A_r^{pq} and B_t^s are tensors then prove that $A_r^{pq} B_t^s$ is also a tensor.
3. If A^{ij} is a contravariant tensor and B_i is covariant vector then prove that $A^{ij} B_k$ is a tensor of rank three and $A^{ij} B_j$ is a tensor of rank one.
4. If A^i is an arbitrary contravariant vector and $C_{ij} A^i A^j$ is an invariant show that $C_{ij} + C_{ji}$ is a covariant tensor of the second order.
5. Show that every tensor can be expressed in the terms of symmetric and skew-symmetric tensor.
6. Prove that in n-dimensional space, symmetric and skew-symmetric tensor have $\frac{n}{2}(n+1)$ and $\frac{n}{2}(n-1)$ independent components respectively.
7. If $U_{ij} \neq 0$ are components of a tensor of the type (0, 2) and if the equation $fU_{ij} + gU_{ji} = 0$ holds w.r.t to a basis then prove that either $f = g$ and U_{ij} is skew-symmetric or $f = -g$ and U_{ij} is symmetric.
8. If A_{ij} is skew-symmetric then $(B_j^i B_l^k + B_l^i B_j^k) A_{ik} = 0$.
9. Explain the process of contraction of tensors. Show that $a_{ij} a^{ij} = \delta_j^i$.

10. If A_r^{pq} is a tensor of rank three. Show that A_r^{pr} is a contravariant tensor of rank one.

11. If $a_k^{ij}\lambda_i\mu_j\gamma^k$ is a scalar or invariant, $\lambda_i, \mu_j, \gamma^k$ are vectors then a_k^{ij} is a mixed tensor of type (2, 1).

12. Show that if $a_{hijk}\lambda^h\mu^i\lambda^h\mu^k = 0$ where λ^i and μ^i are components of two arbitrary vectors then

$$a_{hijk} + a_{hkji} + a_{jihk} + a_{jkhi} = 0$$

13. Prove that $A_{ij}B^iC^j$ is invariant if B^i and C^j are vector and A_{ij} is tensor of rank two.

14. If $A(r, s, t)$ be a function of the coordinates in n-dimensional space such that for an arbitrary vector B^r of the type indicated by the index a $A(r, s, t)B^r$ is equal to the component C^{st} of a contravariant tensor of order two. Prove that $A(r, s, t)$ are the components of a tensor of the form A_r^{st}.

15. If A^{ij} and A_{ij} are components of symmetric relative tensors of weight w. show that

$$\left|\overline{A}^{ij}\right| = \left|A^{ij}\right|\left|\frac{\partial x}{\partial \overline{x}}\right|^{w-2} \quad \text{and} \quad \left|\overline{A}_{ij}\right| = \left|A_{ij}\right|\left|\frac{\partial x}{\partial \overline{x}}\right|^{w+2}$$

16. Prove that the scalar product of a relative covariant vector of weight w and a relative contravariant vector of weight w' is a relative scalar of weight $w + w'$.

CHAPTER – 3

METRIC TENSOR AND RIEMANNIAN METRIC

3.1 THE METRIC TENSOR

In rectangular cartesian coordinates, the distance between two neighbouring point are (x, y, z) and $(x+dx, y+dy, z+dz)$ is given by $ds^2 = dx^2 + dy^2 + dz^2$.

In n-dimensional space, Riemann defined the distance ds between two neighbouring points x^i and $x^i + dx^i (i = 1,2,...n)$ by quadratic differential form

$$\begin{aligned} ds^2 &= g_{11}(dx^1)^2 + g_{12}dx^1dx^2 + \cdots + g_{1n}dx^1dx^n \\ &\quad + g_{12}(dx^2)dx^1 + g_{22}(dx^2)^2 + \cdots + g_{2n}dx^2dx^n \\ &\quad + \cdots\cdots\cdots\cdots\cdots + \\ &\quad + g_{n1}dx^n dx^1 + g_{n2}dx^n dx^2 + \cdots + g_{nn}(dx^n)^2 \end{aligned}$$

$$ds^2 = g_{ij}dx^i dx^j \ (i, j = 1,2,..n) \quad ...(1)$$

using summation convention.

Where g_{ij} are the functions of the coordinates x^i such that

$$g = |g_{ij}| \neq 0$$

The quadratic differential form (1) is called the Riemannian Metric or Metric or line element for n-dimensional space and such n-dimensional space is called *Riemannian space* and denoted by V_n and g_{ij} is called Metric Tensor or Fundamental tensor.

The geometry based on Riemannian Metric is called the Riemannian Geometry.

THEOREM 3.1 *The Metric tensor g_{ij} is a covariant symmetry tensor of rank two.*

Proof: The metric is given by

$$ds^2 = g_{ij}dx^i dx^j \quad ...(1)$$

Let x^i be the coordinates in X-coordinate system and $\bar{x}^i$ be the coordinates in Y-coordinate system. Then metric $ds^2 = g_{ij}\, dx^i dx^j$ transforms to $ds^2 = \bar{g}_{ij} d\bar{x}^i d\bar{x}^j$

Since distance being scalar quantity.

So, $$ds^2 = g_{ij} dx^i dx^j = \bar{g}_{ij} d\bar{x}^i d\bar{x}^j \qquad ...(2)$$

The theorem will be proved in three steps.

(*i*) To show that dx^i is a contravariant vector.

If $$\bar{x}^i = \bar{x}^i (x^1, x^2, ... x^n)$$

$$d\bar{x}^i = \frac{\partial \bar{x}^i}{\partial x^i} dx^1 + \frac{\partial \bar{x}^i}{\partial x^2} dx^2 + \cdots + \frac{\partial \bar{x}^i}{\partial x^n} dx^n$$

$$d\bar{x}^i = \frac{\partial \bar{x}^i}{\partial x^k} dx^k$$

It is law of transformation of contravariant vector. So, dx^i is contravariant vector.

(*ii*) To show that g_{ij} is a covariant tensor of rank two. Since

$$d\bar{x}^i = \frac{\partial \bar{x}^i}{\partial x^k} dx^k \text{ and } d\bar{x}^j = \frac{\partial \bar{x}^j}{\partial x^l} \partial x^l$$

from equation (2)

$$g_{ij} dx^i dx^j = \bar{g}_{ij} \frac{\partial \bar{x}^i}{\partial x^k} dx^k \frac{\partial \bar{x}^j}{\partial x^l} dx^l$$

$$g_{ij} dx^i dx^j = \bar{g}_{ij} \frac{\partial \bar{x}^i}{\partial x^k} \frac{\partial \bar{x}^j}{\partial x^l} \partial x^k dx^l$$

$$g_{kl} dx^k dx^l = \bar{g}_{ij} \frac{\partial \bar{x}^i}{\partial x^k} \frac{\partial \bar{x}^j}{\partial x^l} dx^k dx^l$$

Since $g_{ij} dx^i dx^j = g_{kl} dx^k dx^l$ (*i, j* are dummy indices).

$$\left(g_{kl} - \bar{g}_{ij} \frac{\partial \bar{x}^i}{\partial x^k} \frac{\partial \bar{x}^j}{\partial x^l} \right) dx^k dx^l = 0$$

or $$g_{kl} - \bar{g}_{ij} \frac{\partial \bar{x}^i}{\partial x^k} \frac{\partial \bar{x}^j}{\partial x^l} = 0 \text{ as } dx^k \text{ and } dx^l \text{ are arbitrary.}$$

$$g_{kl} = \bar{g}_{ij} \frac{\partial \bar{x}^j}{\partial x^k} \frac{\partial \bar{x}^j}{\partial x^l}$$

or $$\bar{g}_{ij} = g_{kl} \frac{\partial x^k}{\partial \bar{x}^i} \frac{\partial x^l}{\partial \bar{x}^j}$$

So, g_{ij} is covariant tensor of rank two.

(*iii*) To show that g_{ij} is symmetric. Then g_{ij} can be written as

$$g_{ij} = \frac{1}{2}(g_{ij} + g_{ji}) + \frac{1}{2}(g_{ij} - g_{ji})$$

$$g_{ij} = A_{ij} + B_{ij}$$

where $$A_{ij} = \frac{1}{2}(g_{ij} + g_{ji}) = \text{symmetric}$$

$$B_{ij} = \frac{1}{2}(g_{ij} - g_{ji}) = \text{Skew-symmetric}$$

Now, $$g_{ij}dx^i dx^j = (A_{ij} + B_{ij})dx^i dx^j \text{ from (3)}$$

$$(g_{ij} - A_{ij})dx^i dx^j = B_{ij}dx^i dx^j \tag{4}$$

Interchanging the dummy indices in $B_{ij}\, dx^i dx^j$, we have

$$B_{ij}dx^i dx^j = B_{ji}dx^i dx^j$$

$$B_{ij}dx^i dx^j = -B_{ij}dx^i dx^j$$

Since B_{ij} is Skew-symmetric *i.e.*, $B_{ij} = -B_{ji}$

$$B_{ij}dx^i dx^j + B_{ij}dx^i dx^j = 0$$

$$2B_{ij}dx^i dx^j = 0$$

$\Rightarrow$ $$B_{ij}dx^i dx^j = 0$$

So, from (4),

$$(g_{ij} - A_{ij})dx^i dx^j = 0$$

$\Rightarrow$ $$g_{ij} = A_{ij} \text{ as } dx^i, dx^j \text{ are arbitrary.}$$

So, g_{ij} is symmetric since A_{ij} is symmetric. Hence g_{ij} is a covariant symmetric tensor of rank two. This is called fundamental Covariant Tensor.

THEOREM 3.2 *To show that $g_{ij}dx^i dx^j$ is an invariant.*

Proof: Let x^i be coordinates of a point in *X*-coordinate system and $\overline{x}^i$ be coordinates of a same point in *Y*-coordinate system.

Since g_{ij} is a Covariant tensor of rank two.

Then, $$\overline{g}_{ij} = g_{kl}\frac{\partial x^k}{\partial \overline{x}^i}\frac{\partial x^1}{\partial \overline{x}^j}$$

$$\Rightarrow \qquad \bar{g}_{ij} - g_{kl}\frac{\partial x^k}{\partial \bar{x}^i}\frac{\partial x^l}{\partial \bar{x}^j} = 0$$

$$\left(\bar{g}_{ij} - g_{kl}\frac{\partial x^k}{\partial \bar{x}^i}\frac{\partial x^l}{\partial \bar{x}^j}\right)d\bar{x}^i d\bar{x}^j = 0$$

$$(\bar{g}_{ij} d\bar{x}^i d\bar{x}^j) = g_{kl}\frac{\partial x^k}{\partial \bar{x}^i}\frac{\partial x^l}{\partial \bar{x}^j}d\bar{x}^i d\bar{x}^j$$

$$= g_{kl}\frac{\partial x^k}{\partial \bar{x}^i}d\bar{x}^i \frac{\partial x^l}{\partial \bar{x}^j}d\bar{x}^j$$

$$\bar{g}_{ij} d\bar{x}^i d\bar{x}^j = g_{kl} dx^k dx^l$$

So, $g_{ij} dx^i dx^j$ is an ivariant.

3.2 CONJUGATE METRIC TENSOR: (CONTRAVARIANT TENSOR)

The conjugate Metric Tensor to g_{ij}, which is written as g^{ij}, is defined by

$$g^{ij} = \frac{B_{ij}}{g} \quad \text{(by Art.2.16, Chapter 2)}$$

where B_{ij} is the cofactor of g_{ij} in the determinant $g = |g_{ij}| \neq 0$.

By theorem on page 26

$$A_{ij} A^{kj} = \delta_i^k$$

So,

$$g_{ij} g^{kj} = \delta_i^k$$

Note (*i*) Tensors g_{ij} and g^{ij} are Metric Tensor or Fundamental Tensors.

(*ii*) g_{ij} is called first fundamental Tensor and g^{ij} second fundamental Tensors.

EXAMPLE 1

Find the Metric and component of first and second fundamental tensor is cylindrical coordinates.

Solution

Let (x^1, x^2, x^3) be the Cartesian coordinates and $(\bar{x}^1, \bar{x}^2, \bar{x}^3)$ be the cylindrical coordinates of a point. The cylindrical coordinates are given by

$x = r\cos\theta, \; y = r\sin\theta, \; z = z$

So that

$$x^1 = x, x^2 = y, x^3 = z \text{ and } \bar{x}^1 = r, \bar{x}^2 = \theta, \bar{x}^3 = z \qquad \text{...(1)}$$

Let g_{ij} and $\bar{g}_{ij}$ be the metric tensors in Cartesian coordinates and cylindrical coordinates respectively.

The metric in Cartesian coordinate is given by

$$ds^2 = dx^2 + dy^2 + dz^2$$

$$ds^2 = (dx^1)^2 + (dx^2)^2 + (dx^3)^2 \quad ...(2)$$

But

$$ds^2 = g_{ij}dx^i dx^j$$

$$= g_{11}(dx^1)^2 + g_{12}dx^1dx^2 + g_{13}dx^1dx^3 + g_{21}dx^2dx^1$$

$$+ g_{22}(dx^2)^2 + g_{23}dx^2dx^3 + g_{31}dx^3dx^1$$

$$+ g_{32}dx^3dx^2 + g_{33}(dx^3)^3 \quad ...(3)$$

Comparing (2) and (3), we have

$g_{11} = g_{22} = g_{33} = 1$ and $g_{12} = g_{13} = g_{21} = g_{23} = g_{31} = g_{32} = 0$

On transformation

$\bar{g}_{ij} = g_{ij}\dfrac{\partial x^i}{\partial \bar{x}^i}\dfrac{\partial x^j}{\partial \bar{x}^j}$, since g_{ij} is Covariant Tensor of rank two. ($i, j = 1, 2, 3$)

for $i = j = 1$.

$$\bar{g}_{11} = g_{11}\left(\frac{\partial x^1}{\partial \bar{x}^1}\right)^2 + g_{22}\left(\frac{\partial x^2}{\partial \bar{x}^1}\right) + g_{33}\left(\frac{\partial x^3}{\partial \bar{x}^1}\right)^2$$

since $g_{12} = g_{13} = \cdots = g_{32} = 0$.

$$\bar{g}_{11} = g_{11}\left(\frac{\partial x}{\partial r}\right)^2 + g_{22}\left(\frac{\partial y}{\partial r}\right)^2 + g_{33}\left(\frac{\partial z}{\partial r}\right)^2$$

Since $x = r\cos\theta,\ y = r\sin\theta,\ z = z$

$$\frac{\partial x}{\partial r} = \cos\theta, \quad \frac{\partial y}{\partial r} = \sin\theta, \quad \frac{\partial z}{\partial r} = 0$$

and $g_{11} = g_{22} = g_{33} = 1$.

$$\bar{g}_{11} = \cos^2\theta + \sin^2\theta + 0$$

$$\bar{g}_{11} = 1$$

Put $i = j = 2$.

$$\bar{g}_{22} = g_{11}\left(\frac{\partial x^1}{\partial \bar{x}^2}\right)^2 + g_{22}\left(\frac{\partial x^2}{\partial \bar{x}^2}\right)^2 + g_{33}\left(\frac{\partial x^3}{\partial \bar{x}^2}\right)^2$$

$$\bar{g}_{22} = g_{11}\left(\frac{\partial x}{\partial \theta}\right)^2 + g_{22}\left(\frac{\partial y}{\partial \theta}\right)^2 + g_{33}\left(\frac{\partial z}{\partial \theta}\right)^2$$

Since $g_{11} = g_{22} = g_{33} = 1$

$$\frac{\partial x}{\partial \theta} = -r\sin\theta, \quad \frac{\partial y}{\partial \theta} = r\cos\theta, \quad \frac{\partial z}{\partial \theta} = 0$$

$$\overline{g}_{22} = (-r\sin\theta)^2 + (r\cos\theta)^2 + 0$$

$$= r^2\sin^2\theta + r^2\cos^2\theta$$

$$\overline{g}_{22} = r^2$$

Put $i = j = 3$.

$$\overline{g}_{33} = g_{11}\left(\frac{\partial x^1}{\partial \overline{x}^3}\right)^2 + g_{22}\left(\frac{\partial x^2}{\partial \overline{x}^3}\right) + g_{33}\left(\frac{\partial x^3}{\partial \overline{x}^3}\right)$$

$$= g_{11}\left(\frac{\partial x}{\partial z}\right)^2 + g_{22}\left(\frac{\partial y}{\partial z}\right) + g_{33}\left(\frac{\partial z}{\partial z}\right)$$

Since $\frac{\partial x}{\partial z} = 0, \ \frac{\partial y}{\partial z} = 0, \ \frac{\partial z}{\partial z} = 1$. So, $\overline{g}_{33} = 1$.

So, $\quad \overline{g}_{11} = 1, \ \overline{g}_{22} = r^2, \ \overline{g}_{33} = 1$

and $\quad \overline{g}_{12} = \overline{g}_{13} = \overline{g}_{21} = \overline{g}_{23} = \overline{g}_{31} = \overline{g}_{32} = 0$

(*i*) The metric in cylindrical coordinates

$$ds^2 = \overline{g}_{ij}d\overline{x}^i d\overline{x}^j \quad i, j = 1,2,3.$$

$$ds^2 = \overline{g}_{11}(d\overline{x}^1)^2 + \overline{g}_{22}(d\overline{x}^2)^2 + \overline{g}_{33}(d\overline{x}^3)^2$$

since $\overline{g}_{12} = \overline{g}_{13} = \cdots = \overline{g}_{32} = 0$

$$ds^2 = dr^2 + r^2(d\theta)^2 + d\phi^2$$

(*ii*) The first fundamental tensor is

$$\overline{g}_{ij} = \begin{bmatrix} \overline{g}_{11} & \overline{g}_{12} & \overline{g}_{13} \\ \overline{g}_{21} & \overline{g}_{22} & \overline{g}_{23} \\ \overline{g}_{31} & \overline{g}_{32} & \overline{g}_{33} \end{bmatrix} = \begin{bmatrix} 1 & 0 & 0 \\ 0 & r^2 & 0 \\ 0 & 0 & 1 \end{bmatrix}$$

since $$\mathrm{g} = |\overline{g}_{ij}| = \begin{vmatrix} 1 & 0 & 0 \\ 0 & r^2 & 0 \\ 0 & 0 & 1 \end{vmatrix}$$

$$\mathrm{g} = r^2$$

(*iii*) The cofactor of g are given by

$$B_{11} = r^2, \quad B_{22} = 1, \quad B_{33} = r^2$$

and $$B_{12} = B_{21} = B_{13} = B_{23} = B_{31} = B_{32} = 0$$

The second fundamental tensor or conjugate tensor is $g^{ij} = \dfrac{B_{ij}}{g}$.

$$g^{11} = \frac{\text{cofactor of } g_{11} \text{ in } g}{g}$$

$$g^{11} = \frac{B_{11}}{g} = \frac{r^2}{r^2} = 1$$

$$g^{22} = \frac{B_{12}}{g} = \frac{1}{r^2}$$

$$g^{33} = \frac{B_{33}}{g} = \frac{r^2}{r^2} = 1$$

and $g^{12} = g^{13} = g^{21} = g^{23} = g^{31} = g^{32} = 0$

Hence the second fundamental tensor in matrix form is $\begin{bmatrix} 1 & 0 & 0 \\ 0 & \frac{1}{r^2} & 0 \\ 0 & 0 & 1 \end{bmatrix}$.

EXAMPLE 2

Find the matrix and component of first and second fundamental tensors in spherical coordinates.

Solution

Let (x^1, x^2, x^3) be the cartesian coordinates and $(\bar{x}^1, \bar{x}^2, \bar{x}^3)$ be the spherical coordinates of a point. The spherical coordinates are given by

$$x = r \sin\theta \cos\phi, \quad y = r \sin\theta \sin\phi, \quad z = r\cos\theta$$

So that $x^1 = x, x^2 = y, x^3 = z$ and $\bar{x}^1 = r, \bar{x}^2 = \theta, \bar{x}^3 = \phi$

Let g_{ij} and $\bar{g}_{ij}$ be the metric tensors in cartesian and spherical coordinates respectively.

The metric in cartesian coordinates is given by

$$ds^2 = dx^2 + dy^2 + dz^2$$

$$ds^2 = (dx^1)^2 + (dx^2)^2 + (dx^3)^2$$

But $$ds^2 = g_{ij} dx^i dx^j; \quad (i, j = 1,2,3)$$

$\Rightarrow g_{11} = g_{22} = g_{33} = 1$ and $g_{12} = g_{23} = g_{13} = g_{21} = g_{31} = g_{32} = 0$

On transformation

$$\bar{g}_{ij} = g_{ij} \frac{\partial x^i}{\partial \bar{x}^i} \frac{\partial x^j}{\partial \bar{x}^j}$$

(since g_{ij} is covariant tensor of rank two) (where $i, j = 1,2,3$).

$$\bar{g}_{ij} = g_{11} \frac{\partial x^1}{\partial \bar{x}^i} \frac{\partial x^1}{\partial \bar{x}^j} + g_{22} \frac{\partial x^2}{\partial \bar{x}^i} \frac{\partial x^2}{\partial \bar{x}^j} + g_{33} \frac{\partial x^3}{\partial \bar{x}^i} \frac{\partial x^3}{\partial \bar{x}^1}$$

since i, j are dummy indices.

Put $i = j = 1$

$$\bar{g}_{11} = g_{11} \left(\frac{\partial x^1}{\partial \bar{x}^1} \right)^2 + g_{22} \left(\frac{\partial x^2}{\partial \bar{x}^1} \right)^2 + g_{33} \left(\frac{\partial x^3}{\partial \bar{x}^1} \right)^2$$

$$\bar{g}_{11} = g_{11} \left(\frac{\partial x}{\partial r} \right)^2 + g_{22} \left(\frac{\partial y}{\partial r} \right)^2 + g_{33} \left(\frac{\partial z}{\partial r} \right)^2$$

Since $x = r \sin\theta \cos\phi, \quad y = r \sin\theta \sin\phi, \quad z = r\cos\theta$

$$\frac{\partial x}{\partial r} = \sin\theta\cos\phi, \quad \frac{\partial y}{\partial r} = \sin\theta\sin\phi, \quad \frac{\partial z}{\partial r} = \cos\theta$$

and $g_{11} = g_{22} = g_{33} = 1$.

So,

$$\bar{g}_{11} = (\sin\theta\cos\phi)^2 + (\sin\theta\sin\phi)^2 + \cos^2\theta$$

$$\bar{g}_{11} = 1$$

put $i = j = 2$

$$\bar{g}_{22} = g_{11} \left(\frac{\partial x^1}{\partial \bar{x}^2} \right)^2 + g_{22} \left(\frac{\partial x^2}{\partial \bar{x}^2} \right)^2 + g_{33} \left(\frac{\partial x^3}{\partial \bar{x}^2} \right)^2$$

$$\bar{g}_{22} = g_{11} \left(\frac{\partial x}{\partial \theta} \right)^2 + g_{22} \left(\frac{\partial y}{\partial \theta} \right)^2 + g_{33} \left(\frac{\partial z}{\partial \theta} \right)^2$$

since $g_{11} = g_{22} = g_{33} = 1$

$$\frac{\partial x}{\partial \theta} = r\cos\theta\cos\phi, \quad \frac{\partial y}{\partial \theta} = r\cos\theta\sin\phi, \quad \frac{\partial z}{\partial \theta} = -r\sin\theta$$

$$\bar{g}_{22} = (r\cos\theta\cos\phi)^2 + (r\cos\theta\sin\phi)^2 + (-r\sin\theta)^2$$

$$\bar{g}_{22} = r^2$$

Put $i = j = 3$

$$\overline{g}_{33} = g_{11}\left(\frac{\partial x^1}{\partial \overline{x}^3}\right)^2 + g_{22}\left(\frac{\partial x^2}{\partial \overline{x}^3}\right)^2 + g_{33}\left(\frac{\partial x^3}{\partial \overline{x}^3}\right)^2$$

$$\overline{g}_{33} = g_{11}\left(\frac{\partial x}{\partial \phi}\right)^2 + g_{22}\left(\frac{\partial y}{\partial \phi}\right)^2 + g_{33}\left(\frac{\partial z}{\partial \phi}\right)^2$$

since $g_{11} = g_{22} = g_{33} = 1$

and
$$\frac{\partial x}{\partial \phi} = -r\sin\theta\sin\phi, \frac{\partial y}{\partial \phi} = r\sin\theta\cos\phi, \frac{\partial z}{\partial \phi} = 0$$

$$\overline{g}_{33} = (-r\sin\theta\sin\phi)^2 + (r\sin\theta\cos\phi)^2 + 0$$

$$\overline{g}_{33} = r^2\sin^2\theta$$

So, we have

$$\overline{g}_{11} = 1, \ \overline{g}_{22} = r^2, \ \overline{g}_{33} = r^2\sin^2\theta$$

and
$$\overline{g}_{12} = \overline{g}_{13} = \overline{g}_{21} = \overline{g}_{23} = \overline{g}_{31} = \overline{g}_{32} = 0$$

(*i*) The Metric in spherical coordinates is

$$ds^2 = \overline{g}_{ij}d\overline{x}^i d\overline{x}^j; \ i, j = 1,2,3$$

$$ds^2 = \overline{g}_{11}(d\overline{x}^1)^2 + \overline{g}_{22}(d\overline{x}^2)^2 + \overline{g}_{33}(d\overline{x}^3)^2$$

$$ds^2 = dr^2 + r^2 d\theta^2 + r^2\sin^2\theta d\phi^2$$

(*ii*) The Metric tensor or first fundamental tensor is

$$\overline{g}_{ij} = \begin{bmatrix} \overline{g}_{11} & \overline{g}_{12} & \overline{g}_{13} \\ \overline{g}_{21} & \overline{g}_{22} & \overline{g}_{23} \\ \overline{g}_{31} & \overline{g}_{32} & \overline{g}_{33} \end{bmatrix} = \begin{bmatrix} 1 & 0 & 0 \\ 0 & r^2 & 0 \\ 0 & 0 & r^2\sin^2\theta \end{bmatrix}$$

and

$$g = |\overline{g}_{ij}| = \begin{vmatrix} 1 & 0 & 0 \\ 0 & r^2 & 0 \\ 0 & 0 & r^2\sin^2\theta \end{vmatrix} = r^4\sin^2\theta$$

(*iii*) The cofactor of g are given by $B_{11} = 1$, $B_{22} = r^2$, $B_{33} = r^2\sin^2\theta$ and $B_{12} = B_{21} =$ $B_{31} = B_{13} = B_{23} = B_{32} = 0$

The second fundamental tensor or conjugate tensor is $g^{ij} = \dfrac{B_{ij}}{g}$.

$$g^{11} = \frac{\text{cofactor of } g_{11} \text{ in } g}{g} = \frac{B_{11}}{g}$$

$$= \frac{r^4 \sin^2\theta}{r^4 \sin^2\theta}$$

$$g^{11} = 1$$

$$g^{22} = \frac{B_{22}}{g} = \frac{r^2 \sin^2\theta}{r^4 \sin^2\theta}$$

$$g^{22} = \frac{1}{r^2}$$

$$g^{33} = \frac{B_{33}}{g} = \frac{r^2}{r^4 \sin^2\theta}$$

$$g^{33} = \frac{1}{r^2 \sin^2\theta}$$

and $g^{12} = g^{13} = g^{21} = g^{31} = g^{32} = 0$

Hence the fundamental tensor in matrix form is

$$g^{ij} = \begin{bmatrix} \overline{g}^{11} & \overline{g}^{12} & \overline{g}^{13} \\ \overline{g}^{21} & \overline{g}^{22} & \overline{g}^{23} \\ \overline{g}^{31} & \overline{g}^{32} & \overline{g}^{33} \end{bmatrix} = \begin{bmatrix} 0 & 0 & 0 \\ 0 & \frac{1}{r^2} & 0 \\ 0 & 0 & \frac{1}{r^2 \sin^2\theta} \end{bmatrix}$$

EXAMPLE 3

If the metric is given by

$$ds^2 = 5(dx^1)^2 + 3(dx^2)^2 + 4(dx^3)^2 - 6dx^1dx^2 + 4dx^2dx^3$$

Evaluate *(i)* g and *(ii)* g^{ij}.

Solution

The metric is $ds^2 = g_{ij}dx^i dx^j$; $(i, j = 1, 2, 3)$

$$ds^2 = g_{11}(dx^1)^2 + g_{12}dx^1dx^2 + g_{13}dx^1dx^3 + g_{21}dx^2dx^1$$
$$+ g_{22}(dx^2)^2 + g_{23}dx^2dx^3 + g_{31}dx^3dx^1 + g_{32}dx^3dx^2 + g_{33}(dx^3)^2$$

Since g_{ij} is symmetric $\Rightarrow g_{ij} = g_{ji}$

i.e., $g_{12} = g_{21},\ g_{23} = g_{32}, g_{13} = g_{31}$

So,
$$ds^2 = g_{11}(dx^1)^2 + g_{22}(dx^2)^2 + g_{33}(dx^3)^2 + 2g_{12}dx^1dx^2 + 2g_{23}dx^2dx^3 + 2g_{13}dx^1dx^3 \quad ...(1)$$

Now, the given metric is
$$ds^2 = 5(dx^1)^2 + 3(dx^2)^2 + 4(dx^3)^2 - 6dx^1dx^2 + 4dx^2dx^3 \quad ...(2)$$

Comparing (1) and (2) we have

$g_{11} = 5,\ g_{22} = 3,\ g_{33} = 4,\ 2g_{12} = -6 \Rightarrow g_{12} = -3 = g_{21}$

$2g_{23} = 4 \Rightarrow g_{23} = 2 = g_{32}, g_{13} = 0 = g_{31}$.

$$g = |g_{ij}| = \begin{vmatrix} g_{11} & g_{12} & g_{13} \\ g_{21} & g_{22} & g_{23} \\ g_{31} & g_{32} & g_{33} \end{vmatrix} = \begin{vmatrix} 5 & -3 & 0 \\ -3 & 3 & 2 \\ 0 & 2 & 4 \end{vmatrix} = 4$$

(*ii*) Let B_{ij} be the cofactor of g_{ij} in g.

Then

$$B_{11} = \text{Cofactor of } g_{11} = \begin{vmatrix} 3 & 2 \\ 2 & 4 \end{vmatrix} = 8$$

$$B_{22} = \text{Cofactor of } g_{22} = \begin{vmatrix} 5 & 0 \\ 0 & 4 \end{vmatrix} = 20$$

$$B_{33} = \text{Cofactor of } g_{33} = \begin{vmatrix} 5 & -3 \\ -3 & 3 \end{vmatrix} = 6$$

$$B_{12} = \text{Cofactor of } g_{12} = -\begin{vmatrix} -3 & 2 \\ 0 & 4 \end{vmatrix} = 12 = B_{21}$$

$$B_{13} = \text{Cofactor of } g_{13} = \begin{vmatrix} -3 & 3 \\ 0 & 2 \end{vmatrix} = -6 = B_{31}$$

$$B_{23} = \text{Cofactor of } g_{23} = -\begin{vmatrix} 5 & -3 \\ 0 & 2 \end{vmatrix} = -10 = B_{32}$$

Since
$$g^{ij} = \frac{B_{ij}}{g}$$

We have

$$g^{11} = \frac{B_{11}}{g} = \frac{8}{4} = 2;\ g^{22} = 5,\ g^{33} = \frac{3}{2},\ g^{12} = g^{21} = 3,\ g^{13} = g^{31} = -\frac{3}{2},\ g^{23} = g^{32} = -\frac{5}{2}$$

Hence,

$$g^{ij} = \begin{bmatrix} 2 & 3 & -\frac{3}{2} \\ 3 & 5 & -\frac{5}{2} \\ -\frac{3}{2} & -\frac{5}{2} & \frac{3}{2} \end{bmatrix}$$

3.3 LENGTH OF A CURVE

Consider a continuous curve in a Riemannian V_n *i.e.*, a curve such that the coordinate x^i of any current point on it are expressible as functions of some parameter, say t.

The equation of such curve can be expressed as

$$x^i = x^i(t)$$

The length ds of the arc between the points whose coordinate s are x^i and $x^i + dx^i$ given by

$$ds^2 = g_{ij}dx^i dx^j$$

If s be arc length of the curve between the points P_1 and P_2 on the curve which correspond to the two values t_1 and t_2 of the parameter t.

$$s = \int_{P_1}^{P_2} ds = \int_{t_1}^{t_2} \left(g_{ij} \frac{dx^i}{dt} \frac{dx^j}{dt} \right)^{1/2} dt$$

NULL CURVE

If $g_{ij} \frac{dx^i}{dt} \frac{dx^j}{dt} = 0$ along a curve. Then $s = 0$. Then the points P_1 and P_2 are at zero distance, despite of the fact that they are not coincident. Such a curve is called minimal curve or null curve.

EXAMPLE 4

A curve is in spherical coordinate x^i is given by

$$x^1 = t, \ x^2 = \sin^{-1}\left(\frac{1}{t}\right) \text{ and } x^3 = 2\sqrt{t^2 - 1}$$

Find length of arc $1 \le t \le 2$.

Solution

In spherical coordinate, the metric is given by

$$ds^2 = (dx^1)^2 + (x^1)^2(dx^2)^2 + (x^1 \sin x^2)^2(dx^3)^2$$

given $$x^1 = t,\ x^2 = \sin^{-1}\frac{1}{t},\ x^3 = 2\sqrt{t^2-1}$$

$$dx^1 = dt\ ,\ dx^2 = \frac{1}{\sqrt{1-\left(\frac{1}{t}\right)^2}}\left(-\frac{1}{t^2}\right)dt,\ dx^3 = 2\cdot\frac{1}{2}\left(t^2-1\right)^{-1/2}2t\,dt$$

$$dx^2 = -\frac{dt}{t\sqrt{t^2-1}},\ dx^3 = \frac{2t}{\sqrt{t^2-1}}dt$$

$$ds^2 = (dt)^2 + t^2\left(-\frac{dt}{t\sqrt{t^2-1}}\right)^2 + \left[t\left(\sin\sin^{-1}\frac{1}{t}\right)\right]^2\left(\frac{2t}{\sqrt{t^2-1}}dt\right)^2$$

$$ds^2 = dt^2 + \frac{dt^2}{t^2-1} + \frac{4t^2}{t^2-1}(dt)^2$$

$$ds^2 = \frac{5t^2}{t^2-1}dt^2$$

$$ds = \sqrt{5}\frac{t}{\sqrt{t^2-1}}dt$$

Now, the length of arc, $1 \le t \le 2$, is

$$\int_{t_1}^{t_2} ds = \sqrt{5}\int_1^2 \frac{t}{\sqrt{t^2-1}}dt = \frac{\sqrt{5}}{2}\left[\frac{\sqrt{t^2-1}}{1/2}\right]_1^2 = \sqrt{15}\ \text{units}$$

3.4 ASSOCIATED TENSOR

A tensor obtained by the process of inner product of any tensor $A^{i_1 i_2 \ldots i_r}_{j_1 j_2 \ldots j_s}$ with either of the fundamental tensor g_{ij} or g^{ij} is called associated tensor of given tensor.

e.g. Consider a tensor A_{ijk} and form the following inner product

$$g^{\alpha i}A_{ijk} = A^{\alpha}_{jk};\ g^{\alpha j}A_{ijk} = A^{\alpha}_{ik};g^{\alpha k}A_{ijk} = A^{\alpha}_{ij}$$

All these tensors are called Associated tensor of A_{ijk}.

Associated Vector

Consider a covariant vector A_i. Then $g^{ik}A_i = A^k$ is called associated vector of A_i. Consider a contravariant vector A^j. Then $g_{jk}A^j = A_k$ is called associated vector of A^j.

3.5 MAGNITUDE OF VECTOR

The magnitude or length A of contravariant vector A^i. Then A is defined by

$$A = \sqrt{g_{ij} A^i A^j}$$

or
$$A^2 = g_{ij} A^i A^j$$

Also, $A^2 = A_j A^j$ as $g_{ij} A^i = A_j$

i.e., square of the magnitude is equal to scalar product of the vector and its associate.

The magnitude or length A of covariant vector A_i. Then A is defined by

$$A = \sqrt{g^{ij} A_i A_j}$$

or
$$A^2 = g^{ij} A_i A_j$$

A vector of magnitude one is called Unit vector. A vector of magnitude zero is called zero vector or Null vector.

3.6 SCALAR PRODUCT OF TWO VECTORS

Let $\vec{A}$ and $\vec{B}$ be two vectors. Their scalar product is written as $\vec{A} \cdot \vec{B}$ and defined by

$$\vec{A} \cdot \vec{B} = A^i B_i$$

Also,
$$\vec{A} \cdot \vec{B} = A^i B_i = g_{ij} A^i B^j \text{ since } B_i = g_{ij} B^j$$

$$\vec{A} \cdot \vec{B} = A_i B^i = g^{ij} A_i B_j \text{ since } B^i = g^{ij} B_j$$

Thus
$$\vec{A} \cdot \vec{A} = A^i A_i = g_{ij} A^i A^j = A^2$$

i.e.,
$$A = |\vec{A}| = \sqrt{g_{ij} A^i A^j}$$

Angle between two vectors

Let $\vec{A}$ and $\vec{B}$ be two vectors. Then

$$\vec{A} \cdot \vec{B} = |\vec{A}||\vec{B}| \cos\theta$$

$\Rightarrow$
$$\cos\theta = \frac{\vec{A} \cdot \vec{B}}{|\vec{A}||\vec{B}|} = \frac{g_{ij} A^i B^j}{\sqrt{g_{ij} A^i A^j}\sqrt{g_{ij} B^i B^j}}$$

since
$$|\vec{A}| = \sqrt{g_{ij} A^i A^j}\ ; \ |\vec{B}| = \sqrt{g_{ij} B^i B^j}$$

This is required formula for $\cos\theta$.

Definition

The inner product of two contravariant vectors $\vec{A}$ (or A^i) and $\vec{B}$ (or B^i) associated with a symmetric tensor g_{ij} is defined as $g_{ij} A^i B^j$. It is denoted by

$$g(\vec{A}, \vec{B}) = g_{ij} A^i B^j$$

THEOREM 3.3 *The necessary and sufficient condition that the two vectors $\vec{A}$ and $\vec{B}$ at 0 be orthogonal if* $g(\vec{A},\vec{B})=0$

Proof: Let θ be angle between the vectors $\vec{A}$ and $\vec{B}$ then

$$\vec{A}\cdot\vec{B} = |\vec{A}||\vec{B}|\cos\theta$$

or

$$\vec{A}\cdot\vec{B} = AB\cos\theta$$

$$g_{ij}A^iB^j = AB\cos\theta$$

$$\Rightarrow \qquad \cos\theta = \frac{g_{ij}A^iB^j}{AB} \qquad ...(1)$$

If $\vec{A}$ and $\vec{B}$ are orthogonal then $\theta=\frac{\pi}{2}\Rightarrow\cos\theta=0$ then from (1)

$$g_{ij}A^iB^j = 0$$

$$\Rightarrow \qquad g(\vec{A},\vec{B}) = 0 \text{ since } g(\vec{A},\vec{B}) = g_{ij}A^iB^j$$

Conversely if $g_{ij}A^iB^j = 0$ then from (1)

$$\cos\theta = 0 \Rightarrow \theta = \frac{\pi}{2}.$$

So, two vectors $\vec{A}$ & $\vec{B}$ are orthogonal. **Proved.**

Note: (*i*) If $\vec{A}$ and $\vec{B}$ be unit vectors. Then $|\vec{A}|=|\vec{B}|=1$. Then

$$\cos\theta = \vec{A}\cdot\vec{B} = g_{ij}A^iB^j$$

(*ii*) Two vectors $\vec{A}$ and $\vec{B}$ will be orthogonal if angle between them is $\frac{\pi}{2}$ *i.e.*, $\theta=\frac{\pi}{2}$ then

$$\cos\theta = \cos\theta = \frac{\pi}{2} = 0$$

3.7 ANGLE BETWEEN TWO VECTORS

THEOREM 3.4 *To show that the definition of the angle between two vectors is consistent with the requirement* $\cos^2\theta \le 1$.

OR

To justify the definition of the angle between two vectors.

OR

To show that the angle between the contravariant vectors is real when the Riemannian Metric is positive definition.

Proof: Let θ be the angle between unit vectors $\vec{A}$ and $\vec{B}$ then

$$\cos\theta = g_{ij}A^iB^j = A_jB^j = A_jB^{ij}B_i = g^{ij}A_jB_i = A^iB_i$$

To show that θ is real *i.e.*, $|\cos\theta| \leq 1$.

Consider the vector $lA^i + mB^i$ when l and m are scalars. The square of the magnitude of

$$lA^i + mB^i = g_{ij}(lA^i + mB^i)(lA^j + mB^j)$$

$$= g_{ij}l^2 A^i A^j + g_{ij} lm A^i B^j + ml g_{ij} B^i A^j + m^2 g_{ij} B^i B^j$$

$$= l^2 + 2lm\cos\theta + m^2$$

Since

$$g_{ij}A^i A^j = A^2 = 1; \; g_{ij}B^i B^j = B^2 = 1.$$

and

$$g_{ij}A^i B^j = \cos\theta; \text{ as } \vec{A} \;\&\; \vec{B} \text{ are unit vector } i.e., \left|\vec{A}\right| = 1 \Rightarrow A^2 = 1.$$

Since square of magnitude of any vector ≥ 0.

So, the square of the magnitude of $lA^i + mB^i \geq 0$.

or $\quad l^2 + 2lm\cos\theta + m^2 \geq 0$

$$(l + m\cos\theta)^2 + m^2 - m^2\cos^2\theta \geq 0$$

$$(l + m\cos\theta)^2 + m^2(1 - \cos^2\theta) \geq 0$$

This inequality holds for the real values of l & m.

if $\quad 1 - \cos^2\theta \geq 0$

$\Rightarrow \quad \cos^2\theta \leq 1$

$$\cos\theta \leq 1$$

Proved.

THEOREM 3.5 *The magnitude of two associated vectors are equal.*

Proof: Let A and B be magnitudes of associate vectors A^i and A_i respectively. Then

$$A^2 = g_{ij}A^i A^j \qquad ...(1)$$

and

$$B^2 = g^{ij}A_i A_j \qquad ...(2)$$

From equation (1)

$$A^2 = (g_{ij}A^i)A^j$$

$$A^2 = A_j A^j \qquad ...(3)$$

since $g_{ij}A^i = A^j$ (Associate vector)

From equation (2)

$$B^2 = (g^{ij}A_i)A_j$$

$$B^2 = A^j A_j \qquad ...(4)$$

since $g^{ij}A_i = A^j$

from (3) and (4)

$$A^2 = B^2$$

$\Rightarrow$ $$A = B$$

So, magnitude of A_i and A^i are equal.

3.8 ANGLE BETWEEN TWO COORDINATE CURVES

Let a V_n referred to coordinate $x^i, (i = 1, 2, \ldots n)$. For a coordinate curve of parameter x^l, the coordinate x^l alone varies. Thus the coordinate curve of parameter x^l is defined as

$$x^i = c^i, \ \forall i \text{ except } i = l \qquad \ldots(1)$$

where C^i's are constants.

Differentiating it, we get

$$dx^i = 0, \ \forall i, \text{ except } i = l \text{ and } dx^l \neq 0$$

Let A^i and B^i be the tangent vectors to a coordinate curve of parameters x^p and x^q respectively. Then

$$A^i = dx^i = (0, \ldots 0, x^p, 0 \ldots 0) \qquad \ldots(2)$$

$$B^i = dx^i = (0, \ldots 0, x^q, 0 \ldots 0) \qquad \ldots(3)$$

If θ is required angle then

$$\cos\theta = \frac{g_{ij}A^iB^j}{\sqrt{g_{ij}A^iA^j}\sqrt{g_{ij}B^iB^j}}$$

$$= \frac{g_{pq}A^pB^q}{\sqrt{g_{pp}A^pA^p}\sqrt{g_{qq}B^qB^q}}$$

$$= \frac{g_{pq}A^pB^q}{\sqrt{g_{pp}g_{qq}}\,A^pB^q}$$

$$\cos\theta = \frac{g_{pq}}{\sqrt{g_{pp}g_{qq}}} \qquad \ldots(4)$$

which is required formula for θ.

The angle w_{ij} between the coordinate curves of parameters x^i and x^j is given by

$$\cos w_{ij} = \frac{g_{ij}}{\sqrt{g_{ii}g_{jj}}}$$

If these curves are orthogonal then

$$\cos w_{ij} = \cos\frac{\pi}{2} = 0$$

$$\Rightarrow \qquad g_{ij} = 0$$

Hence the x^i coordinate curve and x^j coordinate curve are orthogonal if $g_{ij} = 0$.

3.9 HYPERSURFACE

The n equations $x^i = x^i(u^1)$ represent a subspace of V_n. If we eliminate the parameter u^1, we get $(n-1)$ equations in x^j's which represent one dimensional curve.

Similarly the n equations $x^i = x^i(u^1, u^2)$ represent two dimensional subspace of V_n. If we eliminating the parameters u^1, u^2, we get $n-2$ equations in x^i's which represent two dimensional curve V_n. This two dimensional curve define a subspace, denoted by V_2 of V_n.

Then n equations $x^i = x^i(u^1, u^2, \ldots u^{n-1})$ represent $n-1$ dimensional subspace V_{n-1} of V_n. If we eliminating the parameters u^1, u^2, ...u^{n-1}, we get only one equation in x^i's which represent $n-1$ dimensional curve in V_n. This particular curve is called *hypersurface* of V_n.

Let ϕ be a scalar function of coordinates x^i. Then $\phi(x^i)$ = constant determines a family of hypersurface of V_n.

3.10 ANGLE BETWEEN TWO COORDINATE HYPERSURFACE

Let

$$\phi(x^i) = \text{constant} \qquad \text{...(1)}$$

and

$$\psi(x^i) = \text{constant} \qquad \text{...(2)}$$

represents two families of hypersurfaces.

Differentiating equation (1), we get

$$\frac{\partial \phi}{\partial x^i} dx^i = 0 \qquad \text{...(3)}$$

This shows that $\frac{\partial \phi}{\partial x^i}$ is orthogonal to dx^i. Hence $\frac{\partial \phi}{\partial x^i}$ is normal to ϕ = constant, since dx^i is tangential to hypersurface (1).

Similarly $\frac{\partial \psi}{\partial x^i}$ is normal to the hypersurface (2). If ω is the angle between the hypersurface (1) and (2) then ω is also defined as the angle between their respective normals. Hence required angle ω is given by

$$\cos\omega = \frac{g^{ij} \frac{\partial \phi}{\partial x^i} \frac{\partial \psi}{\partial x^j}}{\sqrt{g^{ij} \frac{\partial \phi}{\partial x^i} \frac{\partial \phi}{\partial x^j}} \sqrt{g^{ij} \frac{\partial \psi}{\partial x^i} \frac{\partial \psi}{\partial x^j}}} \qquad \text{...(4)}$$

If we take

$$\phi = x^p = \text{constant} \qquad ...(5)$$

and

$$\psi = x^q = \text{constant} \qquad ...(6)$$

The angle ω between (5) and (6) is given by

$$\cos\omega = \frac{g^{ij}\dfrac{\partial x^p}{\partial x^i}\dfrac{\partial x^q}{\partial x^j}}{\sqrt{g^{ij}\dfrac{\partial x^p}{\partial x^i}\dfrac{\partial x^p}{\partial x^j}}\sqrt{g^{ij}\dfrac{\partial x^q}{\partial x^i}\dfrac{\partial x^q}{\partial x^j}}}$$

$$= \frac{g^{ij}\delta_i^p\delta_j^q}{\sqrt{g^{ij}\delta_i^p\delta_j^q}\sqrt{g^{ij}\delta_i^q\delta_j^q}}$$

$$\cos\omega = \frac{g^{pq}}{\sqrt{g^{pp}g^{qq}}} \qquad ...(7)$$

The angle ω_{ij} between the coordinate hypersurfaces of parameters x^i and x^j is given by

$$\cos\omega_{ij} = \frac{g^{ij}}{\sqrt{g^{ii}g^{jj}}} \qquad ...(8)$$

If the coordinate hypersurfaces of parameters x^i and x^j are orthogonal then

$$\omega_{ij} = \frac{\pi}{2}$$

$$\Rightarrow \qquad \cos\omega_{ij} = 0$$

from (8), we have $g^{ij} = 0$.

Hence the coordinate hypersurfaces of parameters x^i and x^j are orthogonal if $g^{ij} = 0$.

3.11 *n*-PLY ORTHOGONAL SYSTEM OF HYPERSURFACES

If in a V_n there are n families of hypersurfaces such that, at every point, each hypersurface is orthogonal to the $n-1$ hypersurface of the other families which pass through that point, they are said to form as n-ply orthogonal system of hypersurfaces.

3.12 CONGRUENCE OF CURVES

A family of curves one of which passes through each point of V_n is called a congruence of curves.

3.13 ORTHOGONAL ENNUPLE

An orthogonal ennuple in a Riemannian V_n consists of n mutually orthogonal congruence of curves.

THEOREM 3.6 *To find the fundamental tensors g_{ij} and g^{ij} in terms of the components of the unit tangent $e_{h|}$ $(h = 1, 2,...n)$ to an orthogonal ennuple.*

Proof: Consider n unit tangents $e^i_{h|}$ $(h = 1, 2,...n)$ to conguence $e_{h|}$ $(h = 1, 2,...n)$ of an orthogonal ennuple in a Riemannian V_n. The subscript h followed by an upright bar simply distinguishes one congruence from other. It does not denote tensor suffix.

The contravariant and covariant components of $e_{h|}$ are denoted by $e_{h|}$ and $e_{h|i}$ respectively.

Suppose any two congruences of orthogonal ennuple are $e_{h|}$ and $e_{k|}$ so that

$$g_{ij} e^i_{h|} e^j_{k|} = \delta^h_k \qquad ...(1)$$

$$e^i_{h|} e_{k|i} = \delta^h_k$$

from (1),

$$g_{ij}\ e^i_{h|}\ e^j_{k|} = 0$$

and

$$g_{ij} e^i_{h|}\ e^j_{h|} = 1$$

We define

$$e^i_{h|} = \frac{\text{cofactor of } e_{h|i} \text{ in determinant } |e_{h|i}|}{|e_{h|i}|}$$

Also, from the determinant property, we get

$$\sum_{h=1}^{n} e^i_{h|} e_{h|j} = \delta^i_j \qquad ...(2)$$

Multiplying by e^{jk}

$$\sum_{h=1}^{n} e^i_{h|} e_{h|j}\ g^{jk} = \delta^i_j g^{jk}$$

or

$$\sum_{h=1}^{n} e^i_{h|} e^k_{h|} = g^{ik} \qquad ...(3)$$

Again multiplying (2) by g_{ik}.

$$\sum_{h=1}^{n} e^i_{h|} e_{h|j} g_{ik} = \delta^i_j g_{ik}$$

or

$$g_{jk} = \sum e_{h|k} e_{h|j} \qquad ...(4)$$

from (3) and (4)

$$g_{ij} = \sum_{h=1}^{n} e_{h|i}\ e_{h|j} \qquad ...(5)$$

$$g^{ij} = \sum_{h=1}^{n} e^i_{h|} e^j_{h|} \quad \text{...(6)}$$

This is the required results.

Corollary: To find the magnitude of any vector u is zero if the projections of u on $e_{h|}$ are all zero.

Proof: Let

$$u^i = \sum_{h=1}^{n} C_h e^i_{h|} \quad \text{...(7)}$$

Then

$$u^i e_{k|i} = \sum_{h=1}^{n} C_h e^i_{h|} e_{k|i} = \sum_{h=1}^{n} C_h \delta^h_k = C_k$$

or $$C_k = u^i e_{k|i} \quad \text{...(8)}$$

i.e., $$C_k = \text{projection of } u^i \text{ on } e_{k|i}$$

Using (8), equation (7) becomes

$$u^i = \sum_{h=1}^{n} u^j e_{h|j} e^i_{k|}$$

Now,

$$u^2 = u^i u_i = \left(\sum_h C_h e^i_{h|}\right)\left(\sum_k C_k e_{k|i}\right) \text{ from (7)}$$

$$= \sum_{h,k} C_h C_k e^i_{h|} e_{k|i}$$

$$= \sum_{h,k} C_h C_k \delta^h_k$$

$$= \sum_h C_h C_h$$

$$u^2 = \sum_{h=1}^{n} (C_h)^2$$

This implies that $u = 0$ iff $u^2 = 0$ iff $C_h = 0$.

Hence the magnitude of a vector u is zero iff all the projections of u (i.e. of u^i) on n mutually orthogonal directions $e^i_{h|}$ are zero.

Miscellaneous Examples

1. If p and q are orthogonal unit vectors, show that

$$(g_{hj}g_{ik} - g_{hk}g_{ij})p^h q^i p^j q^k = 1$$

Solution

Since p and q are orthogonal unit vectors. Then

$$g_{ij}p_i q^j = 0, \quad p^2 = q^2 = 1.$$

Now,

$$(g_{hj}g_{ik} - g_{hk}g_{ij})p^h q^i p^j q^k = g_{hj}g_{ik}p^h p^j q^i q^k - g_{hk}g_{ij}p^h q^k q^i p^j$$

$$= (g_{hi}p^h p^j)(g_{ik}q^i q^k) - (g_{hk}p^j q^k)(g_{ij}q^i p^j)$$

$$= p^2.q^2 - 0.0$$

$$= 1 \cdot 1$$

$$= 1 \text{ (since } g_{hi}p^h p^j = 1 \,\&\, g_{hk}p^h q^k = 0)$$

2. If θ is the inclination of two vectors A and B show that

$$\sin^2\theta = \frac{(g_{hi}g_{ik} - g_{hk}g_{ij})A^h A^j B^i B^k}{g_{hj}g_{ik}A^h A^j B^j B^k}$$

Solution

If θ be the angle between the vectors A and B then

$$\cos\theta = \frac{g_{ij}A^j B^i}{\sqrt{g_{ij}A^i A^j}\sqrt{g_{ik}B^i B^k}}$$

But $\sin^2\theta = 1 - \cos^2\theta$

$$\sin^2\theta = 1 - \frac{(g_{ij}B^i A^j)(g_{hk}A^h B^k)}{(g_{hj}A^h A^j)(g_{ik}B^i B^k)}$$

$$= \frac{(g_{hj}g_{ik} - g_{hk}g_{ij})A^h A^j B^i B^k}{g_{hj}g_{ik}A^h A^j B^i B^k}$$

3. If X_{ij} are components of a symmetric covariant tensor and u, v are unit orthogonal to w and satisfying the relations

$$(X_{ij} - \alpha g_{ij})u^i + \gamma w_j = 0$$

$$(X_{ij} - \beta g_{ij})v^i + \delta w_j = 0$$

where $\alpha \neq \beta$ prove that u and v are orthogonal and that

$$X_{ij}u^i v^j = 0$$

Solution

Suppose X_{ij} is a symmetric tensor. Since u^i, v^j are orthogonal to w^i then

$$u^i w_i = 0 \quad ...(1)$$

$$v^i w_i = 0 \quad ...(2)$$

given $\quad (X_{ij} - \alpha g_{ij})u^i + \gamma w_j = 0 \quad ...(3)$

$$(X_{ij} - \beta g_{ij})v^i + \delta w_j = 0 \quad ...(4)$$

where $\alpha \neq \beta$.

Multiply (3) & (4) by v^j, u^j respectively and using (1) and (2), we have

$$(X_{ij} - \alpha g_{ij})u^i v^j = 0 \quad ...(5)$$

$$(X_{ij} - \beta g_{ij})v^i u^j = 0 \quad ...(6)$$

Interchanging the suffixes i & j in the equation (6) and since g_{ij}, X_{ij} are symmetric, we get

$$(X_{ij} - \alpha g_{ij})u^i v^j = 0 \quad ...(7)$$

Subtract (6) & (7) we get

$$(\beta - \alpha)g_{ij}u^i v^j = 0$$

Since $\beta \neq \alpha$ and $\beta - \alpha \neq 0$.

Hence,

$$g_{ij}u^i v^j = 0 \quad ...(8)$$

So, u and v are orthogonal.

Using (8) in equation (5) & (6), we get

$$X_{ij}u^i v^i = 0$$ **Proved.**

4. Prove the invariance of the expression $\sqrt{g}dx^1 dx^2 ... dx^n$ for the element volume.

Solution

Since g_{ij} is a symmetric tensor of rank two. Then

$$\bar{g}_{ij} = \frac{\partial x^k}{\partial \bar{x}^i} \frac{\partial x^l}{\partial \bar{x}^j} g_{kl}$$

Taking determinant of both sides

$$|\bar{g}_{ij}| = \left|\frac{\partial x^k}{\partial \bar{x}^i}\right| \left|\frac{\partial x^l}{\partial \bar{x}^j}\right| |g_{kl}|$$

Since $\left|\frac{\partial x}{\partial \bar{x}}\right| = J$ (Jacobian)

$$|g_{kl}| = g \quad \& \quad |\bar{g}_{ij}| = \bar{g}$$

So,

$$\bar{g} = gJ^2$$

or

$$J = \sqrt{\frac{\bar{g}}{g}}$$

Now, the transformation of coordinates from x^l to $\bar{x}^i$, we get

$$dx^1 dx^2 \dots dx^n = \left|\frac{\partial x}{\partial \bar{x}}\right| d\bar{x}^1 d\bar{x}^2 \dots d\bar{x}^n$$

$$= J d\bar{x}^1 d\bar{x}^2 \dots d\bar{x}^n$$

$$dx^1 dx^2 \dots dx^n = \sqrt{\frac{\bar{g}}{g}} d\bar{x}^1 d\bar{x}^2 \dots d\bar{x}^n$$

$$\sqrt{g}\, dx^1 dx^2 \dots dx^n = \sqrt{\bar{g}}\, d\bar{x}^1 d\bar{x}^2 \dots d\bar{x}^n$$

So, the volume element $dv = \sqrt{g}\, dx^1 dx^2 \dots dx^n$ is invariant.

EXERCISES

1. For the Metric tensor g_{ij} defined g^{kl} and prove that it is a contravariant tensor.
2. Calculate the quantities g^{ij} for a V_3 whose fundamental form in coordinates u, v, w, is

$$adu^2 + bdv^2 + cdw^2 + 2fdvdw + 2gdwdu + 2hdudv$$

3. Show that for an orthogonal coordinate system

$$g_{11} = \frac{1}{g_{11}},\ g_{22} = \frac{1}{g_{22}},\ g_{33} = \frac{1}{g_{33}}$$

4. For a V_2 in which $g_{11} = E, g_{12} = F, g_{21} = G$ prove that

$$g = EG - F^2,\ g^{11} = G/g,\ g^{12} = -F/g,\ g^{22} = E/g$$

5. Prove that the number of independent components of the metric g_{ij} cannot exceed $\frac{1}{2}n(n+1)$.
6. If vectors u^i, v^i are defined by $u^i = g^{ij}u_j$, $v^i = g^{ij}v_j$ show that $u_i = g^{ij}u^j$, $u^i v_i = u_i v^i$ and $u^i g_{ij} u^j = u_i g^{ij} u_j$
7. Define magnitude of a unit vector. prove that the relation of a vector and its associate vector is reciprocal.
8. If θ is the angle between the two vectors A^i and B^i at a point, prove that

$$\sin^2\theta = \frac{(g_{hi}g_{ik} - g_{hk}g_{ij})A^h A^i B^j B^k}{g_{hi}g_{jk}A^h A^i B^j B^k}$$

9. Show that the angle between two contravariant vectors is real when the Riemannian metric is positive definite.

CHAPTER – 4

CHRISTOFFEL'S SYMBOLS AND COVARIANT DIFFERENTIATION

4.1 CHRISTOFFEL'S SYMBOLS

The German Mathematician Elwin Bruno Christoffel defined symbols

$$[ij,k] = \frac{1}{2}\left(\frac{\partial g_{ik}}{\partial x^j} + \frac{\partial g_{jk}}{\partial x^i} - \frac{\partial g_{ij}}{\partial x^k}\right), \quad (i, j, k = 1,2,...n) \qquad ...(1)$$

called Christoffel 3-index symbols of the first kind.

and $$\begin{Bmatrix} k \\ i\ j \end{Bmatrix} = g^{kl}[ij,l] \qquad ...(2)$$

called *Christoffel 3-index symbols* of second kind, where g_{ij} are the components of the *metric Tensor or fundamental Tensor.*

There are n distinct Christoffel symbols of each kind for each independent g_{ij}. Since g_{ij} is symmetric tensor of rank two and has $\frac{1}{2}n(n+1)$ independent components. So, the number of independent components of Christoffel's symbols are $n \cdot \frac{1}{2}n(n+1) = \frac{1}{2}n^2(n+1)$.

THEOREM 4.1 *The Christoffel's symbols* $[ij,k]$ *and* $\begin{Bmatrix} k \\ i\ j \end{Bmatrix}$ *are symmetric with respect to the indices i and j.*

Proof: By Christoffel's symbols of first kind

$$[ij,k] = \frac{1}{2}\left(\frac{\partial g_{ik}}{\partial x^j} + \frac{\partial g_{jk}}{\partial x^i} \partial - \frac{\partial g_{ij}}{\partial x^k}\right)$$

Interchanging i and j, we get

$$[ji,k] = \frac{1}{2}\left(\frac{\partial g_{jk}}{\partial x^i} + \frac{\partial g_{ik}}{\partial x^j} - \frac{\partial g_{ji}}{\partial x^k}\right)$$

$$= \frac{1}{2}\left(\frac{\partial g_{ik}}{\partial x^j} + \frac{\partial g_{jk}}{\partial x^i} - \frac{\partial g_{ij}}{\partial x^k}\right) \text{ since } g_{ij} = g_{ji}$$

$$[ji,k] = [ij,k]$$

Also, by Christoffel symbol of second kind

$$\begin{Bmatrix} k \\ i\,j \end{Bmatrix} = g^{kl}[ij,l]$$

$$= g^{kl}[ji,l] \text{ since } [ij,l] = [ji,l]$$

$$\begin{Bmatrix} k \\ i\,j \end{Bmatrix} = \begin{Bmatrix} k \\ j\,i \end{Bmatrix}$$

Proved.

THEOREM 4.2 *To prove that*

(*i*) $$[ij,m] = g_{km}\begin{Bmatrix} k \\ i\,j \end{Bmatrix}$$

(*ii*) $$[ik,j] + [jk,i] = \frac{\partial g_{ij}}{\partial x^k}$$

(*iii*) $$\frac{\partial g_{ij}}{\partial x^k} = -g^{jl}\begin{Bmatrix} i \\ l\;k \end{Bmatrix} - g^{im}\begin{Bmatrix} j \\ m\;k \end{Bmatrix}$$

Proof: (*i*) By Christoffel's symbol of second kind

$$\begin{Bmatrix} k \\ i\,j \end{Bmatrix} = g^{kl}[ij,l]$$

Multiplying this equation by g_{km}, we get

$$g_{km}\begin{Bmatrix} k \\ i\,j \end{Bmatrix} = g_{km}g^{kl}[ij,l]$$

$$= \delta^l_m[ij,l] \text{ as } g_{km}g^{kl} = \delta^l_m$$

$$g_{km}\begin{Bmatrix} k \\ i\,j \end{Bmatrix} = [ij,m]$$

(*ii*) By Christoffel's symbol of first kind

$$[ik,j] = \frac{1}{2}\left(\frac{\partial g_{kj}}{\partial x^i} + \frac{\partial g_{ij}}{\partial x^k} - \frac{\partial g_{ik}}{\partial x^j}\right) \quad ...(1)$$

and

$$[jk,i] = \frac{1}{2}\left(\frac{\partial g_{ki}}{\partial x^j} + \frac{\partial g_{ji}}{\partial x^k} - \frac{\partial g_{jk}}{\partial x^i}\right) \quad ...(2)$$

adding (1) and (2),

$$[ik,j] + [jk,i] = \frac{1}{2}\left(\frac{\partial g_{ij}}{\partial x^k} + \frac{\partial g_{ji}}{\partial x^k}\right) \text{ since } g_{ij} = g_{ji}$$

$$[ik, j] + [jk, i] = \frac{1}{2} \cdot 2 \frac{\partial g_{ij}}{\partial x^k} = \frac{\partial g_{ij}}{\partial x^k}$$

(iii) Since $g^{ij} g_{lj} = \delta_l^i$.

Differentiating it w.r.t. to x^k, we get

$$g^{ij} \frac{\partial g_{lj}}{\partial x^k} + g_{lj} \frac{\partial g^{ij}}{\partial x^k} = 0$$

Multiplying this equation by g^{lm}, we get

$$g^{ij} g^{lm} \frac{\partial g_{lj}}{\partial x^k} + g^{lm} g_{lj} \frac{\partial g^{ij}}{\partial x^k} = 0$$

$$g^{lm} g_{lj} \frac{\partial g^{ij}}{\partial x^k} = -g^{ij} g^{lm} \frac{\partial g_{lj}}{\partial x^k}$$

$$\delta_j^m \frac{\partial g^{ij}}{\partial x^k} = -g^{ij} g^{lm} \{ [lk, j] + [jk, l] \} \text{ since } \frac{\partial g_{lj}}{\partial x^k} = [lk, j] + [jk, l].$$

$$\frac{\partial g^{im}}{\partial x^k} = -g^{lm} \{ g^{ij} [lk, j] \} - g^{ij} \{ g^{lm} [jk, l] \}$$

$$\frac{\partial g^{im}}{\partial x^k} = -g^{lm} \begin{Bmatrix} i \\ l \ k \end{Bmatrix} - g^{ij} \begin{Bmatrix} m \\ j \ k \end{Bmatrix}$$

Interchanging m and j, we get

$$\frac{\partial g^{ij}}{\partial x^k} = -g^{lj} \begin{Bmatrix} i \\ l \ k \end{Bmatrix} - g^{im} \begin{Bmatrix} j \\ m \ k \end{Bmatrix}$$

or

$$\frac{\partial g^{ij}}{\partial x^k} = -g^{lj} \begin{Bmatrix} i \\ l \ k \end{Bmatrix} - g^{im} \begin{Bmatrix} j \\ m \ k \end{Bmatrix} \text{ as } g^{lj} = g^{jl}$$

Proved.

THEOREM 4.3 *To show that* $\begin{Bmatrix} i \\ i \ j \end{Bmatrix} = \frac{\partial \log(\sqrt{g})}{\partial x^j}$.

Proof: The matrix form of g_{ik} is

$$g_{ik} = \begin{bmatrix} g_{11} & g_{12} & \cdots & g_{1n} \\ g_{21} & g_{22} & \cdots & G_{2n} \\ \vdots & & & \\ g_{n1} & g_{n2} & \cdots & g_{nn} \end{bmatrix}$$

and
$$g = |g_{ik}| = \begin{vmatrix} g_{11} & g_{12} & \cdots & g_{1n} \\ g_{21} & g_{22} & \cdots & g_{2n} \\ \vdots & & & \\ g_{n1} & g_{n2} & \cdots & gnn \end{vmatrix}$$

But
$$g_{ik}g^{il} = \delta_k^l$$

Take
$$l = k.$$

$$g_{ik}g^{ik} = \delta_k^k = 1$$

$$\Rightarrow \quad g^{ik} = [g_{ik}]^{-1} = \frac{G_{ik}}{g} \quad \text{(Theorem 2.13 , Pg 25)}$$

where G_{ik} is cofactor of g_{ik} in the determinant $|g_{ik}|$

$$\Rightarrow \quad g = g_{ik}G_{ik} \qquad ...(1)$$

Differentiating w.r.t. g_{ik} partially

$$\frac{\partial g}{\partial g_{ik}} = G_{ik} \text{ since } \frac{\partial g_{ik}}{\partial g_{ik}} = 1$$

Now,

$$\frac{\partial g}{\partial x^j} = \frac{\partial g}{\partial g_{ik}} \frac{\partial g_{ik}}{\partial x^j}$$

$$= G_{ik} \frac{\partial g_{ik}}{\partial x^j}$$

But $G_{ik} = gg^{ik}$

$$\frac{\partial g}{\partial x^j} = gg^{ik} \frac{\partial g_{ik}}{\partial x^j}$$

$$\frac{1}{g}\frac{\partial g}{\partial x^j} = g^{ik} \frac{\partial g_{ik}}{\partial x^j}$$

$$\frac{1}{g}\frac{\partial g}{\partial x^j} = g^{ik}\{[jk,i] + [ij,k]\} \text{ as } \frac{\partial g_{ik}}{\partial x^j} = [jk,i] + [ij,k]$$

$$= g^{ik}[jk,i] + g^{ik}[ij,k]$$

$$= \begin{Bmatrix} k \\ j\ k \end{Bmatrix} + \begin{Bmatrix} i \\ i\ j \end{Bmatrix}$$

$$\frac{1}{g}\frac{\partial g}{\partial x^j} = \left\{ \begin{matrix} i \\ j \; i \end{matrix} \right\} + \left\{ \begin{matrix} i \\ j \; i \end{matrix} \right\} \text{ as } k \text{ is dummy indices}$$

$$\frac{1}{g}\frac{\partial g}{\partial x^j} = 2\left\{ \begin{matrix} i \\ j \; i \end{matrix} \right\}$$

$$\frac{1}{2g}\frac{\partial g}{\partial x^j} = \left\{ \begin{matrix} i \\ j \; i \end{matrix} \right\}$$

$$\frac{\partial \log(\sqrt{g})}{\partial x^j} = \left\{ \begin{matrix} i \\ j \; i \end{matrix} \right\}$$

Proved.

EXAMPLE 1

If $|g_{ij}| \neq 0$ show that

$$g_{\alpha\beta}\frac{\partial}{\partial x^j}\left\{ \begin{matrix} \beta \\ i \; k \end{matrix} \right\} = \frac{\partial}{\partial x^j}[ik,\alpha] - \left\{ \begin{matrix} \beta \\ i \; k \end{matrix} \right\}([\beta j,\alpha] + [\alpha j,\beta])$$

Solution

By Christoffel's symbol of second kind

$$\left\{ \begin{matrix} \beta \\ i \; k \end{matrix} \right\} = g^{\beta\alpha}[ik,\alpha]$$

Multiplying it by $g_{\alpha\beta}$, we get

$$g_{\alpha\beta}\left\{ \begin{matrix} \beta \\ i \; k \end{matrix} \right\} = g_{\alpha\beta}g^{\beta\alpha}[ik,\alpha]$$

$$g_{\alpha\beta}\left\{ \begin{matrix} \beta \\ i \; k \end{matrix} \right\} = [ik,\alpha] \text{ as } g_{\alpha\beta}g^{\beta\alpha} = 1$$

Differentiating it w.r.t. to x^j partially

$$\frac{\partial}{\partial x^j}\left[g_{\alpha\beta}\left\{ \begin{matrix} \beta \\ i \; k \end{matrix} \right\}\right] = \frac{\partial}{\partial x^j}[ik,\alpha]$$

$$g_{\alpha\beta}\frac{\partial}{\partial x^j}\left\{ \begin{matrix} \beta \\ i \; k \end{matrix} \right\} + \left\{ \begin{matrix} \beta \\ i \; k \end{matrix} \right\}\frac{\partial g_{\alpha\beta}}{\partial x^j} = \frac{\partial}{\partial x^j}[ik,\alpha]$$

since $$\frac{\partial g_{\alpha\beta}}{\partial x^j} = [\alpha j,\beta] + [\beta j,\alpha]$$

$$g_{\alpha\beta}\frac{\partial}{\partial x^j}\left\{ \begin{matrix} \beta \\ i \; k \end{matrix} \right\} + \left\{ \begin{matrix} \beta \\ i \; k \end{matrix} \right\}([\alpha j,\beta] + [\beta j,\alpha]) = \frac{\partial}{\partial x^j}[ik,\alpha]$$

$$g_{\alpha\beta}\frac{\partial}{\partial x^j}\begin{Bmatrix}\beta\\ i\ k\end{Bmatrix}=\frac{\partial}{\partial x^j}[ik,\alpha]-\begin{Bmatrix}\beta\\ i\ k\end{Bmatrix}([\alpha i,\beta]+[\beta j,\alpha])$$ **Solved.**

EXAMPLE 2

Show that if $g_{ij}=0$ for $i\neq j$ then (*i*) $\begin{Bmatrix}k\\ i\ j\end{Bmatrix}=0$ whenever i, j and k are distinct.
(*ii*) $\begin{Bmatrix}i\\ i\ i\end{Bmatrix}=\frac{1}{2}\frac{\partial\log g_{ii}}{\partial x^i}$ (*iii*) $\begin{Bmatrix}i\\ i\ j\end{Bmatrix}=\frac{1}{2}\frac{\partial\log g_{ii}}{\partial x^j}$ (*iv*) $\begin{Bmatrix}i\\ j\ j\end{Bmatrix}=-\frac{1}{2g_{ii}}\frac{\partial g_{jj}}{\partial x^i}$

Solution

The Christoffel's symbols of first kind

$$[ij,k]=\frac{1}{2}\left(\frac{\partial g_{jk}}{\partial x^i}+\frac{\partial g_{ik}}{\partial x^j}-\frac{\partial g_{ij}}{\partial x^k}\right) \quad ...(1)$$

(*a*) If $i=j=k$

The equation (1) becomes

$$[ii,i]=\frac{1}{2}\frac{\partial g_{ii}}{\partial x^i}$$

(*b*) If $i=j\neq k$

The equation (1) becomes

$$[ii,k]=\frac{1}{2}\left[\frac{\partial g_{ik}}{\partial x^i}+\frac{\partial g_{ik}}{\partial x^i}-\frac{\partial g_{ii}}{\partial x^k}\right]$$

Since $g_{ik}=0$ as $i\neq k$ (given)

$$[ii,k]=-\frac{1}{2}\frac{\partial g_{ii}}{\partial x^k} \text{ or } [jj,i]=-\frac{1}{2}\frac{\partial g_{jj}}{\partial x^i}$$

(*c*) $i=k\neq j$

$$[ij,i]=\frac{1}{2}\left[\frac{\partial g_{ji}}{\partial x^i}+\frac{\partial g_{ii}}{\partial x^j}-\frac{\partial g_{ij}}{\partial x^i}\right]$$

$$=\frac{1}{2}\frac{\partial g_{ii}}{\partial x^j},\ \text{as } g_{ij}=0,\ i\neq j$$

(*d*) $i\neq j\neq k$

$$[ij,k]=0 \text{ as } g_{ij}=0,\ g_{jk}=0,\ i\neq j\neq k$$

(*i*) as i, j, k are distinct *i.e.*, $i\neq j\neq k$

$$\begin{Bmatrix}k\\ i\ j\end{Bmatrix}=g^{kl}[ij,l] \text{ since } g^{kl}=0,\ k\neq l$$

$$\begin{Bmatrix} k \\ i\ j \end{Bmatrix} = 0$$

(ii)

$$\begin{Bmatrix} i \\ i\,i \end{Bmatrix} = g^{ii}[ii,i]$$

$$= g^{ii} \cdot \frac{1}{2}\frac{\partial g_{ii}}{\partial x^i} \text{ from } (a)$$

$$= \frac{1}{2g_{ii}}\frac{\partial g_{ii}}{\partial x^i} \text{ as } g^{ii} = \frac{1}{g_{ii}}$$

$$\begin{Bmatrix} i \\ i\,i \end{Bmatrix} = \frac{1}{2}\frac{\partial \log g_{ii}}{\partial x^i}$$

(iii) $i = k \neq j$

$$\begin{Bmatrix} i \\ i\ j \end{Bmatrix} = g^{ii}[ij,i]$$

$$= \frac{1}{g_{ii}}[ij,i] \text{ as } g^{ii} = \frac{1}{g_{ii}}$$

$$\begin{Bmatrix} i \\ i\ j \end{Bmatrix} = \frac{1}{2}\frac{\partial \log g_{ii}}{\partial x^j}$$

(iv) $j = k \neq i$

$$\begin{Bmatrix} i \\ j\ j \end{Bmatrix} = g^{ii}[jj,i]$$

$$= -\frac{1}{g_{ii}}\frac{1}{2}\frac{\partial g_{jj}}{\partial x^i} \text{ from } (b)$$

$$\begin{Bmatrix} i \\ j\,j \end{Bmatrix} = \frac{-1}{2g_{ii}}\frac{\partial g_{jj}}{\partial x^i}$$

Solved.

EXAMPLE 3

If $ds^2 = dr^2 + r^2 d\theta^2 + r^2 \sin^2\theta d\phi^2$, find the values of

(*i*) [22, 1] and [13, 3], (*ii*) $\begin{Bmatrix} 1 \\ 2\ 2 \end{Bmatrix}$ and $\begin{Bmatrix} 3 \\ 1\ 3 \end{Bmatrix}$

Solution

The given metric is metric in spherical coordinates, $x^1 = r,\ x^2 = \theta,\ x^3 = \phi$.
Clearly,

$$g_{11} = 1,\ g_{22} = r^2,\ g_{33} = r^2 \sin^2\theta \text{ and } g_{ij} = 0 \text{ for } i \neq j$$

Also, $$g^{11} = 1,\ g^{22} = \frac{1}{r^2},\ g^{33} = \frac{1}{r^2 \sin^2\theta}$$

(See Ex. 2, Pg. 39, and $g^{ij} = 0$, for $i \neq j$.)

(*i*) Christoffel Symbols of first kind are given by

$$[ij,k] = \frac{1}{2}\left[\frac{\partial g_{jk}}{\partial x^i} + \frac{\partial g_{ik}}{\partial x^j} - \frac{\partial g_{ij}}{\partial x^k}\right],\quad i,j,k = 1,2,3 \qquad ...(1)$$

Taking $i = j = 2$ and $k = 1$ in (1)

$$[22,1] = \frac{1}{2}\left[\frac{\partial g_{21}}{\partial x^2} + \frac{\partial g_{21}}{\partial x^2} - \frac{\partial g_{22}}{\partial x^1}\right] \text{ Since } g_{21} = 0.$$

$$= \frac{1}{2}\left[\frac{\partial 0}{\partial x^2} + \frac{\partial 0}{\partial x^2} - \frac{\partial r^2}{\partial x^1}\right]$$

$$= -\frac{1}{2}\frac{\partial r^2}{\partial r} = -r$$

Taking $i = 1, j = k = 3$ in (1)

$$[13,3] = \frac{1}{2}\left[\frac{\partial g_{33}}{\partial x^1} + \frac{\partial g_{13}}{\partial x^3} - \frac{\partial g_{13}}{\partial x^3}\right]$$

$$= \frac{1}{2}\frac{\partial r^2 \sin^2\theta}{\partial r} \quad \text{since } g_{13} = 0$$

$$[13,3] = r\sin^2\theta$$

(*ii*) Christoffel symbols of the second kind are given by

$$\begin{Bmatrix} k \\ i\ j \end{Bmatrix} = g^{kl}[ij,l] = g^{k1}[ij,1] + g^{k2}[ij,2] + g^{k3}[ij,3]$$

Taking $k = 1,\ i = j = 2$.

$$\begin{Bmatrix} 1 \\ 2\ 2 \end{Bmatrix} = g^{11}[22,1] + g^{12}[22,2] + g^{13}[22,3]$$

$$\begin{Bmatrix} 1 \\ 2\ 2 \end{Bmatrix} = 1[22,1] + 0[22,2] + 0[22,3] \text{ Since } g^{12} = g^{13} = 0$$

$$\begin{Bmatrix} 1 \\ 2\ 2 \end{Bmatrix} = -r$$

and

$$\begin{Bmatrix} 3 \\ 1\ 3 \end{Bmatrix} = g^{31}[13,1]+g^{32}[13,2]+g^{33}[13,3]$$

$$= \frac{1}{r^2\sin^2\theta}[13,3] \text{ Since } g^{31}=g^{32}=0$$

$$\begin{Bmatrix} 3 \\ 1\ 3 \end{Bmatrix} = \frac{1}{r^2\sin^2\theta}\cdot r\sin^2\theta = \frac{1}{r}$$

4.2 TRANSFORMATION OF CHRISTOFFEL'S SYMBOLS

The fundamental tensors g_{ij} and g^{ij} are functions of coordinates x^i and $[ij,k]$ is also function of coordinates x^i. Let $\bar{g}_{ij}, g^{-ij}$ and $\overline{[ij,k]}$ in another coordinate system $\bar{x}^i$.

(i) Law of Transformation of Christoffel's Symbol for First Kind

Let $[ij,k]$ is a function of coordinate x^i and $\overline{[ij,k]}$ in another coordinate system $\bar{x}^i$. Then

$$\overline{[ij,k]} = \frac{1}{2}\left[\frac{\partial\bar{g}_{ik}}{\partial\bar{x}^i}+\frac{\partial\bar{g}_{ik}}{\partial\bar{x}^j}-\frac{\partial\bar{g}_{ij}}{\partial\bar{x}^k}\right] \quad ...(1)$$

Since $\bar{g}_{ij}$ is a covariant tensor of rank two. Then

$$\bar{g}_{ij} = \frac{\partial x^p}{\partial\bar{x}^i}\frac{\partial x^2}{\partial\bar{x}^j}g_{pq} \quad ...(2)$$

Differentiating it w.r.t. to $\bar{x}^k$, we get

$$\frac{\partial\bar{g}_{ij}}{\partial\bar{x}^k} = \frac{\partial}{\partial\bar{x}^k}\left(\frac{\partial x^p}{\partial\bar{x}^i}\frac{\partial x^q}{\partial\bar{x}^j}g_{pq}\right)$$

$$= \frac{\partial}{\partial\bar{x}^k}\left(\frac{\partial x^p}{\partial\bar{x}^i}\frac{\partial x^q}{\partial\bar{x}^j}\right)g_{pq}+\frac{\partial x^p}{\partial\bar{x}^i}\frac{\partial x^q}{\partial\bar{x}^j}\frac{\partial g_{pq}}{\partial\bar{x}^k}$$

$$\frac{\partial\bar{g}_{ij}}{\partial\bar{x}^k} = \left(\frac{\partial^2x^p}{\partial\bar{x}^k\partial\bar{x}^i}\frac{\partial x^q}{\partial\bar{x}^j}+\frac{\partial x^p}{\partial\bar{x}^i}\frac{\partial^2x^q}{\partial\bar{x}^k\partial\bar{x}^j}\right)g_{pq}+\frac{\partial x^p}{\partial\bar{x}^i}\frac{\partial x^q}{\partial\bar{x}^j}\frac{\partial g_{pq}}{\partial x^r}\frac{\partial x^r}{\partial\bar{x}^k} \quad ...(3)$$

Interchanging *i*, *k* and also interchanging *p*, *r* in the last term in equation (3)

$$\frac{\partial\bar{g}_{kj}}{\partial\bar{x}^i} = \left(\frac{\partial^2x^p}{\partial\bar{x}^i\partial\bar{x}^k}\frac{\partial x^q}{\partial\bar{x}^j}+\frac{\partial x^p}{\partial\bar{x}^k}\frac{\partial^2x^q}{\partial\bar{x}^i\partial\bar{x}^j}\right)g_{pq}+\frac{\partial x^r}{\partial\bar{x}^k}\frac{\partial x^q}{\partial\bar{x}^j}\frac{\partial g_{rq}}{\partial x^p}\frac{\partial x^p}{\partial\bar{x}^i} \quad ...(4)$$

and interchanging j, k and also interchange q, r in the last term of equation (3)

$$\frac{\partial \bar{g}_{ik}}{\partial \bar{x}^j} = \left(\frac{\partial^2 x^p}{\partial \bar{x}^j \partial \bar{x}^i} \frac{\partial x^q}{\partial \bar{x}^i} + \frac{\partial x^p}{\partial \bar{x}^i} \frac{\partial^2 x^q}{\partial \bar{x}^k \partial \bar{x}^j} \right) g_{pq} + \frac{\partial x^p}{\partial \bar{x}^i} \frac{\partial x^r}{\partial \bar{x}^k} \frac{\partial x^q}{\partial \bar{x}^j} \frac{\partial g_{pr}}{\partial x^q} \qquad \ldots(5)$$

Substituting the values of equations (3), (4) and (5) in equation (1), we get

$$\overline{[ij,k]} = \frac{1}{2} \left[2 \frac{\partial^2 x^p}{\partial \bar{x}^i \partial \bar{x}^j} \frac{\partial x^q}{\partial \bar{x}^k} g_{pq} + \frac{\partial x^p}{\partial \bar{x}^i} \frac{\partial x^q}{\partial \bar{x}^j} \frac{\partial x^r}{\partial \bar{x}^k} \left(\frac{\partial g_{rp}}{\partial x^q} + \frac{\partial g_{qr}}{\partial x^p} - \frac{\partial g_{pq}}{\partial x^r} \right) \right]$$

$$\overline{[ij,k]} = \frac{\partial^2 x^p}{\partial \bar{x}^i \partial \bar{x}^j} \frac{\partial x^q}{\partial \bar{x}^k} g_{pq} + \frac{\partial x^p}{\partial \bar{x}^i} \frac{\partial x^q}{\partial \bar{x}^i} \frac{\partial x^r}{\partial \bar{x}^k} \frac{1}{2} \left(\frac{\partial g_{rp}}{\partial x^q} + \frac{\partial g_{qr}}{\partial x^p} - \frac{\partial g_{pq}}{\partial x^r} \right)$$

$$\overline{[ij,k]} = \frac{\partial^2 x^p}{\partial \bar{x}^i \partial \bar{x}^j} \frac{\partial x^q}{\partial \bar{x}^k} g_{pq} + \frac{\partial x^p}{\partial \bar{x}^i} \frac{\partial x^q}{\partial \bar{x}^i} \frac{\partial x^r}{\partial \bar{x}^k} [pq,r] \qquad \ldots(6)$$

It is law of transformation of Christoffel's symbol of the first kind. But it is not the law of transformation of any tensor due to presence of the first term of equation (6).

So, Christoffel's symbol of first kind is not a tensor.

(ii) Law of Transformation of Christoffel's Symbol of the Second Kind

Let $g^{kl}[ij,l] = \left\{ \begin{matrix} k \\ i \ j \end{matrix} \right\}$ is function of coordinates x^i and $\overline{g}^{kl}\overline{[ij,l]} = \overline{\left\{ \begin{matrix} k \\ i \ j \end{matrix} \right\}}$ in another coordinate system $\bar{x}^i$. Then

$$\overline{[ij,l]} = \frac{\partial^2 x^p}{\partial \bar{x}^i \partial \bar{x}^j} \frac{\partial x^q}{\partial \bar{x}^l} g_{pq} + \frac{\partial x^p}{\partial \bar{x}^i} \frac{\partial x^q}{\partial \bar{x}^j} \frac{\partial x^r}{\partial \bar{x}^l} [pq,r] \text{ from (6)}$$

As g^{kl} is contravariant tensor of rank two.

$$\bar{g}^{kl} = \frac{\partial \bar{x}^k}{\partial x^s} \frac{\partial \bar{x}^l}{\partial x^t} g^{st}$$

Now

$$\bar{g}^{kl} \ \overline{[ij,l]} = \frac{\partial \bar{x}^k}{\partial x^s} \frac{\partial \bar{x}^l}{\partial x^t} g^{st} \frac{\partial^2 x^p}{\partial \bar{x}^i \bar{x}^j} \frac{\partial x^q}{\partial \bar{x}^l} g_{pq} + \frac{\partial \bar{x}^k}{\partial x^s} \frac{\partial \bar{x}^l}{\partial x^t} g^{st} \frac{\partial x^p}{\partial \bar{x}^i} \frac{\partial x^q}{\partial \bar{x}^j} \frac{\partial x^r}{\partial \bar{x}^l} [pq,r]$$

$$= \frac{\partial \bar{x}^k}{\partial x^s} \left(\frac{\partial \bar{x}^l}{\partial x^t} \frac{\partial x^q}{\partial \bar{x}^l} \right) g^{st} \frac{\partial^2 x^p}{\partial \bar{x}^i \bar{x}^j} g_{pq} + \frac{\partial \bar{x}^k}{\partial x^s} \left(\frac{\partial \bar{x}^l}{\partial x^t} \frac{\partial x^r}{\partial \bar{x}^l} \right) \frac{\partial x^p}{\partial \bar{x}^i} \frac{\partial x^q}{\partial \bar{x}^j} g^{st} [pq,r]$$

$$= \frac{\partial \bar{x}^k}{\partial x^s} \delta_t^2 g^{st} \frac{\partial^2 x^p}{\partial \bar{x}^i \bar{x}^j} g_{pq} + \frac{\partial \bar{x}^k}{\partial x^s} \delta_t^r \frac{\partial x^p}{\partial \bar{x}^i} \frac{\partial x^q}{\partial \bar{x}^j} g^{st} [pq,r] \text{ as } \frac{\partial \bar{x}^l}{\partial x^t} \frac{\partial x^q}{\partial \bar{x}^l} = \delta_t^q$$

$$= \frac{\partial \bar{x}^k}{\partial x^s} \frac{\partial^2 x^p}{\partial \bar{x}^i \partial \bar{x}^j} g^{sq} g_{pq} + \frac{\partial \bar{x}^k}{\partial x^s} \frac{\partial x^p}{\partial \bar{x}^i} \frac{\partial x^q}{\partial \bar{x}^j} g^{sr} [pq,r] \text{ as } \delta_t^r g^{st} = g^{sr}$$

$$\bar{g}^{kl}\overline{[ij,l]} = \frac{\partial \bar{x}^p}{\partial x^s}\frac{\partial^2 x^p}{\partial \bar{x}^i \partial \bar{x}^j}\delta^s_p + \frac{\partial \bar{x}^k}{\partial x^s}\frac{\partial x^p}{\partial \bar{x}^i}\frac{\partial x^2}{\partial \bar{x}^j}\begin{Bmatrix} s \\ p\ q \end{Bmatrix}$$

$$\text{Since } g^{sq} g_{pq} = \delta^s_p \text{ and } g^{sr}[pq,r] = \begin{Bmatrix} s \\ p\ q \end{Bmatrix}$$

$$\overline{\begin{Bmatrix} k \\ i\ j \end{Bmatrix}} = \frac{\partial \bar{x}^k}{\partial x^s}\frac{\partial^2 x^s}{\partial \bar{x}^i \partial \bar{x}^j} + \frac{\partial \bar{x}^k}{\partial x^s}\frac{\partial x^p}{\partial \bar{x}^i}\frac{\partial x^q}{\partial \bar{x}^j}\begin{Bmatrix} s \\ p\ q \end{Bmatrix} \qquad \text{....(7)}$$

It is law of transformation of Christoffel's symbol of the second kind. But it is not the law of transformation of any tensor. So, *Christoffel's symbol of the second kind* is not a tensor.

Also, multiply (7) by $\frac{\partial x^s}{\partial \bar{x}^k}$, we get

$$\frac{\partial x^s}{d\bar{x}^k}\overline{\begin{Bmatrix} k \\ i\ j \end{Bmatrix}} = \frac{\partial x^s}{\partial \bar{x}^k}\frac{\partial \bar{x}^k}{\partial x^s}\frac{\partial^2 x^s}{\partial \bar{x}^i \partial \bar{x}^j} + \frac{\partial x^s}{\partial \bar{x}^k}\frac{\partial \bar{x}^k}{\partial x^s}\frac{\partial x^p}{\partial \bar{x}^i}\frac{\partial x^q}{\partial \bar{x}^j}\begin{Bmatrix} s \\ p\ q \end{Bmatrix}$$

$$\text{Since } \frac{\partial x^s}{\partial \bar{x}^k}\frac{\partial \bar{x}^k}{\partial x^s} = \delta^s_s = 1$$

$$\frac{\partial x^s}{\partial \bar{x}^k}\overline{\begin{Bmatrix} k \\ i\ j \end{Bmatrix}} = \frac{\partial^2 x^s}{\partial \bar{x}^i \partial \bar{x}^j} + \frac{\partial x^p}{\partial \bar{x}^i}\frac{\partial x^q}{\partial \bar{x}^j}\begin{Bmatrix} s \\ p\ q \end{Bmatrix}$$

$$\frac{\partial^2 x^s}{\partial \bar{x}^i \partial \bar{x}^j} = \frac{\partial x^s}{\partial \bar{x}^k}\overline{\begin{Bmatrix} k \\ i\ j \end{Bmatrix}} - \frac{\partial x^p}{\partial \bar{x}^i}\frac{\partial x^q}{\partial \bar{x}^j}\begin{Bmatrix} s \\ p\ q \end{Bmatrix} \qquad \text{...(8)}$$

It is second derivative of x^s with respect to $\bar{x}$'s in the terms of Christoffel's symbol of second kind and first derivatives.

THEOREM 4.4 *Prove that the transformation of Christoffel's Symbols form a group i.e., possess the transitive property.*

Proof: Let the coordinates x^i be transformed to the coordinate system $\bar{x}^i$ and $\bar{x}^j$ be transformed to $\bar{\bar{x}}^i$.

When coordinate x^i be transformed to $\bar{x}^i$, the law of transformation of Christoffel's symbols of second kind (equation (7)) is

$$\overline{\begin{Bmatrix} k \\ i\ j \end{Bmatrix}} = \frac{\partial \bar{x}^k}{\partial x^s}\frac{\partial^2 x^s}{\partial \bar{x}^i \partial \bar{x}^j} + \frac{\partial \bar{x}^k}{\partial x^s}\frac{\partial x^p}{\partial \bar{x}^i}\frac{\partial x^q}{\partial \bar{x}^j}\begin{Bmatrix} s \\ p\ q \end{Bmatrix} \qquad \text{...(1)}$$

When coordinate $\bar{x}^i$ be transformed to $\bar{\bar{x}}^i$. Then

$$\begin{Bmatrix} \bar{\bar{r}} \\ u\ v \end{Bmatrix} = \overline{\begin{Bmatrix} k \\ i\ j \end{Bmatrix}}\frac{\partial \bar{x}^i}{\partial \bar{\bar{x}}^u}\frac{\partial \bar{x}^j}{\partial \bar{\bar{x}}^v}\frac{\partial \bar{\bar{x}}^r}{\partial \bar{x}^k} + \frac{\partial^2 \bar{x}^k}{\partial \bar{\bar{x}}^u \partial \bar{\bar{x}}^v}\frac{\partial \bar{\bar{x}}^r}{\partial \bar{x}^k}$$

$$=\begin{Bmatrix} s \\ p\ q \end{Bmatrix} \frac{\partial \bar{x}^k}{\partial x^s}\frac{\partial x^p}{\partial \bar{x}^i}\frac{\partial x^q}{\partial \bar{x}^j}\frac{\partial \bar{x}^i}{\partial \bar{\bar{x}}^u}\frac{\partial \bar{x}^j}{\partial \bar{\bar{x}}^v}\frac{\partial \bar{\bar{x}}^r}{\partial \bar{x}^k}+\frac{\partial^2 x^s}{\partial \bar{x}^i \partial \bar{x}^j}\frac{\partial \bar{x}^k}{\partial x^s}\frac{\partial \bar{x}^i}{\partial \bar{\bar{x}}^u}\frac{\partial \bar{x}^j}{\partial \bar{\bar{x}}^v}\frac{\partial \bar{\bar{x}}^r}{\partial \bar{x}^k}$$

$$+\frac{\partial^2 \bar{x}^k}{\partial \bar{\bar{x}}^u \partial \bar{\bar{x}}^v}\frac{\partial \bar{\bar{x}}^r}{\partial \bar{x}^k}$$

$$\begin{Bmatrix} \bar{\bar{r}} \\ u\ v \end{Bmatrix}=\begin{Bmatrix} s \\ p\ q \end{Bmatrix}\frac{\partial x^p}{\partial \bar{\bar{x}}^u}\frac{\partial x^q}{\partial \bar{\bar{x}}^v}\frac{\partial \bar{\bar{x}}^r}{\partial x^s}+\frac{\partial^2 \bar{x}^k}{\partial \bar{\bar{x}}^u \partial \bar{\bar{x}}^v}\frac{\partial \bar{\bar{x}}^r}{\partial \bar{x}^k}+$$

$$\frac{\partial^2 x^s}{\partial \bar{x}^i \partial \bar{x}^j}\frac{\partial \bar{\bar{x}}^r}{\partial x^s}\frac{\partial \bar{x}^i}{\partial \bar{\bar{x}}^u}\frac{\partial \bar{x}^j}{\partial \bar{\bar{x}}^v} \qquad ...(2)$$

as $\frac{\partial x^p}{\partial \bar{x}^i}\frac{\partial \bar{x}^i}{\partial \bar{\bar{x}}^u}=\frac{\partial x^p}{\partial \bar{\bar{x}}^u}$

Since we know that

$$\frac{\partial x^s}{\partial \bar{x}^i}\frac{\partial \bar{x}^i}{\partial \bar{\bar{x}}^u}=\frac{\partial x^s}{\partial \bar{\bar{x}}^u} \qquad ...(3)$$

Differentiating (3) w.r.t. to $\bar{\bar{x}}^v$, we get

$$\frac{\partial}{\partial \bar{\bar{x}}^v}\left(\frac{\partial x^s}{\partial \bar{x}^i}\right)\frac{\partial \bar{x}^i}{\partial \bar{\bar{x}}^u}+\frac{\partial x^s}{\partial \bar{x}^i}\frac{\partial}{\partial \bar{\bar{x}}^v}\left(\frac{\partial \bar{x}^i}{\partial \bar{\bar{x}}^u}\right)=\frac{\partial^2 x^s}{\partial \bar{\bar{x}}^u \partial \bar{\bar{x}}^v}$$

$$\frac{\partial^2 x^s}{\partial \bar{x}^i \partial \bar{x}^j}\frac{\partial \bar{x}^j}{\partial \bar{\bar{x}}^v}\frac{\partial \bar{x}^i}{\partial \bar{\bar{x}}^u}+\frac{\partial x^s}{\partial \bar{x}^i}\frac{\partial^2 \bar{x}^i}{\partial \bar{\bar{x}}^u \partial \bar{\bar{x}}^v}=\frac{\partial^2 x^s}{\partial \bar{\bar{x}}^u \partial \bar{\bar{x}}^v} \qquad ...(4)$$

Mutiply (5) by $\frac{\partial \bar{\bar{x}}^r}{\partial x^s}$.

$$\frac{\partial^2 x^s}{\partial \bar{x}^i \partial \bar{x}^j}\frac{\partial \bar{x}^j}{\partial \bar{\bar{x}}^v}\frac{\partial \bar{x}^i}{\partial \bar{\bar{x}}^u}\frac{\partial \bar{\bar{x}}^r}{\partial x^s}+\frac{\partial^2 \bar{x}^i}{\partial \bar{\bar{x}}^u \partial \bar{\bar{x}}^v}\frac{\partial x^s}{\partial \bar{x}^i}\frac{\partial \bar{\bar{x}}^r}{\partial x^s}=\frac{\partial^2 x^s}{\partial \bar{\bar{x}}^u \partial \bar{\bar{x}}^v}\frac{\partial \bar{\bar{x}}^r}{\partial x^s}$$

Replace dummy index *i* by *k* in second term on L.H.S.

$$\frac{\partial^2 x^s}{\partial \bar{x}^i \partial \bar{x}^j}\frac{\partial \bar{x}^j}{\partial \bar{\bar{x}}^v}\frac{\partial \bar{x}^i}{\partial \bar{\bar{x}}^u}\frac{\partial \bar{\bar{x}}^r}{\partial x^s}+\frac{\partial^2 \bar{x}^k}{\partial \bar{\bar{x}}^u \partial \bar{\bar{x}}^v}\frac{\partial \bar{\bar{x}}^r}{\partial \bar{x}^k}=\frac{\partial^2 x^s}{\partial \bar{\bar{x}}^u \partial \bar{\bar{x}}^v}\frac{\partial \bar{\bar{x}}^r}{\partial x^s} \qquad ...(5)$$

Using (5) in equation (2), we get

$$\begin{Bmatrix} \bar{\bar{r}} \\ u\ v \end{Bmatrix}=\begin{Bmatrix} s \\ p\ q \end{Bmatrix}\frac{\partial x^p}{\partial \bar{\bar{x}}^u}\frac{\partial x^q}{\partial \bar{\bar{x}}^v}\frac{\partial \bar{\bar{x}}^r}{\partial x^s}+\frac{\partial^2 x^s}{\partial \bar{\bar{x}}^u \partial \bar{\bar{x}}^v}\frac{\partial \bar{\bar{x}}^r}{\partial x^s} \qquad ...(6)$$

The equation (6) is same as the equation (1). This shows that if we make direct transformation from x^i to $\overline{\overline{x}}^i$ we get same law of transformation. This property is called that transformation of Christoffel's symbols form a group.

4.3 COVARIANT DIFFERENTIATION OF A COVARIANT VECTOR

Let A_i and $\overline{A}_i$ be the components of a covariant vector in coordinate systems x^i and $\overline{x}^i$ respectively. Then

$$\overline{A}_i = \frac{\partial x^p}{\partial \overline{x}^i} A_p \qquad \text{...(1)}$$

Differentiating (1) partially w.r.t. to $\overline{x}^j$,

$$\frac{\partial \overline{A}_i}{\partial \overline{x}^j} = \frac{\partial^2 x^p}{\partial \overline{x}^j \partial \overline{x}^i} A_p + \frac{\partial x^p}{\partial \overline{x}^i} \frac{\partial A_p}{\partial \overline{x}^j}$$

$$\frac{\partial \overline{A}_i}{\partial \overline{x}^j} = \frac{\partial^2 x^p}{\partial \overline{x}^j \partial \overline{x}^i} A_p + \frac{\partial x^p}{\partial \overline{x}^i} \frac{\partial A_p}{\partial x^q} \frac{\partial x^q}{\partial \overline{x}^j} \qquad \text{...(2)}$$

It is not a tensor due to presence of the first term on the R.H.S. of equation (2).
Now, replace dummy index p by s in the first term on R.H.S. of (2), we have

$$\frac{\partial \overline{A}_i}{\partial \overline{x}^j} = \frac{\partial^2 x^s}{\partial \overline{x}^j \partial \overline{x}^i} A_s + \frac{\partial x^p}{\partial \overline{x}^i} \frac{\partial A_p}{\partial x^q} \frac{\partial x^q}{\partial \overline{x}^j} \qquad \text{...(3)}$$

Since we know that from equation (8), page 65,

$$\frac{\partial^2 x^s}{\partial \overline{x}^i \partial \overline{x}^j} = \frac{\partial x^s}{\partial \overline{x}^k} \overline{\begin{Bmatrix} k \\ i\ j \end{Bmatrix}} - \frac{\partial x^p}{\partial \overline{x}^i} \frac{\partial x^q}{\partial \overline{x}^j} \begin{Bmatrix} s \\ p\ q \end{Bmatrix}$$

Substituting the value of $\dfrac{\partial^2 x^s}{\partial \overline{x}^i \partial \overline{x}^j}$ in equation (3), we have

$$\frac{\partial \overline{A}_i}{\partial \overline{x}^j} = \left(\frac{\partial x^s}{\partial \overline{x}^k} \overline{\begin{Bmatrix} k \\ i\ j \end{Bmatrix}} - \frac{\partial x^p}{\partial \overline{x}^i} \frac{\partial x^q}{\partial \overline{x}^j} \begin{Bmatrix} s \\ p\ q \end{Bmatrix} \right) A_s + \frac{\partial x^p}{\partial \overline{x}^i} \frac{\partial A_p}{\partial x^q} \frac{\partial x^q}{\partial \overline{x}^j}$$

$$= \overline{\begin{Bmatrix} k \\ i\ j \end{Bmatrix}} \frac{\partial x^s}{\partial \overline{x}^k} A_s - \frac{\partial x^p}{\partial \overline{x}^i} \frac{\partial x^q}{\partial \overline{x}^j} \begin{Bmatrix} s \\ p\ q \end{Bmatrix} A_s + \frac{\partial x^p}{\partial \overline{x}^i} \frac{\partial x^q}{\partial \overline{x}^j} \frac{\partial A_p}{\partial x^q}$$

$$\frac{\partial \overline{A}_i}{\partial \overline{x}^j} = \overline{\begin{Bmatrix} k \\ i\ j \end{Bmatrix}} \overline{A}_k + \frac{\partial x^p}{\partial \overline{x}^i} \frac{\partial x^q}{\partial \overline{x}^j} \left(\frac{\partial A_p}{\partial x^q} - A_s \begin{Bmatrix} s \\ p\ q \end{Bmatrix} \right)$$

$$\frac{\partial \overline{A}_i}{\partial \overline{x}^j} - \overline{A}_k \overline{\begin{Bmatrix} k \\ i \;\; j \end{Bmatrix}} = \frac{\partial x^p}{\partial \overline{x}^i}\frac{\partial x^q}{\partial \overline{x}^j}\left(\frac{\partial A_p}{\partial x^q} - A_s \begin{Bmatrix} s \\ p \;\; q \end{Bmatrix}\right) \qquad ...(4)$$

Now, we introduce the comma notation

$$A_{i,j} = \frac{\partial A_i}{\partial x^j} - A_k \begin{Bmatrix} k \\ i \;\; j \end{Bmatrix} \qquad ...(5)$$

Using (5), the equation (4) can be expressed as

$$\overline{A}_{i,j} = \frac{\partial x^p}{\partial \overline{x}^i}\frac{\partial x^q}{\partial \overline{x}^j} A_{p,q} \qquad ..(6)$$

It is law of transformation of a covariant tensor of rank two. Thus, $A_{i,j}$ is a covariant tensor of rank two.

So, $A_{i,j}$ is called *covariant derivative* of A_i with respect to x^j.

4.4 COVARIANT DIFFERENTIATION OF A CONTRAVARIANT VECTOR

Let A^i and $\overline{A}^i$ be the component of contravariant vector in coordinate systems x^i and $\overline{x}^i$ respectively. Then

$$\overline{A}^i = \frac{\partial \overline{x}^i}{\partial x^s} A^s$$

or

$$A^s = \frac{\partial x^s}{\partial \overline{x}^i} \overline{A}^i$$

Differentiating it partially w.r.t. to $\overline{x}^j$, we get

$$\frac{\partial A^s}{\partial \overline{x}^j} = \frac{\partial^2 x^s}{\partial \overline{x}^j \partial x^i} \overline{A}^i + \frac{\partial x^s}{\partial \overline{x}^i}\frac{\partial \overline{A}^i}{\partial \overline{x}^j} \qquad ...(1)$$

Since from equation (8) on page 65,

$$\frac{\partial^2 x^s}{\partial \overline{x}^j \partial \overline{x}^i} = \frac{\partial x^s}{\partial \overline{x}^k}\begin{Bmatrix} k \\ i \;\; j \end{Bmatrix}\overline{A}^i - \frac{\partial x^p}{\partial \overline{x}^i}\frac{\partial x^q}{\partial \overline{x}^j}\begin{Bmatrix} s \\ p \;\; q \end{Bmatrix}$$

substituting the value of $\frac{\partial^2 x^s}{\partial \overline{x}^j \partial \overline{x}^i}$ in the equation (1), we get

$$\frac{\partial A^s}{\partial \overline{A}^j} = \frac{\partial x^s}{\partial \overline{x}^k}\overline{\begin{Bmatrix} k \\ i \;\; j \end{Bmatrix}}\overline{A}^i - \frac{\partial x^p}{\partial \overline{x}^i}\frac{\partial x^q}{\partial \overline{x}^j}\begin{Bmatrix} s \\ p \;\; q \end{Bmatrix}\overline{A}^i + \frac{\partial x^s}{\partial \overline{x}^i}\frac{\partial \overline{A}^i}{\partial \overline{x}^j}$$

$$\frac{\partial A^s}{\partial A^q}\frac{\partial A^q}{\partial \overline{x}^j} = \frac{\partial x^s}{\partial \overline{x}^k}\overline{\begin{Bmatrix} k \\ i \;\; j \end{Bmatrix}}\overline{A}^i - \frac{\partial x^p}{\partial \overline{x}^i}\overline{A}^i\frac{\partial x^q}{\partial \overline{x}^j}\begin{Bmatrix} s \\ p \;\; q \end{Bmatrix} + \frac{\partial x^s}{\partial \overline{x}^i}\frac{\partial \overline{A}^i}{\partial \overline{x}^j}$$

Interchanging the dummy indices i and k in the first term on R.H.S. and put $\frac{\partial x^p}{\partial \overline{x}^i}\overline{A}^i = A^p$ we get

$$\frac{\partial A^s}{\partial x^q}\frac{\partial x^q}{\partial \overline{x}^j} = \frac{\partial x^s}{\partial \overline{x}^i}\overline{\begin{Bmatrix} i \\ k\ j \end{Bmatrix}}\overline{A}^k - \frac{\partial x^q}{\partial \overline{x}^j}A^p\begin{Bmatrix} s \\ p\ q \end{Bmatrix} + \frac{\partial x^s}{\partial \overline{x}^i}\frac{\partial \overline{A}^i}{\partial \overline{x}^j}$$

$$\frac{\partial x^q}{\partial \overline{x}^j}\left(\frac{\partial A^s}{\partial x^q} + A^p\begin{Bmatrix} s \\ p\ q \end{Bmatrix}\right) = \frac{\partial x^s}{\partial \overline{x}^i}\left(\overline{\begin{Bmatrix} i \\ k\ j \end{Bmatrix}}\overline{A}^k + \frac{\partial \overline{A}^i}{\partial \overline{x}^j}\right)$$

$$\frac{\partial \overline{A}^i}{\partial \overline{x}^j} + \overline{A}^k\overline{\begin{Bmatrix} i \\ k\ j \end{Bmatrix}} = \frac{\partial \overline{x}^i}{\partial x^s}\frac{\partial x^q}{\partial \overline{x}^j}\left(\frac{\partial A^s}{\partial x^q} + A^p\begin{Bmatrix} s \\ p\ q \end{Bmatrix}\right) \quad ...(2)$$

Now, we introduce the comma notation

$$A^i_{,j} = \frac{\partial A_i}{\partial x^j} + A^k\begin{Bmatrix} i \\ k\ j \end{Bmatrix} \quad ...(3)$$

Using (3), the equation (2) can be expressed as

$$\overline{A}^i_{,j} = \frac{\partial x^p}{\partial \overline{x}^i}\frac{\partial x^q}{\partial \overline{x}^j}A_{p,q} \quad ...(4)$$

It is law of transformation of a mixed tensor of rank two. Thus, A^i, j is a mixed tensor of rank two. A^i, j is called covariant derivative of A^i with respect to x^j.

4.5 COVARIANT DIFFERENTIATION OF TENSORS

Covariant derivative of a covariant tensor of rank two.

Let A_{ij} and $\overline{A}_{ij}$ be the components of a covariant tensor of rank two in coordinate system x^i and $\overline{x}^i$ respectively then

$$\overline{A}_{ij} = \frac{\partial x^p}{\partial \overline{x}^i}\frac{\partial x^q}{\partial \overline{x}^j}A_{pq} \quad ...(1)$$

Differentiating (1) partially w.r.t. to $\overline{x}^k$

$$\frac{\partial \overline{A}_{ij}}{\partial \overline{x}^k} = \frac{\partial x^p}{\partial \overline{x}^i}\frac{\partial x^q}{\partial \overline{x}^j}\frac{\partial A_{pq}}{\partial \overline{x}^k} + \frac{\partial}{\partial \overline{x}^k}\left(\frac{\partial x^p}{\partial \overline{x}^i}\frac{\partial x^q}{\partial \overline{x}^j}\right)A_{pq}$$

$$\frac{\partial \overline{A}_{ij}}{\partial \overline{x}^k} = \frac{\partial x^p}{\partial \overline{x}^i}\frac{\partial x^q}{\partial \overline{x}^j}\frac{\partial A_{pq}}{\partial x^r}\frac{\partial x^r}{\partial \overline{x}^k} + \frac{\partial^2 x^p}{\partial \overline{x}^k \partial \overline{x}^i}\frac{\partial x^q}{\partial \overline{x}^j}A_{pq} + \frac{\partial x^p}{\partial \overline{x}^i}\frac{\partial^2 x^q}{\partial \overline{x}^k \partial \overline{x}^j}A_{pq} \quad ...(2)$$

as $\frac{\partial A_{pq}}{\partial \overline{x}^k} = \frac{\partial A_{pq}}{\partial x^r}\frac{\partial x^r}{\partial \overline{x}^k}$ (since A_{pq} components in x^i coordinate)

$$A_{pq}\frac{\partial^2 x^p}{\partial \bar{x}^i \partial \bar{x}^k}\frac{\partial x^q}{\partial \bar{x}^j} = A_{lq}\frac{\partial^2 x^l}{\partial \bar{x}^i \partial \bar{x}^k}\frac{\partial x^q}{\partial \bar{x}^j}$$

$$A_{pq}\frac{\partial^2 x^p}{\partial \bar{x}^i \partial \bar{x}^k}\frac{\partial x^q}{\partial \bar{x}^j} = A_{lq}\frac{\partial x^q}{\partial \bar{x}^j}\left[\overline{\begin{Bmatrix} h \\ i\ k \end{Bmatrix}}\frac{\partial x^l}{\partial \bar{x}^h} - \begin{Bmatrix} l \\ p\ r \end{Bmatrix}\frac{\partial x^p}{\partial \bar{x}^i}\frac{\partial x^r}{\partial \bar{x}^k}\right]$$

Since we know that from equation (8) on page 65.

$$\frac{\partial^2 x^l}{\partial \bar{x}^i \partial \bar{x}^k} = \overline{\begin{Bmatrix} h \\ i\ k \end{Bmatrix}}\frac{\partial x^l}{\partial \bar{x}^h} - \begin{Bmatrix} l \\ p\ r \end{Bmatrix}\frac{\partial x^p}{\partial \bar{x}^i}\frac{\partial x^r}{\partial \bar{x}^k}$$

$$A_{pq}\frac{\partial^2 x^p}{\partial \bar{x}^i \partial \bar{x}^k}\frac{\partial x^q}{\partial \bar{x}^j} = \overline{\begin{Bmatrix} h \\ i\ k \end{Bmatrix}} A_{lq}\frac{\partial x^l}{\partial \bar{x}^h}\frac{\partial x^2}{\partial \bar{x}^j} - \begin{Bmatrix} l \\ p\ r \end{Bmatrix} A_{lq}\frac{\partial x^p}{\partial \bar{x}^i}\frac{\partial x^q}{\partial \bar{x}^j}\frac{\partial x^r}{\partial \bar{x}^k}$$

$$A_{pq}\frac{\partial^2 x^p}{\partial \bar{x}^i \partial \bar{x}^k}\frac{\partial x^q}{\partial \bar{x}^j} = \overline{\begin{Bmatrix} h \\ i\ k \end{Bmatrix}}\bar{A}_{hj} - \begin{Bmatrix} l \\ p\ r \end{Bmatrix} A_{lq}\frac{\partial x^p}{\partial \bar{x}^i}\frac{\partial x^q}{\partial \bar{x}^j}\frac{\partial x^r}{\partial \bar{x}^k} \qquad ...(3)$$

$$\text{as } \bar{A}_{hj} = A_{lq}\frac{\partial x^l}{\partial \bar{x}^h}\frac{\partial x^q}{\partial \bar{x}^j} \text{ by equation (1)}$$

and

$$A_{pq}\frac{\partial x^p}{\partial \bar{x}^i}\frac{\partial^2 x^q}{\partial \bar{x}^j \partial \bar{x}^k} = A_{pl}\frac{\partial x^p}{\partial \bar{x}^i}\frac{\partial^2 x^l}{\partial \bar{x}^j \partial \bar{x}^k}$$

$$= A_{pl}\frac{\partial x^p}{\partial \bar{x}^i}\left[\overline{\begin{Bmatrix} h \\ j\ k \end{Bmatrix}}\frac{\partial x^l}{\partial \bar{x}^h} - \begin{Bmatrix} l \\ q\ r \end{Bmatrix}\frac{\partial x^q}{\partial \bar{x}^j}\frac{\partial x^r}{\partial \bar{x}^k}\right]$$

$$= \overline{\begin{Bmatrix} h \\ j\ k \end{Bmatrix}} A_{pl}\frac{\partial x^p}{\partial \bar{x}^i}\frac{\partial x^l}{\partial \bar{x}^h} - \begin{Bmatrix} l \\ q\ r \end{Bmatrix}\frac{\partial x^q}{\partial \bar{x}^j}\frac{\partial x^r}{\partial \bar{x}^k}\frac{\partial x^p}{\partial \bar{x}^i} A_{pl}$$

$$A_{pq}\frac{\partial x^p}{\partial \bar{x}^i}\frac{\partial^2 x^q}{\partial \bar{x}^j \partial \bar{x}^k} = \overline{\begin{Bmatrix} h \\ j\ k \end{Bmatrix}}\bar{A}_{ih} - \begin{Bmatrix} l \\ q\ r \end{Bmatrix}\frac{\partial x^p}{\partial \bar{x}^i}\frac{\partial x^q}{\partial \bar{x}^j}\frac{\partial x^r}{\partial \bar{x}^k} A_{pl} \qquad ...(4)$$

Substituting the value of equations (3) and (4) in equation (2) we get,

$$\frac{\partial \bar{A}_{ij}}{\partial \bar{x}^k} = \left[\frac{\partial A_{pq}}{\partial x^r} - A_{lq}\begin{Bmatrix} l \\ p\ r \end{Bmatrix} - A_{pl}\begin{Bmatrix} l \\ q\ r \end{Bmatrix}\right]\frac{\partial x^p}{\partial \bar{x}^i}\frac{\partial x^q}{\partial \bar{x}^j}\frac{\partial x^r}{\partial \bar{x}^k} + \overline{\begin{Bmatrix} h \\ i\ k \end{Bmatrix}}\bar{A}_{hj} + \overline{\begin{Bmatrix} h \\ j\ k \end{Bmatrix}}\bar{A}_{ih}$$

$$\frac{\partial \bar{A}_{ij}}{\partial \bar{x}^k} - \overline{\begin{Bmatrix} h \\ j\ k \end{Bmatrix}}\bar{A}_{ih} - \overline{\begin{Bmatrix} h \\ i\ k \end{Bmatrix}}\bar{A}_{hj} = \left[\frac{\partial A_{pq}}{\partial x^r} - A_{lq}\begin{Bmatrix} l \\ p\ r \end{Bmatrix} - A_{pl}\begin{Bmatrix} l \\ q\ r \end{Bmatrix}\right]\frac{\partial x^p}{\partial \bar{x}^i}\frac{\partial x^q}{\partial \bar{x}^j}\frac{\partial x^r}{\partial \bar{x}^k}$$

$$A_{ij,k} = \frac{\partial A_{ij}}{\partial x^k} - \begin{Bmatrix} h \\ j\ k \end{Bmatrix} A_{ih} - \begin{Bmatrix} h \\ i\ k \end{Bmatrix} A_{hj}\text{, then}$$

$$\overline{A}_{ij,k} = A_{pq,r} \frac{\partial x^p}{\partial \overline{x}^i} \frac{\partial x^q}{\partial \overline{x}^j} \frac{\partial x^r}{\partial \overline{x}^k}$$

It is law of transformation of a covariant tensor of rank three. Thus, $A_{ij,k}$ is a covariant tensor of rank three.

So, $A_{ij,k}$ is called *covariant derivative of* A_{ij} w.r.t. to x^k.

Similarly we define the covariant derivation x^k of a tensors A^{ij} and A^i_j by the formula

$$A^{ij},k = \frac{\partial A^{ij}}{\partial x^k} + A^{lj} \begin{Bmatrix} i \\ l\ k \end{Bmatrix} + A^{il} \begin{Bmatrix} j \\ l\ k \end{Bmatrix}$$

and
$$A^i_{j,k} = \frac{\partial A^i_j}{\partial x^k} + A^l_j \begin{Bmatrix} i \\ l\ k \end{Bmatrix} - A^i_l \begin{Bmatrix} l \\ j\ k \end{Bmatrix}$$

In general, we define the covariant deriavative x^k of a mixed tensor $A^{ij..l}_{ab...c}$ by the formula

$$A^{ij..l}_{ab...c,k} = \frac{\partial A^{ij..l}_{ab...c}}{\partial x^k} + A^{pj..l}_{ab...c} \begin{Bmatrix} i \\ p\ k \end{Bmatrix} + A^{ip..l}_{ab...c} \begin{Bmatrix} j \\ p\ k \end{Bmatrix} + \cdots + A^{ij...p}_{ab...c} \begin{Bmatrix} l \\ p\ k \end{Bmatrix}$$
$$- A^{ij..l}_{pb...c} \begin{Bmatrix} p \\ a\ k \end{Bmatrix} - A^{ij..l}_{ap...c} \begin{Bmatrix} p \\ b\ k \end{Bmatrix} - \cdots - A^{ij..l}_{ab...p} \begin{Bmatrix} p \\ c\ k \end{Bmatrix}$$

Note: $A_{i,k}$ is also written as $A_{i,k} = \nabla_k A_i$.

4.6 RICCI'S THEOREM

The covariant derivative of Kronecker delta and the fundamental tensors g_{ij} and g^{ij} is zero.

Proof: The covariant derivative x^p of Kronecker delta is

$$\delta^i_{j,k} = \frac{\partial \delta^i_j}{\partial x^k} + \delta^l_j \begin{Bmatrix} i \\ l\ k \end{Bmatrix} - \delta^i_l \begin{Bmatrix} l \\ j\ k \end{Bmatrix}$$

$$= 0 + \begin{Bmatrix} i \\ j\ k \end{Bmatrix} - \begin{Bmatrix} i \\ j\ k \end{Bmatrix}$$

$$\delta^i_{j,k} = 0 \text{ as } \frac{\partial \delta^i_j}{\partial x^k} = 0;\ \delta^l_j \begin{Bmatrix} i \\ l\ k \end{Bmatrix} = \begin{Bmatrix} i \\ j\ k \end{Bmatrix}$$

Also, consider first the tensor g_{ij} and the covariant derivative of g_{ij} is

$$g_{ij,k} = \frac{\partial g_{ij}}{\partial x^k} - g_{mj} \begin{Bmatrix} m \\ i\ k \end{Bmatrix} - g_{im} \begin{Bmatrix} m \\ j\ k \end{Bmatrix}$$

$$g_{ij,k} = \frac{\partial g_{ij}}{\partial x^k} - [ik,j] - [jk,i] \text{ as } g_{mj}\begin{Bmatrix} m \\ i\ k \end{Bmatrix} = [ik,j]$$

But
$$\frac{\partial g_{ij}}{\partial x^k} = [ik,j] + [jk,i]$$

So,
$$g_{ij,k} = \frac{\partial g_{ij}}{\partial x^k} - \frac{\partial g_{ij}}{\partial x^k}$$

$$g_{ij,k} = 0$$

We can perform a similar calculation for the tensor g^{ij}.

Since we know that $g^{im} g_{mj} = \delta^i_j$. Similarly taking covariant derivative, we get

$$g^{im}_{,k} g_{mj} + g^{im} g_{mj,k} = \delta^i_{j,k}$$

But $g_{mj,k} = 0$ and $\delta^i_{j,k} = 0$. So, $g^{im}_{,k} = 0$ as $|g_{mj}| \neq 0$

EXAMPLE 4

Prove that if A^{ij} is a symmetric tensor then

$$A^j_{i,j} = \frac{1}{\sqrt{g}} \frac{\partial}{\partial x^j}\left(A^j_i \sqrt{g}\right) - \frac{1}{2} A^{jk} \frac{\partial g_{jk}}{\partial x^i}$$

Solution

Given that A^{ij} be a symmetric tensor. Then

$$A^{ij} = A^{ji} \quad \text{...(1)}$$

We know that

$$A^j_{i,k} = \frac{\partial A^j_i}{\partial x^k} + A^l_i \begin{Bmatrix} j \\ l\ k \end{Bmatrix} - A^j_l \begin{Bmatrix} l \\ i\ k \end{Bmatrix}$$

Put $k = j$, we get

$$A^j_{i,j} = \frac{\partial A^j_i}{\partial x^j} + A^l_i \begin{Bmatrix} j \\ l\ j \end{Bmatrix} - A^j_l \begin{Bmatrix} l \\ i\ j \end{Bmatrix} \quad \text{...(2)}$$

$$= \frac{\partial A^j_i}{\partial x^j} + A^l_i \frac{\partial\left(\log\sqrt{g}\right)}{\partial x^l} - A^j_l g^{hl}[ij,h]$$

$$= \frac{\partial A^j_i}{\partial x^j} + \frac{A^j_i}{\sqrt{g}} \frac{\partial\sqrt{g}}{\partial x^j} - A^{jh}[ij,h] \text{ since } A^{ij} \text{ is symmetric.}$$

$$A^j_{i,j} = \frac{1}{\sqrt{g}} \frac{\partial\left(A^j_i\sqrt{g}\right)}{\partial x^j} - A^{jk}[ij,k] \quad \text{...(3)}$$

But

$$A^{jk}[ij,k] = \frac{1}{2}A^{jk}\left(\frac{\partial g_{jk}}{\partial x^i}+\frac{\partial g_{ki}}{\partial x^j}-\frac{\partial g_{ij}}{\partial x^k}\right)$$

$$A^{jk}[ij,k] = \frac{1}{2}\left(A^{jk}\frac{\partial g_{jk}}{\partial x^i}+A^{jk}\frac{\partial g_{ki}}{\partial x^j}-A^{jk}\frac{\partial g_{ij}}{\partial x^k}\right)$$

$$A^{jk}\frac{\partial g_{ki}}{\partial x^j} = A^{kj}\frac{\partial g_{ji}}{\partial x^k} \qquad \text{...(4)}$$

On Interchanging the dummy indices j & k.

$$A^{jk}\frac{\partial g_{ki}}{\partial x^j} = A^{jk}\frac{\partial g_{ji}}{\partial x^k} \text{ since } A^{ij} = A^{ji}$$

$$\Rightarrow \qquad A^{jk}\frac{\partial g_{ki}}{\partial x^j} - A^{jk}\frac{\partial g_{ij}}{\partial x^k} = 0 \text{ as } g_{ij} = g_{ji} \qquad \text{...(5)}$$

Using (5), equation (4) becomes

$$A^{jk}[ij,k] = \frac{1}{2}A^{jk}\frac{\partial g_{jk}}{\partial x^i}$$

Put the value of $A^{jk}[ij,k]$ in equation (3), we get

$$A^j_{i,j} = \frac{1}{\sqrt{g}}\frac{\partial\left(A^j_i\sqrt{g}\right)}{\partial x^j} - \frac{1}{2}A^{jk}\frac{\partial g_{jk}}{\partial x^i}.$$

Proved.

EXAMPLE 5

Prove that $\begin{Bmatrix} k \\ i\ j \end{Bmatrix}_a - \begin{Bmatrix} k \\ i\ j \end{Bmatrix}_b$ are components of a tensor of rank three where $\begin{Bmatrix} k \\ i\ j \end{Bmatrix}_a$ and $\begin{Bmatrix} k \\ i\ j \end{Bmatrix}_b$ are the Christoffel symbols formed from the symmetric tensors a_{ij} and b_{ij}.

Solution

Since we know that from equation (8), page 65.

$$\frac{\partial^2 x^s}{\partial\bar{x}^i\partial\bar{x}^j} = \frac{\partial x^s}{\partial\bar{x}^k}\overline{\begin{Bmatrix} k \\ i\ j \end{Bmatrix}} - \frac{\partial x^p}{\partial\bar{x}^i}\frac{\partial x^q}{\partial\bar{x}^j}\begin{Bmatrix} s \\ p\ q \end{Bmatrix}$$

$$\frac{\partial x^s}{\partial\bar{x}^k}\overline{\begin{Bmatrix} k \\ i\ j \end{Bmatrix}} = \frac{\partial^2 x^s}{\partial\bar{x}^i\partial\bar{x}^j} + \frac{\partial x^p}{\partial\bar{x}^i}\frac{\partial x^q}{\partial\bar{x}^j}\begin{Bmatrix} s \\ p\ q \end{Bmatrix}$$

or

$$\overline{\begin{Bmatrix} k \\ i\ j \end{Bmatrix}} = \left[\frac{\partial^2 x^s}{\partial\bar{x}^i\partial\bar{x}^j} + \frac{\partial x^p}{\partial\bar{x}^i}\frac{\partial x^q}{\partial\bar{x}^j}\begin{Bmatrix} s \\ p\ q \end{Bmatrix}\right]\frac{\partial\bar{x}^k}{\partial x^s}$$

Using this equation, we can write

$$\overline{\begin{Bmatrix} k \\ i\ j \end{Bmatrix}}_a = \left[\frac{\partial^2 x^s}{\partial \bar{x}^i \partial \bar{x}^j} + \frac{\partial x^p}{\partial \bar{x}^i}\frac{\partial x^q}{\partial \bar{x}^j}\begin{Bmatrix} s \\ p\ q \end{Bmatrix}_a\right]\frac{\partial \bar{x}^k}{\partial x^s}$$

and
$$\overline{\begin{Bmatrix} k \\ i\ j \end{Bmatrix}}_b = \left[\frac{\partial^2 x^s}{\partial \bar{x}^i \partial \bar{x}^j} + \frac{\partial x^p}{\partial \bar{x}^i}\frac{\partial x^q}{\partial \bar{x}^j}\begin{Bmatrix} s \\ p\ q \end{Bmatrix}_b\right]\frac{\partial \bar{x}^k}{\partial x^s}$$

Subtracting, we obtain

$$\overline{\begin{Bmatrix} k \\ i\ j \end{Bmatrix}}_a - \overline{\begin{Bmatrix} k \\ i\ j \end{Bmatrix}}_b = \left[\begin{Bmatrix} s \\ p\ q \end{Bmatrix}_a - \begin{Bmatrix} s \\ p\ q \end{Bmatrix}_b\right]\frac{\partial x^p}{\partial \bar{x}^i}\frac{\partial x^q}{\partial \bar{x}^j}\frac{\partial \bar{x}^k}{\partial x^s}$$

Put

$$\begin{Bmatrix} s \\ p\ q \end{Bmatrix}_a - \begin{Bmatrix} s \\ p\ q \end{Bmatrix}_b = A^s_{pq}$$

Then above equation can written as

$$\bar{A}^k_{ij} = A^s_{pq}\frac{\partial x^p}{\partial \bar{x}^i}\frac{\partial x^q}{\partial \bar{x}^j}\frac{\partial \bar{x}^k}{\partial x^s}$$

It is law of transformation of tensor of rank three.

So, $\begin{Bmatrix} k \\ i\ j \end{Bmatrix}_a - \begin{Bmatrix} k \\ i\ j \end{Bmatrix}_b$ are components of a tensor of rank three.

EXAMPLE 6

If a specified point, the derivatives of g_{ij} w.r.t. to x^k are all zero. Prove that the components of covariant derivatives at that point are the same as ordinary derivatives.

Solution

Given that

$$\frac{\partial g_{ii}}{\partial x^k} = 0, \ \forall\ i, j, k \ \text{ at } P_0 \qquad ...(1)$$

Let A^i_j be tensor.

Now, we have to prove that $A^i_{j,k} = \frac{\partial A^i_j}{\partial x^k}$ at P_0.

$$A^i_{j,k} = \frac{\partial A^i_j}{\partial x^k} + A^\alpha_j\begin{Bmatrix} i \\ \alpha\ k \end{Bmatrix} - A^i_\alpha\begin{Bmatrix} \alpha \\ j\ k \end{Bmatrix} \qquad ...(2)$$

Since $\begin{Bmatrix} k \\ i\ j \end{Bmatrix}$ and $[ij,k]$ both contain terms of the type $\dfrac{\partial g_{ij}}{\partial x^k}$ and using equation (1) we get

$$\begin{Bmatrix} k \\ i\ j \end{Bmatrix} = 0 = [ij,k] \text{ at } P_0.$$

So, equation (2) becomes

$$A^i_{j,k} = \frac{\partial A^i_j}{\partial x^k} \text{ at } P_0$$

4.7 GRADIENT, DIVERGENCE AND CURL

(a) *Gradient*

If ϕ be a scalar function of the coordinates, then the gradient of ϕ is denoted by

$$\text{grad } \phi = \frac{\partial \phi}{\partial x^i}$$

which is a covariant vector.

(b) *Divergence*

The divergence of the contravariant vector A^i is defined by

$$\text{div } A^i = \frac{\partial A^i}{\partial x^i} + A^k \begin{Bmatrix} i \\ k\ i \end{Bmatrix}$$

It is also written as $A^i_{,i}$

The divergence of the covariant vector A_i is defined by

$$\text{div } A_i = g^{ik} A_{ik}$$

EXAMPLE 7

Prove that div $A^i = \dfrac{1}{\sqrt{g}} \dfrac{\partial\left(\sqrt{g}A^k\right)}{\partial x^k}$

Solution:

If A^i be components of contravariant vector then

$$\text{div } A^i = A^i_{,i} = \frac{\partial A^i}{\partial x^i} + A^k \begin{Bmatrix} i \\ k\ i \end{Bmatrix}$$

Since

$$\begin{Bmatrix} i \\ k\ i \end{Bmatrix} = \frac{\partial}{\partial x^k}\left(\log\sqrt{g}\right) = \frac{1}{\sqrt{g}} \frac{\partial \sqrt{g}}{\partial x^k}$$

So,

$$\text{div } A^i = \frac{\partial A^i}{\partial x^i} + \frac{1}{\sqrt{g}} \frac{\partial \sqrt{g}}{\partial x^k} A^k$$

Since i is dummy index. Then put $i = k$, we get

$$\text{div } A^i = \frac{\partial A^k}{\partial x^k} + \frac{1}{\sqrt{g}}\frac{\partial \sqrt{g}}{\partial x^k} A^k$$

$$\text{div } A^i = \frac{1}{\sqrt{g}}\frac{\partial\left(\sqrt{g}A^k\right)}{\partial x^k} \quad \text{...(1)}$$

Proved

(c) Curl

Let A_i be a covariant vector then

$$A_{i,j} = \frac{\partial A_i}{\partial x^j} - A_k\begin{Bmatrix} k \\ i\ j \end{Bmatrix}$$

and

$$A_{j,i} = \frac{\partial A_j}{\partial x^i} - A_k\begin{Bmatrix} k \\ j\ i \end{Bmatrix}$$

are covariant tensor.

So, $A_{i,j} - A_{j,i} = \dfrac{\partial A_i}{\partial x^j} - \dfrac{\partial A_j}{\partial x^i}$ is covariant tensor of second order, which is called curl of A_i.

Thus

$$\text{curl } A_i = A_{i,j} - A_{j,i}$$

Note: curl A_i is a skew-symmetric tensor.

Since

$$A_{j,i} - A_{i,j} = -(A_{i,j} - A_{j,i})$$

EXAMPLE 8

If A_{ij} be a skew-symmetric tensor of rank two. Show that

$$A_{ij,k} + A_{jk,i} + A_{ki,j} = \frac{\partial A_{ij}}{\partial x^k} + \frac{\partial A_{jk}}{\partial x^i} + \frac{\partial A_{ki}}{\partial x^j}$$

Solution

Since we know that

$$A_{ij,k} = \frac{\partial A_{ij}}{\partial x^k} - A_{lj}\begin{Bmatrix} l \\ i\ k \end{Bmatrix} - A_{il}\begin{Bmatrix} l \\ j\ k \end{Bmatrix}$$

$$A_{jk,i} = \frac{\partial A_{jk}}{\partial x^i} - A_{lk}\begin{Bmatrix} h \\ j\ i \end{Bmatrix} - A_{jl}\begin{Bmatrix} l \\ j\ k \end{Bmatrix}$$

$$A_{ki,j} = \frac{\partial A_{ki}}{\partial x^j} - A_{li}\begin{Bmatrix} l \\ k\ j \end{Bmatrix} - A_{kl}\begin{Bmatrix} l \\ i\ j \end{Bmatrix}$$

Adding these, we get

$$A_{ij,k} + A_{jk,i} + A_{ki,j} = \frac{\partial A_{ij}}{\partial x^k} + \frac{\partial A_{jk}}{\partial x^i} + \frac{\partial A_{ki}}{\partial x^j} - \left(A_{lj}\begin{Bmatrix} l \\ i\ k \end{Bmatrix} + A_{jl}\begin{Bmatrix} l \\ k\ i \end{Bmatrix} \right)$$

$$- \left(A_{li}\begin{Bmatrix} l \\ k\ j \end{Bmatrix} + A_{il}\begin{Bmatrix} l \\ j\ k \end{Bmatrix} \right) - \left(A_{kl}\begin{Bmatrix} l \\ i\ j \end{Bmatrix} + A_{lk}\begin{Bmatrix} l \\ j\ i \end{Bmatrix} \right)$$

Since $\begin{Bmatrix} l \\ i\ k \end{Bmatrix}$ is symmetric *i.e.*, $\begin{Bmatrix} l \\ i\ k \end{Bmatrix} = \begin{Bmatrix} l \\ k\ i \end{Bmatrix}$ etc.

$$= \frac{\partial A_{ij}}{\partial x^k} + \frac{\partial A_{jk}}{\partial x^i} + \frac{\partial A_{ki}}{\partial x^j} - \begin{Bmatrix} l \\ i\ k \end{Bmatrix} (A_{lj} + A_{jl})$$

$$- \begin{Bmatrix} l \\ k\ j \end{Bmatrix} (A_{li} + A_{il}) - \begin{Bmatrix} l \\ i\ j \end{Bmatrix} (A_{kl} + A_{lk})$$

Since A_{ij} is skew-symmetric. Then $A_{lj} = -A_{jl} \Rightarrow A_{lj} + A_{jl} = 0$. Similarly,

$$A_{li} + A_{il} = 0 \text{ and } A_{kl} + A_{lk} = 0$$

So,

$$A_{ij,k} + A_{jk,i} + A_{ki,j} = \frac{\partial A_{ij}}{\partial x^k} + \frac{\partial A_{jk}}{\partial x^i} + \frac{\partial A_{ki}}{\partial x^j}$$

THEOREM 4.5 *A necessary and sufficient condition that the curl of a vector field vanishes is that the vector field be gradient.*

Proof: Suppose that the curl of a vector A_i vanish so that

$$\text{curl } A_i = A_{i,j} - A_{j,i} = 0 \qquad \text{...(1)}$$

To prove that $A_i = \nabla\phi$, ϕ is scalar.

Since from (1),

$$A_{i,j} - A_{j,i} = 0$$

$$\Rightarrow \quad \frac{\partial A_i}{\partial x^j} - \frac{\partial A_j}{\partial x^i} = 0$$

$$\Rightarrow \quad \frac{\partial A_i}{\partial x^j} = \frac{\partial A_j}{\partial x^i}$$

$$\Rightarrow \quad \frac{\partial A_i}{\partial x^j} dx^j = \frac{\partial A_j}{\partial x^i} dx^j$$

$$\Rightarrow \quad dA_i = \frac{\partial}{\partial x^i}\left(A^j dx^j\right)$$

Integrating it we get

$$A_i = \int \frac{\partial}{\partial x^i}\left(A_j dx^j\right)$$

$$= \frac{\partial}{\partial x^i}\int A_j dx^j$$

$$A_i = \frac{\partial \phi}{\partial x^i}, \text{ where } \phi = \int A_j dx^j$$

or $$A_i = \nabla\phi.$$

Conversely suppose that a vector A_i is such that

$$A_i = \nabla\phi, \ \phi \text{ is scalar.}$$

To prove curl $A_i = 0$

Now,

$$A_i = \nabla\phi = \frac{\partial \phi}{\partial x^i}$$

$$\frac{\partial A_i}{\partial x^j} = \frac{\partial^2 \phi}{\partial x^j \partial x^i}$$

and $$\frac{\partial A_j}{\partial x^i} = \frac{\partial^2 \phi}{\partial x^i \partial x^j}$$

So, $$\frac{\partial A_i}{\partial x^j} - \frac{\partial A_j}{\partial x^i} = 0$$

So, $$\text{curl } A_i = A_{i,j} - A_{j,i} = \frac{\partial A_i}{\partial x^j} - \frac{\partial A_j}{\partial x^i} = 0$$

So, $$\text{curl } A_i = 0$$ **Proved.**

THEOREM 4.6 *Let ϕ and ψ be scalar functions of coordinates x^i. Let A be an arbitrary vector then*

(*i*) $div\,(\phi A) = \phi\; div\, A + A \cdot \nabla\phi$

(*ii*) $\nabla(\phi\psi) = \phi\nabla\psi + \psi\nabla\phi$

(*iii*) $\nabla^2(\phi\psi) = \phi\nabla^2\psi + \psi\nabla^2\psi + 2\nabla\phi\cdot\nabla\psi$

(*iv*) $div\,(\psi\nabla\phi) = \psi\nabla^2\phi + \nabla\phi\cdot\nabla\psi$

Proof: (*i*) Since we know that

$$\text{div } A^i = \frac{1}{\sqrt{g}}\frac{\partial\left(\sqrt{g}A^i\right)}{\partial x^i} \qquad ...(1)$$

replace A^i by ϕA^i, we get

$$\text{div}(\phi A^i) = \frac{1}{\sqrt{g}}\frac{\partial\left(\sqrt{g}\phi A^i\right)}{\partial x^i}$$

$$= \frac{1}{\sqrt{g}}\left[\frac{\partial(\sqrt{g}A^i)}{\partial x^i}\phi + \sqrt{g}A^i \cdot \frac{\partial \phi}{\partial x^i}\right]$$

$$= \frac{\partial \phi}{\partial x^i} \cdot A^i + \phi \frac{1}{\sqrt{g}} \frac{\partial(\sqrt{g}A^i)}{\partial x^i}$$

$$= \nabla\phi \cdot A^i + \phi \operatorname{div} A^i$$

Thus

$$\operatorname{div}(\phi A) = \nabla\phi \cdot A + \phi \operatorname{div} A \quad ...(2)$$

(*ii*) By definition of gradient,

$$\nabla(\phi\psi) = \frac{\partial(\phi\psi)}{\partial x^i}$$

$$= \phi\frac{\partial \psi}{\partial x^i} + \psi\frac{\partial \phi}{\partial x^i}$$

Thus $$\nabla(\phi\psi) = \phi\nabla\psi + \psi\nabla\phi \quad ...(3)$$

(*iii*) Taking divergence of both sides in equation (3), we get

$$\operatorname{div}(\nabla\phi\psi) = \operatorname{div}[\phi\nabla\psi + \psi\nabla\phi]$$

$$\nabla^2(\phi\psi) = \operatorname{div}(\phi\nabla\psi) + \operatorname{div}(\psi\nabla\phi)$$

$$= \nabla\phi \cdot \nabla\psi + \phi \operatorname{div}(\nabla\psi) + \nabla\psi \cdot \nabla\phi + \psi\operatorname{div}(\nabla\phi)$$

$$\nabla^2(\phi\psi) = \phi\operatorname{div}(\nabla\psi) + \psi\operatorname{div}(\nabla\phi) + 2\nabla\phi \cdot \nabla\psi$$

Thus,

$$\nabla^2(\phi\psi) = \phi\nabla^2\psi + \psi\nabla^2\phi + 2\nabla\phi \cdot \nabla\psi$$

(*iv*) Replace A by $\nabla\psi$ in equation (3), we get

$$\operatorname{div}(\phi\nabla\psi) = \nabla\phi \cdot \nabla\psi + \phi\operatorname{div}(\nabla\psi)$$

$$\operatorname{div}(\phi\nabla\psi) = \nabla\phi \cdot \nabla\psi + \phi\nabla^2\psi$$

THEOREM 4.7 *Let A_i be a covariant vector and ϕ a scalar function. Then*

(*i*) *curl* $(\phi A) = A \times \nabla\phi + \phi$ *curlA*

(*ii*) *curl* $(\psi\nabla\phi) = \nabla\phi \times \nabla\psi$

Proof: (*i*) Let A_i be a covariant vector then

$$\operatorname{curl} A = \operatorname{curl} A_i = A_{i,j} - A_{j,i}$$

Replacing A_i by ϕA_i, we get

$$\operatorname{curl}(\phi A_i) = (\phi A_i),j - (\phi A_j),i$$

$$= \phi,jA_i + \phi A_{i,j} - \phi,_i A_j - \phi A_{j,i}$$

$$= (A_i\phi,j - A_j\phi,i) + \phi(A_{i,j} - A_{j,i})$$

$$= A_i \times \nabla\phi + \phi\text{curl } A_i$$

So,

$$\text{curl } (\phi A) = A \times \nabla\phi + \phi\text{curl}A \quad \text{...(1)}$$

(*ii*) Replacing A by $\nabla\psi$ in equation (1), we get

$$\text{curl}(\phi\nabla\psi) = \phi\text{curl}(\nabla\psi) + \nabla\psi \times \nabla\phi$$

Interchange of ϕ and ψ, we get

$$\text{curl}(\psi\nabla\phi) = \psi\text{curl}(\nabla\phi) + \nabla\phi \times \nabla\psi.$$

Since curl $(\nabla\phi) = 0.$

So,

$$\text{curl}(\psi\nabla\phi) = \nabla\phi \times \nabla\psi.$$

Proved.

4.8 THE LAPLACIAN OPERATOR

The operator ∇^2 is called *Laplacian operator* read as "del square".

THEOREM 4.8 *If* ϕ *is a scalar function of coordinates* x^i *then*

$$\nabla^2\phi = \frac{1}{\sqrt{g}} \frac{\partial}{\partial x^k}\left(\sqrt{g} g^{kr} \frac{\partial\phi}{\partial x^r}\right)$$

Proof: Since

$$\nabla^2\phi = \text{div grad}\,\phi \quad \text{...(1)}$$

and

$$\text{grad}\,\phi = \frac{\partial\phi}{\partial x^r},$$

which is covariant vector.

But we know that any contravariant vector A^k associated with A_r (covariant vector) is

$$A^k = g^{kr}A_r \text{ (Sec Art. 3.4, Pg 43)}$$

Now, the contravariant vector A^k associated with $\frac{\partial\phi}{\partial x^r}$ (Covariant vector) is

$$A^k = g^{kr}\frac{\partial\phi}{\partial x^r}$$

Since

$$\text{div } A^i = \frac{1}{\sqrt{g}} \frac{\partial\left(\sqrt{g}A^k\right)}{\partial x^k}, \text{ (Sec Ex. 7, Pg 75)}$$

So, from (1)

$$\nabla^2\phi = \text{div}\left(g^{kr}\frac{\partial\phi}{\partial x^r}\right) = \frac{1}{\sqrt{g}}\frac{\partial\left(\sqrt{g}g^{kr}\frac{\partial\phi}{\partial x^r}\right)}{\partial x^k}$$ **Proved.**

EXAMPLE 9

Show that, in the cylindrical coordinates,

$$\nabla^2 V = \frac{1}{r}\frac{\partial}{\partial r}\left(r\frac{\partial V}{\partial r}\right) + \frac{1}{r^2}\frac{\partial^2 V}{\partial\theta^2} + \frac{\partial^2 V}{\partial z^2}$$

by Tensor method

Solution

The cylindrical coordinates are (r,θ,z). If V is a scalar function of (r,θ,z).

Now,

$$\nabla^2 V = \nabla\cdot\nabla V$$

Since

$$\nabla V = i\frac{\partial V}{\partial r} + j\frac{\partial V}{\partial\theta} + k\frac{\partial V}{\partial z}$$

Let

$$A_1 = \frac{\partial V}{\partial r},\ A_2 = \frac{\partial V}{\partial\theta},\ A_3 = \frac{\partial V}{\partial z} \qquad ...(1)$$

Then $\nabla V = iA_1 i + jA_2 j + kA_3 k$, since ∇V is covariant tensor. The metric in cylindrical coordinates is

$$ds^2 = dr^2 + r^2 d\theta^2 + dz^2$$

here, $x^1 = r,\ x^2 = \theta,\ x^3 = z$.

Since $ds^2 = g_{ij}dx^i dx^j$

$$g_{11} = 1,\ g_{22} = r^2,\ g_{33} = 1$$

and others are zero.

$$g = |g_{ij}| = \begin{vmatrix} g_{11} & g_{12} & g_{13} \\ g_{21} & g_{22} & g_{23} \\ g_{31} & g_{32} & g_{33} \end{vmatrix} = \begin{vmatrix} 1 & 0 & 0 \\ 0 & r^2 & 0 \\ 0 & 0 & 1 \end{vmatrix} = r^2$$

$\sqrt{g} = r$ (See Pg. 34, Example 1)

Now,

$$\text{div}(\nabla V) = \text{div}\,A^i = \frac{1}{\sqrt{g}}\frac{\partial\left(\sqrt{g}A^k\right)}{\partial x^k}$$

$$= \frac{1}{\sqrt{g}} \frac{\partial(\sqrt{g} A^1)}{\partial x^1} + \frac{1}{\sqrt{g}} \frac{\partial(\sqrt{g} A^2)}{\partial x^2} + \frac{\partial(\sqrt{g} A^3)}{\partial x^3}$$

$$\text{div}(\nabla V) = \frac{1}{r}\left[\frac{\partial(rA^1)}{\partial r} + \frac{\partial(rA^2)}{\partial \theta} + \frac{\partial(rA^3)}{\partial z}\right] \qquad \text{...(2)}$$

We can write

$$A^k = g^{kq} A_q \text{ (Associated tensor)}$$

$$A^k = g^{k1} A_1 + g^{k2} A_2 + g^{k3} A_3$$

Put $k = 1$

$$A^1 = g^{11} A_1 + g^{12} A_2 + g^{13} A_3$$

$$A^1 = g^{11} A_1 \text{ as } g^{12} = g^{13} = 0$$

Similarly,

$$A^2 = g^{22} A_2$$

$$A^3 = g^{33} A_3$$

and

$$g^{11} = \frac{\text{Cofactor of } g_{11} \text{ in } g}{g} = \frac{r^2}{r^2} = 1$$

$$g^{22} = \frac{\text{Cofactor of } g_{22} \text{ in } g}{g} = \frac{1}{r^2}$$

$$g^{33} = \frac{\text{Cofactor of } g_{33} \text{ in } g}{g} = \frac{r^2}{r^2} = 1 \text{ (See. Pg. 34, Ex.1)}$$

So,

$$A^1 = g^{11} A_1 = A_1$$

$$A^2 = g^{22} A_2 = \frac{1}{r^2} A_2$$

$$A^3 = g^{33} A_3 = A_3.$$

or $A^1 = A_1,\ A^2 = \frac{1}{r^2} A_2,\ A^3 = A_3.$

from (1), we get

$$A^1 = \frac{\partial V}{\partial r},\ A^2 = \frac{1}{r^2}\frac{\partial V}{\partial \theta},\ A^3 = \frac{\partial V}{\partial z}$$

from (2),

So,

$$\nabla^2 V = \text{div } A^i = \frac{1}{r}\left[\frac{\partial}{\partial r}\left(r\frac{\partial V}{\partial r}\right)+\frac{\partial}{\partial \theta}\left(r\frac{1}{r^2}\frac{\partial V}{\partial \theta}\right)+\frac{\partial}{\partial z}\left(r\frac{\partial V}{\partial z}\right)\right]$$

$$\text{div}(\nabla V) = \frac{1}{r}\left[\frac{\partial}{\partial r}\left(r\frac{\partial V}{\partial r}\right)+\frac{\partial}{\partial \theta}\left(\frac{1}{r}\frac{\partial V}{\partial \theta}\right)+\frac{\partial}{\partial z}\left(r\frac{\partial V}{\partial z}\right)\right]$$

$$= \frac{1}{r}\left[\frac{\partial}{\partial r}\left(r\frac{\partial V}{\partial r}\right)+\frac{1}{r}\frac{\partial^2 V}{\partial \theta^2}+r\frac{\partial^2 V}{\partial z^2}\right]$$

$$\text{div}(\nabla V) = \frac{1}{r}\frac{\partial}{\partial r}\left(r\frac{\partial V}{\partial r}\right)+\frac{1}{r^2}\frac{\partial^2 V}{\partial \theta^2}+\frac{\partial^2 V}{\partial z^2}$$

$$\nabla^2 V = \frac{1}{r}\frac{\partial}{\partial r}\left(r\frac{\partial V}{\partial r}\right)+\frac{1}{r^2}\frac{\partial^2 V}{\partial \theta^2}+\frac{\partial^2 V}{\partial z^2}$$

EXERCISES

1. Prove that the expressions are tensors

(*a*) $A_{ij},l = \dfrac{\partial A_{ij}}{\partial x^l} - \begin{Bmatrix}\alpha \\ i\,l\end{Bmatrix} A_{\alpha j} - \begin{Bmatrix}\alpha \\ j\,l\end{Bmatrix} A_{i\alpha}$

(*b*) $A^r_{i\,jk,l} = \dfrac{\partial A^r_{i\,jk}}{\partial x^l} - \begin{Bmatrix}\alpha \\ i\;l\end{Bmatrix} A^r_{\alpha jk} - \begin{Bmatrix}\alpha \\ j\;l\end{Bmatrix} A^r_{i\alpha k} - \begin{Bmatrix}\alpha \\ k\;l\end{Bmatrix} A^r_{ij\alpha} + \begin{Bmatrix}r \\ \alpha\;l\end{Bmatrix} A^\alpha_{ijk}$

2. Prove that

$$A^j_{i,j} = \frac{1}{\sqrt{g}}\frac{\partial(A^j_i\sqrt{g})}{\partial x^j} - A^j_k\begin{Bmatrix}k \\ i\;j\end{Bmatrix}$$

3. If A^{ijk} is a skew-symmetric tensor show that $\dfrac{1}{\sqrt{g}}\dfrac{\partial}{\partial x^k}(\sqrt{g}A^{i\,jk})$ is a tensor.

4. Prove that the necessary and sufficient condition that all the Christoffel symbols vanish at a point is that g_{ij} are constant.

5. Evaluate the Christoffel symbols in cylindrical coordinates.

6. Define covariant differentiation of a tensor w.r. to the fundamental tensor g_{ij}. Show that the covariant differentiation of sums and products of tensors obey the same result as ordinary differentiation.

7. Let contravariant and covariant components of the same vector A be A^i and A_i respectively then prove that

$$\text{div } A^i = \text{div } A_i$$

8. If A_{ij} is the curl of a covariant vector. Prove that

$$A_{ij,k} + A_{jk,i} + A_{ki,j} = 0$$

9. To prove that $u \cdot \nabla u = -\, u \operatorname{curl} u$ if u a vector of constant magnitude.
10. A necessary and sufficient condition that the covariant derivative vector be symmetric is that the vector must be gradient.
11. Show that, in spherical coordinates

$$\nabla^2 V = \frac{1}{r^2}\frac{\partial}{\partial r}\left(r^2 \frac{\partial V}{\partial r}\right) + \frac{1}{r^2 \sin\theta}\frac{\partial}{\partial \theta}\left(\sin\theta \frac{\partial V}{\partial \theta}\right) + \frac{1}{r^2 \sin^2\theta}\frac{\partial^2 V}{\partial \phi^2}$$

by tensor method.

CHAPTER – 5

RIEMANN-CHRISTOFFEL TENSOR

5.1 RIEMANN-CHRISTOFFEL TENSOR

If A_i is a covariant tensor then the covariant x^j derivative of A_i is given by

$$A_{i,j} = \frac{\partial A_i}{\partial x^j} - \begin{Bmatrix} \alpha \\ i \ j \end{Bmatrix} A_\alpha \quad \text{...(1)}$$

Differentiating covariantly the equation (1) w.r. to x^k, we get

$$\begin{aligned} A_{i,jk} &= \frac{\partial A_{i,j}}{\partial x^k} - \begin{Bmatrix} \alpha \\ i \ k \end{Bmatrix} A_{\alpha,j} - \begin{Bmatrix} \alpha \\ j \ k \end{Bmatrix} A_{i,\alpha} \\ &= \frac{\partial}{\partial x^k}\left(\frac{\partial A_i}{\partial x^j} - \begin{Bmatrix} \alpha \\ i \ j \end{Bmatrix} A_\alpha\right) - \begin{Bmatrix} \alpha \\ i \ k \end{Bmatrix}\left(\frac{\partial A_\alpha}{\partial x^j} - \begin{Bmatrix} \beta \\ \alpha \ j \end{Bmatrix} A_\beta\right) \\ &\quad - \begin{Bmatrix} \alpha \\ i \ k \end{Bmatrix}\left(\frac{\partial A_i}{\partial x^\alpha} - \begin{Bmatrix} \gamma \\ i \ k \end{Bmatrix} A_\gamma\right) \end{aligned}$$

$$\begin{aligned} A_{i,jk} &= \frac{\partial^2 A_i}{\partial x^k \partial x^j} - \frac{\partial \begin{Bmatrix} \alpha \\ i \ j \end{Bmatrix}}{\partial x^k} A_\alpha - \begin{Bmatrix} \alpha \\ i \ k \end{Bmatrix} \frac{\partial A_\alpha}{\partial x^k} - \begin{Bmatrix} \alpha \\ i \ k \end{Bmatrix} \frac{\partial A_\alpha}{\partial x^j} \\ &\quad + \begin{Bmatrix} \alpha \\ i \ k \end{Bmatrix} \begin{Bmatrix} \beta \\ \alpha \ j \end{Bmatrix} A_\beta - \begin{Bmatrix} \alpha \\ j \ k \end{Bmatrix} \frac{\partial A_i}{\partial x^\alpha} + \begin{Bmatrix} \alpha \\ j \ k \end{Bmatrix} \begin{Bmatrix} \gamma \\ i \ \alpha \end{Bmatrix} A_\gamma \end{aligned} \quad \text{...(2)}$$

Interchanging j and k in equation (2), we get

$$A_{i,kj} = \frac{\partial^2 A_i}{\partial x^j \partial x^k} - \frac{\partial \begin{Bmatrix} \alpha \\ i \ k \end{Bmatrix}}{\partial x^j} A_\alpha - \begin{Bmatrix} \alpha \\ i \ k \end{Bmatrix} \frac{\partial A_\alpha}{\partial x^j} - \begin{Bmatrix} \alpha \\ i \ j \end{Bmatrix} \frac{\partial A_\alpha}{\partial x^k}$$

$$+\begin{Bmatrix}\alpha\\ i\ \ j\end{Bmatrix}\begin{Bmatrix}\beta\\ \alpha\ \ k\end{Bmatrix}A_\beta-\begin{Bmatrix}\alpha\\ k\ \ i\end{Bmatrix}\frac{\partial A_i}{\partial x^\alpha}+\begin{Bmatrix}\alpha\\ k\ \ j\end{Bmatrix}\begin{Bmatrix}\gamma\\ i\ \ \alpha\end{Bmatrix}A_\gamma \quad ...(3)$$

Subtract equation (3) from (2), we get

$$A_{i,jk}-A_{i,kj}=\begin{Bmatrix}\alpha\\ i\ \ k\end{Bmatrix}\begin{Bmatrix}\beta\\ \alpha\ \ j\end{Bmatrix}A_\beta-\frac{\partial\begin{Bmatrix}\alpha\\ i\ \ j\end{Bmatrix}}{\partial x^k}A_\alpha-\begin{Bmatrix}\alpha\\ i\ \ j\end{Bmatrix}\begin{Bmatrix}\beta\\ \alpha\ \ k\end{Bmatrix}A_\beta+\frac{\partial\begin{Bmatrix}\alpha\\ i\ \ k\end{Bmatrix}}{\partial x^j}A_\alpha$$

Interchanging of α and β in the first and third term of above equation, we get

$$A_{i,jk}-A_{i,kj}=\left[\frac{\partial\begin{Bmatrix}\alpha\\ i\ \ k\end{Bmatrix}}{\partial x^j}-\frac{\partial\begin{Bmatrix}\alpha\\ i\ \ j\end{Bmatrix}}{\partial x^k}+\begin{Bmatrix}\beta\\ i\ \ k\end{Bmatrix}\begin{Bmatrix}\alpha\\ \beta\ \ j\end{Bmatrix}-\begin{Bmatrix}\beta\\ i\ \ j\end{Bmatrix}\begin{Bmatrix}\alpha\\ \beta\ \ k\end{Bmatrix}\right]A_\alpha \quad ...(4)$$

$$A_{i,jk}-A_{i,kj}=A_\alpha R^\alpha_{ijk} \quad ...(5)$$

where

$$R^\alpha_{ijk}=\frac{\partial\begin{Bmatrix}\alpha\\ i\ \ k\end{Bmatrix}}{\partial x^j}-\frac{\partial\begin{Bmatrix}\alpha\\ i\ \ j\end{Bmatrix}}{\partial x^k}+\begin{Bmatrix}\beta\\ i\ \ k\end{Bmatrix}\begin{Bmatrix}\alpha\\ \beta\ \ j\end{Bmatrix}-\begin{Bmatrix}\beta\\ i\ \ j\end{Bmatrix}\begin{Bmatrix}\alpha\\ \beta\ \ k\end{Bmatrix} \quad ...(6)$$

Since A_i is an arbitrary covariant tensor of rank one and difference of two tensors $A_{i,jk}-A_{i,kj}$ is a covariant tensor of rank three. Hence it follows from quotient law that R^α_{ijk} is a mixed tensor of rank four. The tensor R^α_{ijk} is called *Riemann Christoffel tensor or Curvature tensor* for the metric $g_{ij}dx^i dx^j$. The symbol R^α_{ijk} is called Riemann's symbol of second kind.

Now, if the left hand side of equation (4) is to vanish *i.e.,* if the order of covariant differentiation is to be immaterial then

$$R^\alpha_{ijk}=0$$

Since A_α is arbitrary. In general $R^\alpha_{ijk}\neq 0$, so that the order of covariant differentiation is not immaterial, It is clear from the equation (4) that "a necessary and sufficient condition for the validity of inversion of the order of covariant differentiation is that the tensor R^α_{ijk} vanishes identically.

Remark

The tensor

$$R^i_{jkl}=\begin{vmatrix}\dfrac{\partial}{\partial x^k} & \dfrac{\partial}{\partial x^l}\\ \begin{Bmatrix}i\\ j\ \ k\end{Bmatrix} & \begin{Bmatrix}i\\ j\ \ l\end{Bmatrix}\end{vmatrix}+\begin{vmatrix}\begin{Bmatrix}i\\ \alpha\ \ k\end{Bmatrix} & \begin{Bmatrix}i\\ \alpha\ \ l\end{Bmatrix}\\ \begin{Bmatrix}\alpha\\ j\ \ k\end{Bmatrix} & \begin{Bmatrix}\alpha\\ j\ \ l\end{Bmatrix}\end{vmatrix} \quad ...(7)$$

THEOREM 5.1 *Curvature tensor $R^{\alpha}_{i\,jk}$ is anti symmetric w.r.t. indices j and k.*

Proof: We know that from (7) curvature tensor

$$R^{\alpha}_{i\,jk} = \begin{vmatrix} \dfrac{\partial}{\partial x^{j}} & \dfrac{\partial}{\partial x^{k}} \\ \begin{Bmatrix} \alpha \\ i\ \ j \end{Bmatrix} & \begin{Bmatrix} \alpha \\ i\ \ k \end{Bmatrix} \end{vmatrix} + \begin{vmatrix} \begin{Bmatrix} \alpha \\ \beta\ \ j \end{Bmatrix} & \begin{Bmatrix} \alpha \\ \beta\ \ k \end{Bmatrix} \\ \begin{Bmatrix} \beta \\ i\ \ j \end{Bmatrix} & \begin{Bmatrix} \beta \\ i\ \ k \end{Bmatrix} \end{vmatrix}$$

$$R^{\alpha}_{i\,jk} = \frac{\partial \begin{Bmatrix} \alpha \\ i\ \ k \end{Bmatrix}}{\partial x^{j}} - \frac{\partial \begin{Bmatrix} \alpha \\ i\ \ j \end{Bmatrix}}{\partial x^{k}} + \begin{Bmatrix} \alpha \\ \beta\ \ j \end{Bmatrix} \begin{Bmatrix} \beta \\ i\ \ k \end{Bmatrix} - \begin{Bmatrix} \alpha \\ \beta\ \ k \end{Bmatrix} \begin{Bmatrix} \beta \\ i\ \ j \end{Bmatrix}$$

Interchanging j and k, we get

$$R^{\alpha}_{i\,kj} = \frac{\partial \begin{Bmatrix} \alpha \\ i\ \ j \end{Bmatrix}}{\partial x^{k}} - \frac{\partial \begin{Bmatrix} \alpha \\ i\ \ k \end{Bmatrix}}{\partial x^{j}} + \begin{Bmatrix} \alpha \\ \beta\ \ k \end{Bmatrix} \begin{Bmatrix} \beta \\ i\ \ j \end{Bmatrix} - \begin{Bmatrix} \alpha \\ \beta\ \ j \end{Bmatrix} \begin{Bmatrix} \beta \\ i\ \ k \end{Bmatrix}$$

$$= -\left[\frac{\partial \begin{Bmatrix} \alpha \\ i\ \ k \end{Bmatrix}}{\partial x^{j}} - \frac{\partial \begin{Bmatrix} \alpha \\ i\ \ j \end{Bmatrix}}{\partial x^{k}} + \begin{Bmatrix} \alpha \\ \beta\ \ j \end{Bmatrix} \begin{Bmatrix} \beta \\ i\ \ k \end{Bmatrix} - \begin{Bmatrix} \alpha \\ \beta\ \ k \end{Bmatrix} \begin{Bmatrix} \beta \\ i\ \ j \end{Bmatrix} \right]$$

$$R^{\alpha}_{ik\,j} = -R^{\alpha}_{i\,jk}$$

So, $R^{\alpha}_{i\,jk}$ antisymmetric w.r.t. indices j and k.

Theorem 5.2 *To prove that*

$$R^{\alpha}_{i\,jk} + R^{\alpha}_{jk\,i} + R^{\alpha}_{ki\,j} = 0$$

Proof: Since we know that

$$R^{\alpha}_{ijk} = \begin{vmatrix} \dfrac{\partial}{\partial x^{j}} & \dfrac{\partial}{\partial x^{k}} \\ \begin{Bmatrix} \alpha \\ i\ \ j \end{Bmatrix} & \begin{Bmatrix} \alpha \\ i\ \ k \end{Bmatrix} \end{vmatrix} + \begin{vmatrix} \begin{Bmatrix} \alpha \\ \beta\ \ j \end{Bmatrix} & \begin{Bmatrix} \alpha \\ \beta\ \ k \end{Bmatrix} \\ \begin{Bmatrix} \beta \\ i\ \ j \end{Bmatrix} & \begin{Bmatrix} \beta \\ i\ \ k \end{Bmatrix} \end{vmatrix}$$

$$R^{\alpha}_{ij\,k} = \frac{\partial \begin{Bmatrix} \alpha \\ i\ \ k \end{Bmatrix}}{\partial x^{j}} - \frac{\partial \begin{Bmatrix} \alpha \\ i\ \ j \end{Bmatrix}}{\partial x^{k}} + \begin{Bmatrix} \alpha \\ \beta\ \ j \end{Bmatrix} \begin{Bmatrix} \beta \\ i\ \ k \end{Bmatrix} - \begin{Bmatrix} \beta \\ i\ \ j \end{Bmatrix} \begin{Bmatrix} \alpha \\ \beta\ \ k \end{Bmatrix} \qquad \text{...(1)}$$

Similarly

$$R^{\alpha}_{jki} = \frac{\partial \begin{Bmatrix} \alpha \\ j\ i \end{Bmatrix}}{\partial x^k} - \frac{\partial \begin{Bmatrix} \alpha \\ j\ k \end{Bmatrix}}{\partial x^i} + \begin{Bmatrix} \alpha \\ \beta\ k \end{Bmatrix}\begin{Bmatrix} \beta \\ j\ i \end{Bmatrix} - \begin{Bmatrix} \beta \\ j\ k \end{Bmatrix}\begin{Bmatrix} \alpha \\ \beta\ i \end{Bmatrix} \quad \text{...(2)}$$

and

$$R^{\alpha}_{kij} = \frac{\partial \begin{Bmatrix} \alpha \\ k\ j \end{Bmatrix}}{\partial x^i} - \frac{\partial \begin{Bmatrix} \alpha \\ k\ i \end{Bmatrix}}{\partial x^j} + \begin{Bmatrix} \alpha \\ \beta\ i \end{Bmatrix}\begin{Bmatrix} \beta \\ k\ j \end{Bmatrix} - \begin{Bmatrix} \beta \\ k\ i \end{Bmatrix}\begin{Bmatrix} \alpha \\ \beta\ j \end{Bmatrix} \quad \text{...(3)}$$

On adding (1), (2) and (3), we get

$$R^{\alpha}_{ijk} + R^{\alpha}_{jki} + R^{\alpha}_{kij} = 0$$

This is called cyclic property.

5.2 RICCI TENSOR

The curvature tensor $R^{\alpha}_{i\,jk}$ can be contracted in three ways with respect to the index α and any one of its lower indices

$$R^{\alpha}_{\alpha jk},\ R^{\alpha}_{i\alpha k},\ R^{\alpha}_{ij\alpha}$$

Now, from equation (7), art. 5.1,

$$R^{\alpha}_{\alpha jk} = \begin{vmatrix} \dfrac{\partial}{\partial x^j} & \dfrac{\partial}{\partial x^k} \\ \begin{Bmatrix} \alpha \\ \alpha\ j \end{Bmatrix} & \begin{Bmatrix} \alpha \\ \alpha\ k \end{Bmatrix} \end{vmatrix} + \begin{vmatrix} \begin{Bmatrix} \alpha \\ \beta\ j \end{Bmatrix} & \begin{Bmatrix} \alpha \\ \beta\ k \end{Bmatrix} \\ \begin{Bmatrix} \beta \\ \alpha\ j \end{Bmatrix} & \begin{Bmatrix} \beta \\ \alpha\ k \end{Bmatrix} \end{vmatrix}$$

$$= \frac{\partial \begin{Bmatrix} \alpha \\ \alpha\ k \end{Bmatrix}}{\partial x^j} - \frac{\partial \begin{Bmatrix} \alpha \\ \alpha\ j \end{Bmatrix}}{\partial x^k} + \begin{Bmatrix} \alpha \\ \beta\ j \end{Bmatrix}\begin{Bmatrix} \beta \\ \alpha\ k \end{Bmatrix} - \begin{Bmatrix} \alpha \\ \beta\ k \end{Bmatrix}\begin{Bmatrix} \beta \\ \alpha\ j \end{Bmatrix}$$

$$= \frac{\partial^2 \log\sqrt{g}}{\partial x^j \partial x^k} - \frac{\partial^2 \log\sqrt{g}}{\partial x^k \partial x^j}$$

Since $\begin{Bmatrix} \alpha \\ \alpha\ k \end{Bmatrix} = \dfrac{\partial \log\sqrt{g}}{\partial x^k}$ and α and β are free indices $R^{\alpha}_{\alpha jk} = 0$.

Also for $R^{\alpha}_{ij\alpha}$.

Write R_{ij} for $R^{\alpha}_{ij\alpha}$

$$R_{ij} = R^{\alpha}_{ij\alpha} = \begin{vmatrix} \dfrac{\partial}{\partial x^j} & \dfrac{\partial}{\partial x^\alpha} \\ \begin{Bmatrix} \alpha \\ i\ j \end{Bmatrix} & \begin{Bmatrix} \alpha \\ i\ \alpha \end{Bmatrix} \end{vmatrix} + \begin{vmatrix} \begin{Bmatrix} \alpha \\ \beta\ j \end{Bmatrix} & \begin{Bmatrix} \alpha \\ \beta\ \alpha \end{Bmatrix} \\ \begin{Bmatrix} \beta \\ i\ j \end{Bmatrix} & \begin{Bmatrix} \beta \\ i\ \alpha \end{Bmatrix} \end{vmatrix}$$

$$R_{ij} = \frac{\partial \begin{Bmatrix} \alpha \\ i\ \alpha \end{Bmatrix}}{\partial x^j} - \frac{\partial \begin{Bmatrix} \alpha \\ i\ j \end{Bmatrix}}{\partial x^\alpha} + \begin{Bmatrix} \alpha \\ \beta\ j \end{Bmatrix} \begin{Bmatrix} \beta \\ i\ \alpha \end{Bmatrix} - \begin{Bmatrix} \alpha \\ \beta\ \alpha \end{Bmatrix} \begin{Bmatrix} \beta \\ i\ j \end{Bmatrix}$$

$$R_{ij} = \frac{\partial^2 \log \sqrt{g}}{\partial x^j \partial x^i} - \frac{\partial \begin{Bmatrix} \alpha \\ i\ j \end{Bmatrix}}{\partial x^\alpha} + \begin{Bmatrix} \alpha \\ \beta\ j \end{Bmatrix} \begin{Bmatrix} \beta \\ i\ \alpha \end{Bmatrix} - \begin{Bmatrix} \alpha \\ \beta\ \alpha \end{Bmatrix} \begin{Bmatrix} \beta \\ i\ j \end{Bmatrix} \qquad \ldots(1)$$

Interchanging the indices i and j we get

$$R_{ji} = \frac{\partial \begin{Bmatrix} \alpha \\ j\ \alpha \end{Bmatrix}}{\partial x^i} - \frac{\partial \begin{Bmatrix} \alpha \\ j\ i \end{Bmatrix}}{\partial x^\alpha} + \begin{Bmatrix} \alpha \\ \beta\ i \end{Bmatrix} \begin{Bmatrix} \beta \\ j\ \alpha \end{Bmatrix} - \begin{Bmatrix} \alpha \\ \beta\ \alpha \end{Bmatrix} \begin{Bmatrix} \beta \\ j\ i \end{Bmatrix}$$

$$R_{ji} = \frac{\partial^2 \log \sqrt{g}}{\partial x^i \partial x^j} - \frac{\partial \begin{Bmatrix} \alpha \\ i\ j \end{Bmatrix}}{\partial x^\alpha} + \begin{Bmatrix} \beta \\ i\ \alpha \end{Bmatrix} \begin{Bmatrix} \alpha \\ \beta\ j \end{Bmatrix} - \begin{Bmatrix} \alpha \\ \beta\ \alpha \end{Bmatrix} \begin{Bmatrix} \beta \\ i\ j \end{Bmatrix} \qquad \ldots(2)$$

(Since α and β are dummy indices in third term).

Comparing (1) and (2), we get

$$R_{ij} = R_{ji}$$

Thus R_{ij} is a symmetric Tensor and is called Ricci Tensor.

For $R^{\alpha}_{i\alpha k}$:

$$R^{\alpha}_{i\alpha k} = -R^{\alpha}_{ik\alpha} = -R_{ik}$$

5.3 COVARIANT RIEMANN-CHRISTOFFEL TENSOR

The associated tensor

$$R_{ijkl} = g_{i\alpha} R^{\alpha}_{jkl} \qquad \ldots(1)$$

is known as the covariant Riemann-Christoffel tensor or the *Riemann-Christoffel tensor of the first kind.*

Expression for R_{ijkl}

$$R_{ijkl} = g_{i\alpha} R^{\alpha}_{jkl}$$

$$= g_{i\alpha}\left(\frac{\partial}{\partial x^k}\begin{Bmatrix}\alpha\\ j\ l\end{Bmatrix} - \frac{\partial}{\partial x^l}\begin{Bmatrix}\alpha\\ j\ k\end{Bmatrix} + \begin{Bmatrix}\beta\\ j\ l\end{Bmatrix}\begin{Bmatrix}\alpha\\ \beta\ k\end{Bmatrix} - \begin{Bmatrix}\beta\\ j\ k\end{Bmatrix}\begin{Bmatrix}\alpha\\ \beta\ l\end{Bmatrix}\right) \quad ...(2)$$

Now,

$$g_{i\alpha}\frac{\partial}{\partial x^k}\begin{Bmatrix}\alpha\\ j\ l\end{Bmatrix} = \frac{\partial}{\partial x^k}\left[g_{i\alpha}\begin{Bmatrix}\alpha\\ j\ l\end{Bmatrix}\right] - \begin{Bmatrix}\alpha\\ j\ l\end{Bmatrix}\frac{\partial g_{i\alpha}}{\partial x^k}$$

$$= \frac{\partial[jl,i]}{\partial x^k} - \begin{Bmatrix}\alpha\\ j\ l\end{Bmatrix}\frac{\partial g_{i\alpha}}{\partial x^k} \quad ...(3)$$

Similarly,

$$g_{i\alpha}\frac{\partial}{\partial x^l}\begin{Bmatrix}\alpha\\ j\ k\end{Bmatrix} = \frac{\partial[jk,i]}{\partial x^l} - \begin{Bmatrix}\alpha\\ j\ k\end{Bmatrix}\frac{\partial g_{i\alpha}}{\partial x^l} \quad ...(4)$$

$$\left[\text{By the formula } \frac{udv}{dx} = \frac{duv}{dx} - \frac{vdu}{dx}\right]$$

Using (3) and (4) in equation (2), we get

$$R_{ijkl} = \frac{\partial[jl,i]}{\partial x^k} - \begin{Bmatrix}\alpha\\ j\ l\end{Bmatrix}\frac{\partial g_{i\alpha}}{\partial x^k} - \frac{\partial[jk,i]}{\partial x^l} + \begin{Bmatrix}\alpha\\ j\ k\end{Bmatrix}\frac{\partial g_{i\alpha}}{\partial x^l} + g_{i\alpha}\begin{Bmatrix}\beta\\ j\ l\end{Bmatrix}\begin{Bmatrix}\alpha\\ \beta\ k\end{Bmatrix} - g_{i\alpha}\begin{Bmatrix}\beta\\ j\ k\end{Bmatrix}\begin{Bmatrix}\alpha\\ \beta\ l\end{Bmatrix}$$

$$= \frac{\partial[jl,i]}{\partial x^k} - \frac{\partial[jk,i]}{\partial x^l} + \begin{Bmatrix}\alpha\\ j\ k\end{Bmatrix}\frac{\partial g_{i\alpha}}{\partial x^l} - \begin{Bmatrix}\alpha\\ j\ l\end{Bmatrix}\frac{\partial g_{i\alpha}}{\partial x^k} + g_{i\alpha}\begin{Bmatrix}\beta\\ j\ l\end{Bmatrix}\begin{Bmatrix}\alpha\\ \beta\ k\end{Bmatrix} - g_{i\alpha}\begin{Bmatrix}\beta\\ j\ k\end{Bmatrix}\begin{Bmatrix}\alpha\\ \beta\ l\end{Bmatrix}$$

$$R_{ijkl} = \frac{\partial[jl,i]}{\partial x^k} - \frac{\partial[jk,i]}{\partial x^l} + \begin{Bmatrix}\alpha\\ j\ k\end{Bmatrix}([il,\alpha]+[\alpha l,i]) - \begin{Bmatrix}\alpha\\ j\ l\end{Bmatrix}([ik,\alpha]+[\alpha k,i])$$

$$+ \begin{Bmatrix}\beta\\ j\ l\end{Bmatrix}[\beta k,i] - \begin{Bmatrix}\beta\\ j\ k\end{Bmatrix}[\beta l,i]$$

$$R_{ijkl} = \frac{\partial[jl,i]}{\partial x^k} - \frac{\partial[jk,i]}{\partial x^l} + \begin{Bmatrix}\alpha\\ j\ k\end{Bmatrix}[il,\alpha] - \begin{Bmatrix}\alpha\\ j\ l\end{Bmatrix}[ik,\alpha] \quad ...(5)$$

It is also written as

$$R_{ijkl} = \begin{vmatrix}\frac{\partial}{\partial x^k} & \frac{\partial}{\partial x^k}\\ [jk,i] & [jl,i]\end{vmatrix} + \begin{vmatrix}\begin{Bmatrix}\alpha\\ j\ k\end{Bmatrix} & \begin{Bmatrix}\alpha\\ j\ l\end{Bmatrix}\\ [ik,\alpha] & [il,\alpha]\end{vmatrix} \quad ...(6)$$

But we know that

$$[jl,i] = \frac{1}{2}\left(\frac{\partial g_{li}}{\partial x^j} + \frac{\partial g_{ji}}{\partial x^l} - \frac{\partial g_{jl}}{\partial x^i}\right) \quad ...(7)$$

and $$[jk,i] = \frac{1}{2}\left(\frac{\partial g_{ki}}{\partial x^j} + \frac{\partial g_{ji}}{\partial x^k} - \frac{\partial g_{ik}}{\partial x^i}\right) \qquad ...(8)$$

Using (7) and (8), equation (5) becomes

$$R_{ijkl} = \frac{1}{2}\frac{\partial}{\partial x^k}\left(\frac{\partial g_{li}}{\partial x^j} + \frac{\partial g_{ji}}{\partial x^l} - \frac{\partial g_{jl}}{\partial x^i}\right) - \frac{1}{2}\frac{\partial}{\partial x^l}\left(\frac{\partial g_{ki}}{\partial x^j} + \frac{\partial g_{ji}}{\partial x^k} - \frac{\partial g_{jk}}{\partial x^i}\right)$$

$$+\begin{Bmatrix}\alpha \\ j\ k\end{Bmatrix}[il,\alpha] - \begin{Bmatrix}\alpha \\ j\ l\end{Bmatrix}[ik,\alpha]$$

$$R_{ijkl} = \frac{1}{2}\left(\frac{\partial^2 g_{il}}{\partial x^j \partial x^k} + \frac{\partial^2 g_{jk}}{\partial x^i \partial x^l} - \frac{\partial^2 g_{ik}}{\partial x^j \partial x^l} - \frac{\partial^2 g_{jl}}{\partial x^i \partial x^k}\right) +$$

$$\begin{Bmatrix}\alpha \\ j\ k\end{Bmatrix}[il,\alpha] - \begin{Bmatrix}\alpha \\ j\ l\end{Bmatrix}[ik,\alpha] \qquad ...(9)$$

Since

$$\begin{Bmatrix}\alpha \\ j\ k\end{Bmatrix} = g^{\alpha\beta}[jk,\beta] \text{ and } \begin{Bmatrix}\alpha \\ j\ l\end{Bmatrix} = g^{\alpha\beta}[jl,\beta]$$

$$R_{ijkl} = \frac{1}{2}\left(\frac{\partial^2 g_{il}}{\partial x^j \partial x^k} + \frac{\partial^2 g_{jk}}{\partial x^i \partial x^l} - \frac{\partial^2 g_{ik}}{\partial x^j \partial x^l} - \frac{\partial^2 g_{jl}}{\partial x^i \partial x^l}\right)$$

$$+ g^{\alpha\beta}[jk,\beta][il,\alpha] - g^{\alpha\beta}[jl,\beta][ik,\alpha] \qquad ...(10)$$

This is expression for R_{ijkl}.

The equation (9) can also be written as

$$R_{ijkl} = \frac{1}{2}\left(\frac{\partial^2 g_{il}}{\partial x^j \partial x^k} + \frac{\partial^2 g_{jk}}{\partial x^i \partial x^l} - \frac{\partial^2 g_{ik}}{\partial x^j \partial x^l} - \frac{\partial^2 g_{jl}}{\partial x^i \partial x^k}\right)$$

$$+ g^{\alpha\beta}\begin{Bmatrix}\alpha \\ j\ k\end{Bmatrix}\begin{Bmatrix}\beta \\ i\ l\end{Bmatrix} - g^{\alpha\beta}\begin{Bmatrix}\alpha \\ j\ l\end{Bmatrix}\begin{Bmatrix}\beta \\ i\ k\end{Bmatrix}$$

5.4 PROPERTIES OF RIEMANN-CHRISTOFFEL TENSORS OF FIRST KIND R_{ijkl}

(*i*) $R_{jikl} = -R_{ijkl}$

(*ii*) $R_{ijlk} = -R_{ijkl}$

(*iii*) $R_{klij} = R_{ijkl}$

(*iv*) $R_{ijkl} + R_{iklj} + R_{iljk} = 0$

Proof: We know that from equation (9), Pg. 91.

$$R_{ijkl} = \frac{1}{2}\left(\frac{\partial^2 g_{il}}{\partial x^j \partial x^k} + \frac{\partial^2 g_{jk}}{\partial x^i \partial x^l} - \frac{\partial^2 g_{ik}}{\partial^2 x^j \partial x^l} - \frac{\partial^2 g_{jl}}{\partial x^i \partial x^k}\right)$$

$$+ \begin{Bmatrix} \alpha \\ j\ k \end{Bmatrix}[il,\alpha] - \begin{Bmatrix} \alpha \\ j\ l \end{Bmatrix}[ik,\alpha] \qquad ...(1)$$

(*i*) Interchanging of *i* and *j* in (1), we get

$$R_{jikl} = \frac{1}{2}\left(\frac{\partial^2 g_{jl}}{\partial x^i \partial x^k} = \frac{\partial^2 g_{ik}}{\partial x^j \partial x^l} - \frac{\partial^2 g_{jk}}{\partial x^i \partial x^l} - \frac{\partial^2 g_{il}}{\partial x^j \partial x^k}\right) + \begin{Bmatrix} \alpha \\ i\ k \end{Bmatrix}[jl,\alpha] - \begin{Bmatrix} \alpha \\ i\ l \end{Bmatrix}[jk,\alpha]$$

$$= -\frac{1}{2}\left(\frac{\partial^2 g_{il}}{\partial x^j \partial x^k} + \frac{\partial^2 g_{jk}}{\partial x^i \partial x^l} - \frac{\partial^2 g_{jl}}{\partial x^i \partial x^k} - \frac{\partial^2 g_{ik}}{\partial x^j \partial x^l}\right) + \begin{Bmatrix} \alpha \\ i\ k \end{Bmatrix}[jl,\alpha] - \begin{Bmatrix} \alpha \\ i\ l \end{Bmatrix}[jk,\alpha]$$

$$R_{jikl} = -R_{ijkl}$$

or $$R_{ijkl} = -R_{jikl}$$

(*ii*) Interchange *l* and *k* in equation (1) and Proceed as in (*i*)

(*iii*) Interchange *i* and *k* in (1), we get

$$R_{kjil} = \frac{1}{2}\left(\frac{\partial^2 g_{kl}}{\partial x^j \partial x^i} + \frac{\partial^2 g_{ji}}{\partial x^k \partial x^l} - \frac{\partial^2 g_{ki}}{\partial x^j \partial x^l} - \frac{\partial^2 g_{jl}}{\partial x^k \partial x^i}\right) + \begin{Bmatrix} \alpha \\ j\ i \end{Bmatrix}[kl,\alpha] - \begin{Bmatrix} \alpha \\ j\ l \end{Bmatrix}[ki,\alpha]$$

Now interchange *j* and *l*, we get

$$R_{klij} = \frac{1}{2}\left(\frac{\partial^2 g_{kj}}{\partial x^l \partial x^i} + \frac{\partial^2 g_{li}}{\partial x^k \partial x^j} - \frac{\partial^2 g_{ki}}{\partial x^l \partial x^j} - \frac{\partial^2 g_{lj}}{\partial x^k \partial x^i}\right) + \begin{Bmatrix} \alpha \\ l\ i \end{Bmatrix}[kj,\alpha] - \begin{Bmatrix} \alpha \\ l\ j \end{Bmatrix}[ki,\alpha]$$

For

$$\begin{Bmatrix} \alpha \\ l\ i \end{Bmatrix}[kj,\alpha] = \begin{Bmatrix} \alpha \\ l\ i \end{Bmatrix} g^{\alpha\beta} \begin{Bmatrix} \beta \\ k\ \ j \end{Bmatrix}$$

$$= \begin{Bmatrix} \beta \\ k\ \ j \end{Bmatrix} g^{\alpha\beta} \begin{Bmatrix} \alpha \\ l\ i \end{Bmatrix}$$

$$= \begin{Bmatrix} \beta \\ k\ \ j \end{Bmatrix}[li,\beta]$$

$$\begin{Bmatrix} \alpha \\ l\ i \end{Bmatrix}[li,\beta] = \begin{Bmatrix} \alpha \\ k\ j \end{Bmatrix}[li,\alpha]$$

So,

$$R_{klij} = \frac{1}{2}\left(\frac{\partial^2 g_{kj}}{\partial x^l \partial x^i} + \frac{\partial^2 g_{li}}{\partial x^k \partial x^j} - \frac{\partial^2 g_{ki}}{\partial x^l \partial x^j} - \frac{\partial^2 g_{lj}}{\partial x^k \partial x^i}\right) + \begin{Bmatrix} \alpha \\ k \ j \end{Bmatrix}[li,\alpha] - \begin{Bmatrix} \alpha \\ l \ j \end{Bmatrix}[ki,\alpha]$$

$$R_{klij} = R_{ijkl}, \text{ from (1)}$$

from equation (1)

$$R_{ijkl} = \frac{1}{2}\left(\frac{\partial^2 g_{il}}{\partial x^j \partial x^k} + \frac{\partial^2 g_{jk}}{\partial x^i \partial x^l} - \frac{\partial^2 g_{jk}}{\partial x^j \partial x^l} - \frac{\partial^2 g_{jl}}{\partial x^j \partial x^k}\right) + \begin{Bmatrix} \alpha \\ j \ k \end{Bmatrix}[il,\alpha] - \begin{Bmatrix} \alpha \\ j \ l \end{Bmatrix}[ik,\alpha]$$

$$R_{iklj} = \frac{1}{2}\left(\frac{\partial^2 g_{ij}}{\partial x^k \partial x^l} + \frac{\partial^2 g_{kl}}{\partial x^i \partial x^j} - \frac{\partial^2 g_{il}}{\partial x^k \partial x^j} - \frac{\partial^2 g_{kj}}{\partial x^i \partial x^l}\right) + \begin{Bmatrix} \alpha \\ k \ l \end{Bmatrix}[ij,\alpha] - \begin{Bmatrix} \alpha \\ k \ j \end{Bmatrix}[il,\alpha]$$

$$R_{iljk} = \frac{1}{2}\left(\frac{\partial^2 g_{ik}}{\partial x^l \partial x^j} + \frac{\partial^2 g_{lj}}{\partial x^i \partial x^k} - \frac{\partial^2 g_{ij}}{\partial x^l \partial x^k} - \frac{\partial^2 g_{lk}}{\partial x^i \partial x^j}\right) + \begin{Bmatrix} \alpha \\ l \ j \end{Bmatrix}[ik,\alpha] - \begin{Bmatrix} \alpha \\ l \ k \end{Bmatrix}[ij,\alpha]$$

On adding these equations, we get

$$R_{ijkl} + R_{iklj} + R_{iljk} = 0$$

This property of R_{ijkl} is called cyclic property.

Theorem 5.3 *Show that the number of not necessarily independent components of curvature tensor does not exceed* $\frac{1}{12}n^2(n^2-1)$.

Or

Show that number of distinct non-vanishing components of curvature tensor does not exceed $\frac{1}{12}n^2(n^2-1)$.

Proof: The distinct non-vanishing components of R_{ijkl} of three types.

(*i*) Symbols with two distinct indices *i.e.,* R_{ijij}. In this case total number of distinct non-vanishing components of R_{ijkl} are $\frac{1}{2}n(n-1)$.

(*ii*) Symbols with three distinct indices *i.e.,* R_{ijik}. In this case, total number of distinct non-vanishing components R_{ijkl} are $\frac{1}{2}n(n-1)(n-2)$.

(*iii*) Symbols R_{ijkl} with four distinct indices. In this case, total number of distinct non-vanishing components of R_{ijkl} is $\frac{n^2(n^2-1)}{12}$.

Hence the number of distinct non-vanishing components of the curvature tensor R_{ijkl} does not exceed $\frac{1}{12}n^2(n^2-1)$.

Remark

When R_{ijkl} is of the form R_{iiii} *i.e.*, all indices are same

In this case, R_{iiii} has no components.

5.5 BIANCHI IDENTITY

It states that

$$R^i_{jkl,m} + R^i_{jlm,k} + R^i_{jmk,l} = 0$$

and $$R_{hjkl,m} + R_{hjlm,k} + R_{hjmk,l} = 0$$

Proof: Introducing geodesic coordinate[1] in which Christoffel symbols are constant with the pole at P_0.

Since we know that

$$R^i_{jkl} = \begin{vmatrix} \frac{\partial}{\partial x^k} & \frac{\partial}{\partial x^l} \\ \begin{Bmatrix} i \\ jk \end{Bmatrix} & \begin{Bmatrix} i \\ jl \end{Bmatrix} \end{vmatrix} + \begin{vmatrix} \begin{Bmatrix} i \\ mk \end{Bmatrix} & \begin{Bmatrix} i \\ ml \end{Bmatrix} \\ \begin{Bmatrix} m \\ jk \end{Bmatrix} & \begin{Bmatrix} m \\ jl \end{Bmatrix} \end{vmatrix}$$

$$R^i_{jkl} = \frac{\partial \begin{Bmatrix} i \\ jl \end{Bmatrix}}{\partial x^k} - \frac{\partial \begin{Bmatrix} i \\ jk \end{Bmatrix}}{\partial x^l} + \begin{Bmatrix} i \\ mk \end{Bmatrix}\begin{Bmatrix} m \\ jl \end{Bmatrix} - \begin{Bmatrix} m \\ jk \end{Bmatrix}\begin{Bmatrix} i \\ ml \end{Bmatrix}$$

$$R^i_{jkl,m} = \frac{\partial^2 \begin{Bmatrix} i \\ j\,l \end{Bmatrix}}{\partial x^m \partial x^k} - \frac{\partial^2 \begin{Bmatrix} i \\ j\,k \end{Bmatrix}}{\partial x^m \partial x^l} \qquad ...(1)$$

Since $\begin{Bmatrix} i \\ j\,k \end{Bmatrix}, \begin{Bmatrix} m \\ j\,l \end{Bmatrix}$ etc. are constant at pole.

So, their derivatives are zero.

$$R^i_{jlm} = \begin{vmatrix} \frac{\partial}{\partial x^l} & \frac{\partial}{\partial x^m} \\ \begin{Bmatrix} i \\ j\,l \end{Bmatrix} & \begin{Bmatrix} i \\ j\,m \end{Bmatrix} \end{vmatrix} + \begin{vmatrix} \begin{Bmatrix} i \\ \alpha\ l \end{Bmatrix} & \begin{Bmatrix} i \\ \alpha\ m \end{Bmatrix} \\ \begin{Bmatrix} \alpha \\ j\ l \end{Bmatrix} & \begin{Bmatrix} \alpha \\ j\ m \end{Bmatrix} \end{vmatrix}$$

[1] Details of geodesic coordinate given in chapter *curvature in curve*. Geodesic.

$$= \frac{\partial \begin{Bmatrix} i \\ j\ m \end{Bmatrix}}{\partial x^l} - \frac{\partial \begin{Bmatrix} i \\ j\ l \end{Bmatrix}}{\partial x^m} + \begin{Bmatrix} i \\ \alpha\ l \end{Bmatrix}\begin{Bmatrix} \alpha \\ j\ m \end{Bmatrix} - \begin{Bmatrix} i \\ \alpha\ m \end{Bmatrix}\begin{Bmatrix} \alpha \\ j\ l \end{Bmatrix}$$

$$R^i_{jlm,k} = \frac{\partial^2 \begin{Bmatrix} i \\ j\ m \end{Bmatrix}}{\partial x^k \partial x^l} - \frac{\partial^2 \begin{Bmatrix} i \\ j\ l \end{Bmatrix}}{\partial x^k \partial x^m} \qquad \ldots(2)$$

and

$$R^i_{jmk} = \begin{vmatrix} \dfrac{\partial}{\partial x^m} & \dfrac{\partial}{\partial x^k} \\ \begin{Bmatrix} i \\ j\ m \end{Bmatrix} & \begin{Bmatrix} i \\ j\ k \end{Bmatrix} \end{vmatrix} + \begin{vmatrix} \begin{Bmatrix} i \\ \beta\ m \end{Bmatrix} & \begin{Bmatrix} i \\ \beta\ k \end{Bmatrix} \\ \begin{Bmatrix} \beta \\ j\ m \end{Bmatrix} & \begin{Bmatrix} \beta \\ j\ k \end{Bmatrix} \end{vmatrix}$$

$$R^i_{jmk} = \frac{\partial \begin{Bmatrix} i \\ j\ k \end{Bmatrix}}{\partial x^m} - \frac{\partial \begin{Bmatrix} i \\ j\ m \end{Bmatrix}}{\partial x^k} + \begin{Bmatrix} i \\ \beta\ m \end{Bmatrix}\begin{Bmatrix} \beta \\ j\ k \end{Bmatrix} - \begin{Bmatrix} i \\ \beta\ k \end{Bmatrix}\begin{Bmatrix} \alpha \\ j\ m \end{Bmatrix}$$

$$R^i_{jmk} = \frac{\partial^2 \begin{Bmatrix} i \\ j\ k \end{Bmatrix}}{\partial x^m \partial x^l} - \frac{\partial^2 \begin{Bmatrix} i \\ j\ m \end{Bmatrix}}{\partial x^k \partial x^l} \qquad \ldots(3)$$

On adding (1), (2) and (3), we get

$$R^i_{jkl,m} + R^i_{jlm,k} + R^i_{jmk,l} = 0 \qquad \ldots(4)$$

Multiplying $R^i_{jkl,m}$ by g_{hi} *i.e.*,

$$g_{hi} R^i_{jkl,m} = R_{hjkl,m}$$

Then equation (4) becomes

$$R_{hjkl,m} + R_{hjlm,k} + R_{hjmk,l} = 0 \qquad \ldots(5)$$

Since every term of equation (4) and (5) is a tensor. So, equation (4) and (5) are tensor equations and therefore hold in every coordinate system. Further, P_0 is an arbitrary point of V_n. Thus there hold throughout V_n. Hence equation (4) or (5) is called Bianchi identity.

5.6 EINSTEIN TENSOR

Theorem 5.4 *To prove the tensor* $R^i_j - \frac{1}{2}\delta^i_j R$ *is divergence free.*

Proof: We know that from equation (5)

$$R_{hjkl,m} + R_{hjlm,k} + R_{hjmk,l} = 0$$

Multiply it by $g^{hl} g^{jk}$, we get

$$g^{hl} g^{jk} R_{hjkl,m} + g^{hl} g^{jk} R_{hjlm,k} + g^{hl} g^{jk} R_{hjmk,l} = 0$$

$$g^{jk} R_{jk,m} + g^{hl} g^{jk} (-R_{hjml,k}) + g^{hl} g^{jk} (-R_{jhmk,l}) = 0$$

Since $R_{hjlm} = -R_{hjml}$ & $R_{hjmk} = -R_{jhmk}$

$$g^{jk} R_{jk,m} - g^{jk} R_{jm,k} - g^{hl} R_{hm,l} = 0$$

$$R,m - R^k_{m,k} - R^l_{m,l} = 0 \text{ Since } g^{jk} R_{jk} = R$$

$$R,m - R^k_{m,k} - R^k_{m,k} = 0$$

$$R,m - 2R^k_{m,k} = 0$$

$$R^k_{m,k} - \frac{1}{2} R,m = 0$$

$$R^k_{m,k} - \frac{1}{2}\delta^k_m R,k = 0 \quad \text{since } R,m = \delta^k_m R,k$$

$$\left(R^k_m - \frac{1}{2}\delta^k_m R\right)_{,k} = 0$$

$R^k_m - \frac{1}{2}\delta^k_m R$ is divergence free.

The tensor $R^k_m - \frac{1}{2}\delta^k_m R = G^k_m$ or $R^i_j - \frac{1}{2}\delta^i_j R$ is known as *Einstein Tensor*.

5.7 RIEMANN CURVATURE OF A V_n

Consider two unit vectors p^i and q^i at a point P_0 of V_n. These vectors at P_0 determine a pencil of directions deferred by $t^i = \alpha p^i + \beta q^i$. α and β being parameters. One and only one geodesic will pass through P_0 in the direction of p^i. Similarly one and only one geodesic will pass through in the direction q^i. These two geodesics through P_0 determined by the orientation of the unit vectors p^i and q^i. Let this surface is denoted by S.

The Gaussian curvature of S at P_0 is defined to be the Riemannian Curvature of V_n at P_0 for the orientation determined by p^i and q^i.

Let the coordinates y^i of V_n are Riemannian coordinates with origin at P_0. The equation of surfaces S in given by

$$y^i = (p^i\alpha + q^i\beta)s \qquad ...(1)$$

i.e.,

$$y^i = p^i u^1 + q^i u^2 \qquad ...(2)$$

where $\alpha s = u^1$ and $\beta s = u^2$, three parameters namely α, β, s can be reduced to two parameters u^1 and u^2. Here u^1 and u^2 are coordinates of any current point on S.

Let $ds^2 = b_{\alpha\beta} du^\alpha du^\beta$ be the metric for the surface S. where

$$b_{\alpha\beta} = g_{ij} \frac{\partial y^i}{\partial u^\alpha} \frac{\partial y^j}{\partial u^\beta} \quad (\alpha, \beta = 1, 2)$$

Let $\begin{Bmatrix} k \\ i \ j \end{Bmatrix}_g$ and $\begin{Bmatrix} \gamma \\ \alpha \ \beta \end{Bmatrix}$ be the Christoffel symbols corresponding to the coordinates y^i and u^α.

Let $\overline{R}_{\alpha\beta\gamma\delta}$ and R_{hijk} be curvature tensor corresponding to the metrices $b_{\alpha\beta} du^\alpha du^\beta$ and $g_{ij} dy^i dy^j$.

Since the Greek letters $\alpha, \beta, \gamma, \delta$ take values 1,2 and so that the number of independent non-vanishing components of $\overline{R}_{\alpha\beta\gamma\delta}$ is $\frac{1}{12} n^2 (n^2 - 1)$ for $n = 2$, *i.e.*, they are $\frac{1}{12} \cdot 2^2 (2^2 - 1) = 1$. Let us transform the coordinate system u^α to u'^α and suppose that the corresponding value of $\overline{R}_{1212}$ are $\overline{R}'_{1212}$.

Then

$$\overline{R}'_{1212} = \overline{R}_{\alpha\beta\gamma\delta} \frac{\partial u^\alpha}{\partial u'^1} \frac{\partial u^\beta}{\partial u'^2} \frac{\partial u^\gamma}{\partial u'^1} \frac{\partial u^\delta}{\partial u'^2}$$

$$= \overline{R}_{1\beta\gamma\delta} \frac{\partial u^1}{\partial u'^1} \frac{\partial u^\beta}{\partial u'^2} \frac{\partial u^\gamma}{\partial u'^1} \frac{\partial u^\delta}{\partial u'^2} + \overline{R}_{2\beta\gamma\delta} \frac{\partial u^2}{\partial u'^1} \frac{\partial u^\beta}{\partial u'^2} \frac{\partial u^\gamma}{\partial u'^1} \frac{\partial u^\delta}{\partial u'^2}$$

$$= \overline{R}_{12\gamma\delta} \frac{\partial u^1}{\partial u'^1} \frac{\partial u^2}{\partial u'^2} \frac{\partial u^\gamma}{\partial u'^1} \frac{\partial u^\delta}{\partial u'^2} + \overline{R}_{21\gamma\delta} \frac{\partial u^2}{\partial u'^1} \frac{\partial u^1}{\partial u'^2} \frac{\partial u^\gamma}{\partial u'^1} \frac{\partial u^\delta}{\partial u'^2}$$

$$= \overline{R}_{1212} \frac{\partial u^1}{\partial u'^1} \frac{\partial u^2}{\partial u'^2} \frac{\partial u^1}{\partial u'^1} \frac{\partial u^2}{\partial u'^2} + \overline{R}_{1221} \frac{\partial u^1}{\partial u'^1} \frac{\partial u^2}{\partial u'^2} \frac{\partial u^2}{\partial u'^1} \frac{\partial u^1}{\partial u'^2}$$

$$+ \overline{R}_{2112} \frac{\partial u^2}{\partial u'^{11}} \frac{\partial u^1}{\partial u'^2} \frac{\partial u^1}{\partial u'^2} \frac{\partial u^2}{\partial u'^2} + \overline{R}_{2121} \frac{\partial u^2}{\partial u'^1} \frac{\partial u^1}{\partial u'^2} \frac{\partial u^2}{\partial u'^1} \frac{\partial u^1}{\partial u'^2}$$

$$= \overline{R}_{2112} \left[\frac{\partial u^1}{\partial u'^1} \frac{\partial u^2}{\partial u'^2} - 2 \frac{\partial u^1}{\partial u'^1} \frac{\partial u^2}{\partial u'^2} \frac{\partial u^2}{\partial u'^2} \frac{\partial u'}{\partial u'^2} + \left(\frac{\partial u^2}{\partial u'^1} \frac{\partial u^1}{\partial u'^2} \right)^2 \right]$$

$$= \overline{R}_{1212} J^2 \text{ where } J = \begin{vmatrix} \dfrac{\partial u^1}{\partial u'^1} & \dfrac{\partial u^1}{\partial u'^2} \\ \dfrac{\partial u^2}{\partial u'^1} & \dfrac{\partial u^2}{\partial u'^2} \end{vmatrix} = \left| \frac{\partial u}{\partial u'} \right|$$

so,

$$\overline{R}'_{1212} = \overline{R}_{1212} J^2 \qquad ...(3)$$

Again

$$b'_{\alpha\beta} = b_{\alpha\beta}\frac{\partial u^\gamma}{\partial u'^\alpha}\frac{\partial u^\delta}{\partial u'^\beta}$$

$$\Rightarrow \qquad |b'_{\alpha\beta}| = |b_{\alpha\beta}|\left|\frac{\partial u^\gamma}{\partial u'^\alpha}\right|\left|\frac{\partial u^\delta}{\partial u'^\beta}\right|$$

or

$$b' = bJ^2 \qquad \text{...(4)}$$

from (3) and (4), we get

$$\frac{\overline{R}'_{1212}}{b'} = \frac{\overline{R}_{1212}}{b} = K \text{ (say)} \qquad \text{...(5)}$$

This shows that the quantity K is an invariant for transformation of coordinates. The invariant K is defined to be the Guasian curvature of S. Hence K is the Riemannian Curvature of S at P_0.

Since Riemannian Coordinates y^i with the origin at P_0. We have as geodesic coordinates with the pole at P_0.

Therefore

$$\begin{Bmatrix} k \\ i\ j \end{Bmatrix}_g, \begin{Bmatrix} \gamma \\ \alpha\ \beta \end{Bmatrix}_b = 0 \text{ at } P_0$$

Then

$$R_{hijk} = \left(-\frac{\partial[ij,h]_g}{\partial y^k} + \frac{\partial[ik,h]_g}{\partial y^j}\right) \text{ at } P_0.$$

and

$$\overline{R}_{\alpha\beta\gamma\delta} = \left(-\frac{\partial[\beta\gamma,\alpha]_b}{\partial u^\delta} + \frac{\partial[\beta\delta,\alpha]_b}{\partial u^\gamma}\right) \text{ at } P_0.$$

$$\overline{R}_{1212} = -\frac{\partial[21,1]_b}{\partial u^2} + \frac{\partial[22,1]_b}{\partial u^1} \text{ at } P_0 \qquad \text{...(6)}$$

from (5) we get

$$K = \frac{1}{b}\left(-\frac{\partial[21,1]_b}{\partial u^2} + \frac{\partial[22,1]_b}{\partial u^1}\right) \qquad \text{...(7)}$$

This is required expression for Riemannian curvature at P_0.

5.8 FORMULA FOR RIEMANNIAN CURVATURE IN THE TERMS OF COVARIANT CURVATURE TENSOR OF V_n

Let $\begin{Bmatrix} k \\ i\ j \end{Bmatrix}_g$ and $\begin{Bmatrix} \gamma \\ \alpha\ \beta \end{Bmatrix}$ be the Christoffel symbols of second kind relative to the metrices $b_{\alpha\beta}du^\alpha du^\beta$

and $g_{ij}dy^i dy^j$ respectively. We have

$$[\alpha\beta, \gamma]_b = \frac{\partial y^i}{\partial u^\alpha}\frac{\partial y^j}{\partial u^\beta}\frac{\partial y^k}{\partial u^\gamma}[ij, k]_g \qquad ...(8)$$

$$[21, 1]_b = \frac{\partial y^i}{\partial u^2}\frac{\partial y^j}{\partial u^1}\frac{\partial y^k}{\partial u^1}[ij, k]_g$$

$$= q^i p^j p^k [ij, k]_g \text{ using (2)}$$

Now,

$$\frac{\partial[21, 1]_b}{\partial u^2} = \frac{\partial[ij, k]_g}{\partial u^2} q^i p^j p^k$$

$$= q^i p^j p^k \frac{\partial[ij, k]_g}{\partial y^h}\frac{\partial y^h}{\partial u^2}$$

$$= q^i p^j p^k q^h \frac{\partial[ij, k]_g}{\partial y^h}$$

Interchanging h and k, we get

$$\frac{\partial[21, 1]_b}{\partial u^2} = q^i q^k p^j p^h \frac{\partial[ij, h]_g}{\partial y^k} \qquad ...(9)$$

Similarly,

$$\frac{\partial[22, 1]_b}{\partial u^1} = q^i q^k p^j p^h \frac{\partial[ij, h]_g}{\partial y^k} \qquad ...(10)$$

Using (9) and (10), equation (6) becomes

$$\overline{R}_{1212} = p^h q^i p^j q^k \left[-\frac{\partial[ij, h]_g}{\partial y^k} + \frac{\partial[ik, h]_g}{\partial y^j}\right] \text{at } P_0$$

$$\overline{R}_{1212} = p^h q^i p^j q^k R_{hijk} \text{ at } P_0 \qquad ...(11)$$

Since

$$b_{\alpha\beta} = g_{ij}\frac{\partial y^i}{\partial u^\alpha}\frac{\partial y^j}{\partial u^\beta}$$

$$b_{11} = g_{ij}\frac{\partial y^i}{\partial u^1}\frac{\partial y^j}{\partial u^1} = g_{ij}p^i p^j$$

$$b_{11} = g_{hj}p^h p^j$$

Similarly

$$b_{22} = g_{ij}q^i q^j = g_{ik}q^i q^k$$

$$b_{12} = g_{ij}p^i q^j = g_{ji}p^j q^i = g_{hk}p^h q^k$$

$$b = \begin{vmatrix} b_{11} & b_{12} \\ b_{21} & b_{22} \end{vmatrix} = b_{11}b_{22} - b_{12}b_{21}$$

$$b = p^h q^i p^j q^k \left[g_{hi}g_{ik} - g_{ij}g_{hk}\right] \qquad ...(12)$$

Dividing (11) and (12), we get

$$K = \frac{\overline{R}_{1212}}{b} = \frac{p^h q^i q^k R_{hijk}}{p^h q^i p^j q^k (g_{ik}g_{hj} - g_{ij}g_{hk})} \qquad ...(13)$$

This is formula for Riemannian Curvature of V_n at P_0 determined by the orientation of Unit vectors p^i and q^i at P_0.

5.9 SCHUR'S THEOREM

If at each point, the Riemannian curvature of a space is independent of the orientation choosen then it is constant throughout the space.

Proof: If K is the Riemannian curvature of V_n at P for the orientation determined by unit vectors p^i and q^i then it is given by

$$K = \frac{p^h q^i p^j q^k R_{hijk}}{(g_{ik}g_{hj} - g_{ij}g_{hk})p^h q^i p^j q^k} \qquad ..(1)$$

Let K be independent of the orientation choosen. Then equation (1) becomes

$$K = \frac{R_{hijk}}{g_{ik}g_{hj} - g_{ij}g_{hk}}$$

$$R_{hijk} = K(g_{ik}g_{hj} - g_{ij}g_{hk}) \qquad ...(2)$$

We have to prove that K is constant throughout the space V_n.

If $N = 2$, the orientation is the same at every point. So, consider the case of V_n when $n > 2$.

Since g_{ij} are constants with respect to covariant differentation, therefore covariant differentation of (2) gives

$$R_{hijk,l} = (g_{ik}g_{hj} - g_{ij}g_{hk})K_{,l} \qquad ...(3)$$

where $K_{,l}$ is the partial derivative of K.

Taking the sum of (3) and two similar equations obtained by cyclic permutation of the suffices, j, k and l.

$$(g_{hj}g_{ik} - g_{hk}g_{ij})K,_i + (g_{hk}g_{il} - g_{hl}g_{ik})K,_j + (g_{hl}g_{ij} - g_{hj}g_{il})K,_k = R_{hijk,l} + R_{hikl,j} + R_{hilj,k} \quad ...(4)$$

Here $n > 2$ therefore three or more distinct values to indices *j, k, m* can be given .

Multiplying (4) by g^{hj} and using $g^{hj}g_{hl} = \delta_l^j$, we get

$$(ng_{ik} - \delta_i^h g_{hk})k,_l + (g_{il}\delta_k^j - g_{ik}\delta_l^j)K,_j + (\delta_i^h g_h l - ng_{il})K,_k = 0$$

or,

$$(n-1)g_{ik}K_{,l} + 0 + (1-n)g_{il}K_{,k} = 0 \text{ for } \delta_j^i = 0, i \neq j$$

$$g_{ik}K_{,l} - g_{il}K_{,k} = 0$$

Multiplying by g^{ik} and using

$$g^{ik}\delta_{ik} = n, g_{il}g^{ik} = \delta_l^k,$$

we get

$$nK,_l - \delta_l^k K,_k = 0 \text{ or } (n-1)K,_l = 0$$

or $$K,_k = 0 \text{ as } (n-1)K,_l = 0$$

or $$\frac{\partial K}{\partial x^l} = 0.$$

integrating it, we get K = constant. This proves that the partial derivatives of K w.r.t. to x's are all zero. Consequently K is constant at P. But P is an arbitrary point of V_n. Hence K is constant throughout V_n.

5.10 MEAN CURVATURE

The sum of mean curvatures of a V_n for a mutually orthogonal directions at a point, is independent of the ennuple choosen. Obtain the value of this sum.

Or

Prove that the mean curvature (or Riccian Curvature) in the direction e_i at a point of a V_n is the sum of $n - 1$ Riemmanian curvatures along the direction pairs consisting of the direction and $n - 1$ other directions forming with this directions an orthogonal frame.

Proof: Let $e_{h|}^i$ be the components of unit vector in a given direction at a point P of a V_n. Let $e_{k|}^i$ be the components of unit vector forming an orthogonal ennuple.

Let the Riemannian curvature at P of V_n for the orientation determined by $l_{h|}^i$ and $e_{k|}^i (h \neq k)$ be denoted by K_{hk} and given by

$$K_{hk} = \frac{e_{h|}^p e_{k|}^q \lambda_{k|}^r \lambda_{k|}^s R_{pqrs}}{e_{h|}^p e_{k|}^q e_{h|}^r e_{k|}^s (g_{pr} q_{qs} - g_{ps} g_{qr})}$$

$$= \frac{e_{h1}^p e_{k1}^q e_{h1}^r e_{k1}^s R_{pqrs}}{(e_{h|}^p e_{h|}^r g_{pr})(e_{k|}^q e_{k|}^s g_{qs}) - (e_{h|}^p e_{k|}^s g_{ps})(e_{k|}^q e_{h|}^r g_{qr})} \quad ...(1)$$

Since unit vectors $e^i_{h|}, \lambda^i_{k|}$ are orthogonal. Therefore

$$e^p_{h|}e^r_{h|}g_{pr} = 1$$

and $$e^p_{h|}e^s_{h|}g_{ps} = 0 \text{ etc.}$$

Using these in equation (1), then equation (1) becomes

$$K_{hk} = \frac{e^p_{h|}e^q_{k|}e^r_{h|}e^s_{k|}R_{pqrs}}{1\times 1 - 0\times 0}$$

$$K_{hk} = e^p_{h|}e^q_{k|}e^r_{h|}e^s_{k|}R_{pqrs} \qquad ...(2)$$

$$\sum_{k=1}^{n} K_{hk} = \sum_{k=1}^{n} e^p_{h|}e^q_{k|}e^r_{h|}e^s_{k|}R_{pqrs}$$

Put $\sum_{k=1}^{n} K_{hk} = M_h$. Then

$$M_h = e^p_{h|}e^r_{h|}\sum_{k=1}^{n} e^q_{k|}e^s_{k|}R_{pqrs}$$

$$= e^p_{h|}e^r_{h|}g^{qs}R_{pqrs}$$

$$= -e^p_{h|}e^r_{h|}g^{qs}R_{qprs}$$

$$M_h = -e^p_{h|}e^r_{h|}R_{pr} \qquad ...(3)$$

This shows that M_h is independent of $(n-1)$ orthogonal direction choosen to complete an orthogonal ennuple. Here M_h is defined as mean curvature or Riccian curvature of V_n for the direction $e^p_{h|}$.

Summing the equation (1) from $h = 1$ to $h = n$, we get

$$\sum_{h=1}^{n} M_h = -e^p_{h|}e^r_{h|}R_{pr}$$

$$= -g^{pr}R_{pr}$$

$$= -R$$

or $$\sum_{h=1}^{n} M_h = -g^{pr}R_{pr} == -R$$

This proves that the sum of mean curvatures for n mutual orthogonal directions is independent of the directions chosen to complete an orthogonal ennuple and has the value $-R$.

5.11 RICCI'S PRINCIPAL DIRECTIONS

Let $e^i_{h|}$ is not a unit vector and the mean curvature M_h is given by

$$M_h = \frac{R_{ij}e^i_{h|}e^j_{h|}}{g_{ij}e^i_{h|}e^j_{h|}}$$

$$\Rightarrow \qquad (R_{ij} + M_h g_{ij}) e^i_{h|} e^j_{h|} = 0$$

Differentiating it w.r. to $e^i_{h|}$, we get

$$\frac{\partial M_h}{\partial e^i_{h|}} g_{jk} e^j_{h|} e^k_{h|} + 2(R_{ij} + Mg_{ij}) e^j_{h|} = 0 \qquad ...(1)$$

For maximum and minimum value of M_h.

$$\frac{\partial M_h}{\partial e^i_{h|}} = 0$$

Then equation (1) becomes

$$(R_{ij} + Mg_{ij}) e^i_{h|} = 0$$

These are called Ricci's Principal direction of the space as they are principal directions of Ricci tensor R_{ij}.

5.12 EINSTEIN SPACE

A space, which is homogeneous relative to the Ricci tensor R_{ij} is called Einstein space.

If space is homogeneous then we have

$$R_{ij} = \lambda g_{ij} \qquad ...(1)$$

Inner multiplication by g^{ij}, we get

$$R = \lambda n \text{ since } R_{ij} g^{ij} = R \text{ and } g^{ij} g_{ij} = n$$

$$\Rightarrow \qquad \lambda = \frac{1}{n} R$$

from (1)

$$R_{ij} = \frac{R}{n} g_{ij}$$

Hence a space is an Einstein space if $R_{ij} = \frac{R}{n} g_{ij}$ at every point of the space.

Theorem 5.5 *To show that a space of constant Curvature is an Einstein space.*

Proof: Let the Riemannian curvature K at P of V_n for the orientation determined by p^i and q^i, is given by

$$K = \frac{p^h q^i p^j q^k R_{hijk}}{p^h q^i p^j q^k (g_{ik} g_{hj} - g_{ij} g_{hk})}$$

Since K is constant and independent of the orientaion.

$$K = \frac{R_{hijk}}{(g_{ik}g_{hj} - g_{ij}g_{hk})}$$

$$R_{hijk} = K(g_{ik}g_{hj} - g_{ij}g_{hk})$$

Multiplying by g^{hk}

$$Kg^{hk}(g_{ik}g_{hj} - g_{ij}g_{hk}) = g^{hk}R_{hijk}$$

$$K(\delta_i^h g_{hj} - ng_{ij}) = R_{ij}$$

Since $g^{hk}g_{ik} = \delta_i^h$; $g^{hk}g_{hk} = n$ and $g^{hk}R_{hijk} = R_{ij}$

$$K(g_{ij} - ng_{ij}) = R_{ij}$$

$$K(1-n)g_{ij} = R_{ij} \qquad ...(1)$$

Multiplying by g^{ij}, we get

$$Kn(1-n) = R \text{ as } g^{ij}R_{ij} = R \qquad ...(2)$$

from (1) & (2)

$$R_{ij} = (1-n)g_{ij} \cdot \frac{R}{n(1-n)} = \frac{1}{n}Rg_{ij}$$

$$\Rightarrow \qquad R_{ij} = \frac{R}{n}g_{ij}$$

This is necessary and sufficient condition for the space V_n to be Einstein space.

5.13 WEYL TENSOR OR PROJECTIVE CURVATURE TENSOR

Weyl Tensor denoted as W_{hijk} and defined by

$$W_{hijk} = R_{hijk} + \frac{1}{1-n}(g_{kl}R_{hj} - g_{kh}g_{ij})$$

Theorem 5.6 *A necessary and sufficient condition for a Riemannian $V_n(n > 3)$ to be of constant curvature to that the Weyl tensor vanishes identically throughout V_n.*

Proof: *Necessary Condition:*

Let K be Riemannian Curvature of V_n. Let K = constant.

We have to prove that $W_{hijk} = 0$

Since we know that

$$K = \frac{p^h q^i p^j q^k R_{hijk}}{(g_{hj}g_{ik} - g_{ij}g_{hk})p^h q^i p^j q^k} = \text{constant}$$

Since K be independent of the orientation determined by the vector p^i and q^i. Then

$$K = \frac{R_{hijk}}{g_{hj}g_{ik} - g_{ij}g_{hk}} \qquad ...(1)$$

Multiplying by g^{hk}, we get

$$g^{hk}R_{hijk} = g^{hk}K(g_{hj}g_{ik} - g_{ij}g_{hk})$$

$$R_{ij} = K(\delta^k_j g_{ik} - ng_{ij})$$

$$R_{ij} = K(1-n)g_{ij} \qquad ...(2)$$

Multiplying by g^{ij} again, we get

$$g^{ij}R_{ij} = K(1-n)g_{ij}g^{ij}$$

$$R = K(1-n)n \qquad ...(3)$$

Putting the value of K from (3) in (2), we get

$$R_{ij} = \frac{R}{n}g_{ij} \qquad ...(4)$$

The equation (3) shows that R is constant since K is constant.
Now, the W tensor is given by

$$W_{hijk} = R_{hijk} + \frac{1}{1-n}[g_{ik}R_{hj} - g_{hk}g_{ij}]$$

from (5), we get

$$W_{hijk} = R_{hijk} + \frac{1}{1-n}\left[g_{ik}\frac{R}{n}g_{hj} - g_{hk}\frac{R}{n}g_{ij}\right]$$

$$= R_{hijk} + \frac{R}{n(1-n)}[g_{ik}g_{hj} - g_{hk}g_{ij}]$$

$$= R_{hijk} + \frac{R}{n(1-n)}\cdot\frac{R_{hijk}}{K}, \text{ by. eqn. (1)}$$

$$W_{hijk} = R_{hijk} + K\frac{R_{hijk}}{K}, \text{ by equation (3)}$$

$$W_{hijk} = 2R_{hijk}$$

Since K is constant. The equation (5) shows that $W_{hijk} = 0$.

This proves necessary condition.

Sufficient Condition

Let $W_{hijk} = 0$. Then we have to prove that K is constant.

Now, $W_{hijk} = 0$.

$$\Rightarrow \quad R_{hijk} + \frac{1}{1-n}[g_{ik}R_{hj} - g_{hk}R_{ij}] = 0$$

Multiplying by g^{hk}, we get

$$R_{ij} + \frac{1}{1-n}[g^{hk}g_{ik}R_{hj} - g^{hk}g_{hk}R_{ij}] = 0$$

$$R_{ij} + \frac{1}{1-n}[\delta_i^h R_{hj} - nR_{ij}] = 0$$

$$R_{ij} + \frac{R_{ij}}{1-n}(1-n) = 0$$

$$\Rightarrow \quad 2R_{ij} = 0$$

$$\Rightarrow \quad R_{ij} = 0$$

Since $R_{ij} = 0 \Rightarrow g^{hk}R_{hijk} = 0$.

$\Rightarrow \quad g^{hk} = 0$ or $R_{hijk} = 0$

If $R_{hijk} = 0$, then clearly $K = 0$. So, K is constant.

If $g^{hk} = 0$ then

$$K = \frac{R_{hijk}p^h q^i p^j q^k}{(g_{hj}g_{ik} - 0g_{ij})p^h q^i p^j q^k}$$

$$K = \frac{R_{hijk}p^h q^i p^j q^k}{(p^h p^j g_{hj})(q^i q^k g_{ik})} = \frac{R_{hijk}p^h q^i p^j q^k}{p^2 \cdot q^2}$$

$$K = R_{hijk}p^h q^i p^j q^k \text{ since } p^2 = 1, q^2 = 1$$

$$K = \text{constant as } R_{ij} = 0$$

This proves sufficient condition.

EXAMPLE 1

For a V_2 referred to an orthogonal system of parametric curves ($g_{12} = 0$) show that

$$R_{12} = 0, \; R_{11}g_{22} = g_{22}g_{11} = R_{1221}$$

$$R = g^{ij}R_{ij} = \frac{2R_{1221}}{g_{11}g_{22}}$$

Consequently

$$R_{ij} = \frac{1}{2}Rg_{ij}.$$

Solution

Given that $g_{12} = 0$ so that $g^{12} = 0$.

Also,

$$g^{ij} = \frac{1}{g_{ij}} \Rightarrow g^{11} = \frac{1}{g_{11}} \ \& \ g^{22} = \frac{1}{g_{22}}.$$

The metric of V_2 is given by

$$ds^2 = g_{ij}dx^i dx^j; \ (i, j = 1, 2)$$

$$ds^2 = g_{11}(dx^1)^2 + g_{22}(dx^2)^2 \text{ Since } g_{12} = 0.$$

We know that $g_1^{hk} R_{hijk} = R_{ij}$.

and
$$g = |g_{ij}| = \begin{vmatrix} g_{11} & g_{12} \\ g_{21} & g_{22} \end{vmatrix} = \begin{vmatrix} g_{11} & 0 \\ 0 & g_{22} \end{vmatrix} = g_{11}g_{22}$$

(*i*) To prove $R_{12} = 0$

$$R_{12} = g^{hk} R_{h12k} = g^{h1} R_{h|2|} \text{ as } R_{h122} = 0$$

$$= g^{21} R_{2121} \text{ as } R_{1121} = 0$$

$$R_{12} = 0$$

(*ii*) To prove $R_{11}g_{22} = R_{22}g_{11} = R_{1221}$

$$R_{11} = g^{hk} R_{h11k} = g^{2k} R_{211k}$$

$$R_{11} = g^{22} R_{2112} = \frac{R_{2112}}{g_{22}}$$

So,

$$R_{11} = \frac{R_{2112}}{g_{22}} \quad \text{...(1)}$$

and

$$R_{22} = g^{hk} R_{h22k} = g^{1k} R_{122k}$$

$$= g^{11} R_{1221} = \frac{R_{1221}}{g_{11}}$$

So,

$$R_{22} = \frac{R_{1221}}{g_{11}} \quad \text{...(2)}$$

from (1) and (2)

$$R_{11}g_{22} = R_{1221} = R_{22}g_{11} \quad \text{... (3)}$$

(*iii*) To prove

$$R = \frac{2R_{1221}}{g_{11}g_{22}}$$

$$R = g^{ij} R_{ij} = g^{i1} R_{i1} + g^{i2} R_{i2}$$

$$= g^{11}R_{11} + g^{22}R_{22} \text{ for } g^{12} = 0$$

$$= \frac{R_{11}}{g_{11}} + \frac{R_{22}}{g_{22}}$$

$$= \frac{R_{1221}}{g_{22}g_{11}} + \frac{R_{1221}}{g_{11}g_{22}} \quad [\because \text{by eqn. (3)}]$$

$$R = \frac{2R_{1221}}{g_{11}g_{22}}$$

(*iv*) To prove $R_{ij} = \frac{1}{2}Rg_{ij}$

$$R = \frac{2R_{1221}}{g_{11}g_{22}}$$

$$R = \frac{2R_{1221}}{g} \text{ as } g = g_{11}g_{22}$$

$$R_{1221} = \frac{1}{2}Rg$$

The eqn (3) expressed as

$$R_{11}g_{22} = \frac{1}{2}Rg = R_{22}g_{11}.$$

it becomes

$$R_{11} = \frac{Rg}{2g_{22}} = \frac{Rg_{11}g_{22}}{2g_{22}} = \frac{Rg_{11}}{2}$$

$$R_{22} = \frac{Rg}{2g_{11}} = \frac{Rg_{11}g_{22}}{2g_{11}} = \frac{Rg_{22}}{2}$$

So,

$$R_{11} = \frac{1}{2}Rg_{11} \;\&\; R_{22} = \frac{1}{2}g_{22}$$

$$R_{12} = \frac{1}{2}g_{12} \text{ as } R_{12} = 0 = g_{12}$$

This prove that $R_{ij} = \frac{1}{2}R_{ij}$.

EXAMPLE 2

The metric of the V_2 formed by the surface of a sphere of radius r is $ds^2 = r^2 d\theta^2 + r^2 \sin^2\theta d\phi^2$ in spherical polar coordinates. Show that the surface of a sphere is a surface of constant curvature $\frac{1}{r^2}$.

Solution

Given that

$$ds^2 = r^2 d\theta^2 + r^2 \sin^2\theta d\phi^2$$

Since r is radius of curvature then r is constant.

$$g_{11} = r^2, \quad g_{22} = r^2 \sin^2\theta, \; g_{12} = 0, \quad g = g_{11}g_{22} = r^4 \sin^2\theta.$$

We can prove

$$R_{1221} = r^2 \sin^2\theta$$

Now, the Riemannian curvature K of V_n is given by

$$K = \frac{p^h q^i p^j q^k}{p^h q^i p^j q^k (g_{hj}g_{ik} - g_{hk}g_{ij})}.$$

At any point of V_2 there exists only two independent vectors.

Consider two vectors whose components are (1, 0) and (0, 1) respectively in V_2. Then

$$K = \frac{R_{1212}}{g_{11}g_{22}} = \frac{R_{1212}}{g}$$

$$K = \frac{r^2 \sin^2\theta}{r^4 \sin^2\theta} = \frac{1}{r^2} = \text{constant.}$$

EXERCISES

1. Show that

$$R_{ijkl} = \frac{\partial[jl,i]}{\partial x^k} - \frac{\partial[jk,i]}{\partial x^l} + \begin{Bmatrix} \alpha \\ j\ k \end{Bmatrix} [il,\alpha] - \begin{Bmatrix} \alpha \\ j\ l \end{Bmatrix} [ik,\alpha]$$

2. Show that

$$R_{ijkl} = \frac{1}{2}\left(\frac{\partial^2 g_{il}}{\partial x^j \partial x^k} + \frac{\partial^2 g_{jk}}{\partial x^i \partial x^l} - \frac{\partial^2 g_{ik}}{\partial x^j \partial x^l} - \frac{\partial^2 g_{jl}}{\partial x^i \partial x^k} \right) + g^{\alpha\beta}\big([jk,\beta][il,\alpha] - [jl,\beta][ik,\alpha]\big).$$

3. Using the formula of the problem 2. Show that

$$R_{ijkl} = -R_{jikl} = -R_{ijlk} = R_{klij} \text{ and } R_{ijkl} + R_{iklj} + R_{iljk} = 0$$

4. Show that the curvature tensor of a four dimensional Riemannian space has at the most 20 distinct non-vanishing components.

5. (*a*) If prove that the process of contraction applied to the tensor R^h_{ijk} generates only one new tensor R_{ij} which is symmetric in i and j.

   (*b*) If $ds^2 = g_{11}(dx^1)^2 + g_{22}(dx^2)^2 + g_{33}(dx^3)^2$. Prove that $R_{ij} = \frac{1}{g_{hh}} R_{ihhj}$, ($h$, i, j being unequal).

6. Show that when in a V_3 the coordinates can be chosen so that the components of a tensor g_{ij} are zero when i, j, k are unequal then

   (*i*) $R_{hj} = \frac{1}{g_{ii}} R_{hiij}$

(ii) $R_{hh} = \frac{1}{g_{ii}} R_{hiih} + \frac{1}{g_{jj}} R_{hjjh}$

7. Prove that if

$$R_i^\alpha = g^{\alpha j} R_{ij} \text{ then } R_{i,\alpha}^\alpha = \frac{1}{2} \frac{\partial R}{\partial x^i}$$

and hence deduce that when $n > 2$ then scalar curvature of an Einstein space is constant.

8. If the Riemannian curvature K of V_n at every point of a neighbourhood U of V_n is independent of the direction chosen, show that K is constant throughout the neighbourhood U. Provided $n > 2$.
9. Show that a space of constant curvature K_0 is an Einstein space and that $R = K_0 n(1 - n)$.
10. Show that the necessary and sufficient condition that V_n be locally flat in the neighbourhood of 0 is that Riemannian Christoffel tensor is zero.
11. Show that every V_2 is an Einstein space.
12. For two dimensional manifold prove that

$$K = -\frac{R}{2}$$

13. Show that if Riemann-Christoffel curvature tensor vanishes then order of covariant differentiation is commutative.

CHAPTER – 6

THE *e*-SYSTEMS AND THE GENERALIZED KRÖNECKER DELTAS

The concept of symmetry and skew-symmetry with respect to pairs of indices can be extended to cover to pairs of indices can be extended to cover the sets of quantities that are symmetric or skew-symmetric with respect to more than two indices. Now, consider the sets of quantities $A^{i_1 \ldots i_k}$ or $A_{i_1 \ldots i_k}$ depending on k indices written as subscripts or superscripts, although the quantities A may not represent tensor.

6.1 COMPLETELY SYMMETRIC

The system of quantities $A^{i_1 \ldots i_k}$ (or $A_{i_1 \ldots i_k}$) depending on k indices, is said to be completely symmetric if the value of the symbol A is unchanged by any permutation of the indices.

6.2 COMPLETELY SKEW-SYMMETRIC

The systems $A^{i_1 \ldots i_k}$ or ($A_{i_1 \ldots i_k}$) depending on k indices, is said to be completely skew-symmetric if the value of the symbol A is unchanged by any even permutation of the indices and A merely changes the sign after an odd permutation of the indices.

Any permutation of n distinct objects say a permutation of n distinct integers, can be accomplished by a finite number of interchanges of pairs of these objects and that the number of interchanges required to bring about a given permutation form a perscribed order is always even or always odd.

In any skew-symmetric system, the term containing two like indices is necessarily zero. Thus if one has a skew-symmetric system of quantities A_{ijk} where i, j, k assume value 1, 2, 3. Then

$$A_{122} = A_{112} = 0$$

$$A_{123} = -A_{213}, \quad A_{312} = A_{123} \text{ etc.}$$

In general, the components A_{ijk} of a skew-symmetric system satisfy the relations.

$$A_{ijk} = -A_{ikj} = -A_{jik}$$

$$A_{ijk} = A_{jki} = A_{kij}$$

6.3 e-SYSTEM

Consider a skew-symmetric system of quantities $e_{i_1 \dots i_n}$ (or $e_{i_1, \dots, i_n}$) in which the indices $i_1 \dots i_n$ assume values 1,2,...n. The system $e_{i_1 \dots i_n}$ (or $e^{i_1 \dots i_n}$) is said to be the e-system if

$$e_{i_1 \dots i_n} (\text{or } e^{i_1 \dots i_n}) \begin{cases} = +1; \text{ when } i_1, i_2, \dots, i_n \\ \qquad \text{an even permutation of number } 1, 2, \dots, n \\ = -1; \text{ when } i_1, i_2, \dots, i_n \\ \qquad \text{an odd permutation of number } 1, 2, \dots, n \\ = 0 \quad \text{in all other cases} \end{cases}$$

EXAMPLE 1

Find the components of system e_{ij} when i, j takes the value 1,2.

Solution

The components of system e_{ij} are

$$e_{11},\ e_{12},\ e_{21},\ e_{22}.$$

By definition of e-system, we have

$$e_{11} = 0, \qquad \text{indices are same}$$
$$e_{12} = 1, \qquad \text{since } i\,j \text{ has even permutation of } 12$$
$$e_{21} = -e_{12} = -1 \qquad \text{since } i\,j \text{ has odd permutation of } 12$$
$$e_{22} = 0, \qquad \text{indices are same}$$

EXAMPLE 2

Find the components of the system $e_{i\,jk}$.

Solution

By the definition of e-system,

$$e_{123} = e_{231} = e_{321} = 1$$
$$e_{213} = e_{132} = e_{321} = -1$$
$$e_{ijk} = 0 \text{ if any two indices are same.}$$

6.4 GENERALISED KRÖNECKER DELTA

A symbol $\delta^{i_1 \dots i_k}_{j_1 \dots j_k}$ depending on k superscripts and k subscripts each of which take values from 1 to n, is called a generalised Krönecker delta provided that

(a) it is completely skew-symmetric in superscripts and subscripts (*b*) if the superscripts are distinct from each other and the subscripts are the same set of numbers as the superscripts.

The value of symbol

$$\delta^{i_1 \cdots i_k}_{j_1 \cdots j_k} \begin{cases} = 1; & \text{an even number of transposition is required to arrange the superscripts in the same order as subscripts.} \\ = -1; & \text{where odd number of transpositions arrange the superscripts in the same order as subscripts} \\ = 0, & \text{in all other cases the value of the symbol is zero} \end{cases}$$

EXAMPLE 3

Find the values of δ^{ij}_{kl}.

Solution

By definition of generalised Kronecker Delta, $\delta^{ij}_{kl} = 0$ if $i = j$ or $k = l$ or if the set. ij is not the set kl.

i.e., $$\delta^{11}_{pq} = \delta^{22}_{pq} = \delta^{23}_{13} = \cdots = 0$$

$$\delta^{ij}_{kl} = 1 \text{ if } kl \text{ is an even permutation of } ij$$

i.e., $$\delta^{12}_{12} = \delta^{21}_{21} = \delta^{13}_{13} = \delta^{31}_{31} = \delta^{23}_{23} = \cdots = 1$$

and $\delta^{ij}_{kl} = -1$ if kl is an odd permutation of ij.

i.e., $$\delta^{12}_{21} = \delta^{31}_{13} = \delta^{13}_{31} = \delta^{21}_{12} = \cdots = -1$$

Theorem 6.1 *To prove that the direct product* $e^{i_1 i_2 \ldots i_n} e_{j_1 j_2 \ldots j_n}$ *of two systems* $e^{i_1 \ldots i_n}$ *and* $e_{j_1 j_2 \ldots j_n}$ *is the generalized Krönecker delta.*

Proof: By definition of generalized Krönecker delta, the product $e^{i_1 i_2 \ldots i_n} e_{j_1 j_2 \ldots j_n}$ has the following values.

(*i*) Zero if two or more subscripts or superscripts are same.

(*ii*) +1, if the difference in the number of transpositions of $i_1, i_2, \ldots, i_n$ and $j_1, j_2, \ldots, j_n$ from $1, 2, \ldots n$ is an even number.

(*iii*) –1, if the difference in the number of transpositions of $i_1, i_2, \ldots, i_n$ and $j_1, j_2, \ldots j_n$ from $1, 2, \ldots n$ an odd number.

Thus we can write

$$e^{i_1 i_2 \ldots i_n} e_{j_1 j_2 \ldots j_n} = \delta^{i_1 i_2 \ldots i_n}_{j_1 j_2 \ldots j_n}$$

THEOREM 6.2 *To prove that*

(*i*) $e_{i_1 i_2 \ldots i_n} = \delta^{i_1 i_2 \ldots i_n}_{j_1 j_2 \ldots j_n}$

(*ii*) $e_{i_1 i_2 \ldots i_n} = \delta^{i_1 i_2 \ldots i_n}_{j_1 j_2 \ldots j_n}$

Proof: By Definition of e-system, $e^{i_1 i_2 \dots i_n}$ (or $e_{i_1 i_2 \dots i_n}$) has the following values.

(*i*) +1; if $i_1, i_2, \dots, i_n$ is an even permutation of numbers $1, 2, \dots n$.

(*ii*) -1; if $i_1, i_2, \dots, i_n$ is an odd permutation of numbers $1, 2, \dots n$

(*iii*) 0; in all other cases

Hence by Definition of generalized krönecker delta, we can write

(1) $e^{i_1 i_2 \dots i_n} = \delta^{i_1 i_2 \dots i_n}_{1\,2 \dots n}$

and

(2) $e_{i_1 i_2 \dots i_n} = \delta^{1\,2 \dots n}_{i_1 i_2 \dots i_n}$

6.5 CONTRACTION OF $\delta^{ijk}_{\alpha\beta\gamma}$

Let us contract $\delta^{ijk}_{\alpha\beta\gamma}$ on k and γ. For $n = 3$, the result is

$$\delta^{ijk}_{\alpha\beta\gamma} = \delta^{ij1}_{\alpha\beta 1} + \delta^{ij2}_{\alpha\beta 2} + \delta^{ij3}_{\alpha\beta 3} = \delta^{ij}_{\alpha\beta}$$

This expression vanishes if i and j are equal or if α and β are equal.

If $i = 1$, and $j = 2$, we get $\delta^{123}_{\alpha\beta 3}$.

Hence

$$\delta^{12}_{\alpha\beta} = \begin{cases} +1; & \text{if } \alpha\beta \text{ is an even permutation of } 12 \\ -1; & \text{if } \alpha\beta \text{ is an odd permutation of } 12 \\ 0; & \text{if } \alpha\beta \text{ is not permutation of } 12 \end{cases}$$

Similarly results hold for all values of α and β selected from the set of numbers 1, 2, 3.

Hence

$$\delta^{ij}_{\alpha\beta} = \begin{cases} +1; & \text{if } ij \text{ is an even permutation of } \alpha\beta \\ -1; & \text{if } ij \text{ is an odd permutation of } \alpha\beta \\ 0; & \text{if two of the subscripts or superscripts are equal or when the} \\ & \text{subscripts and superscripts are not formed from the same numbers.} \end{cases}$$

If we contract $\delta^{ij}_{\alpha\beta}$. To contract $\delta^{ij}_{\alpha\beta}$ first contract it and the multiply the result by $\frac{1}{2}$. We obtain a system depending on two indices

$$\delta^{i}_{\alpha} = \frac{1}{2}\delta^{ij}_{\alpha j} = \frac{1}{2}(\delta^{i1}_{\alpha 1} + \delta^{i2}_{\alpha 2} + \delta^{i3}_{\alpha 3})$$

It $i = 1$ in δ^i_α then we get $\delta^1_\alpha = \frac{1}{2}\left(\delta^{12}_{\alpha 2} + \delta^{13}_{\alpha 3}\right)$

This vanishes unless $\alpha = 1$ and if $\alpha = 1$ then $\delta^1_1 = 1$.

Similar result can be obtained by setting $i = 2$ or $i = 3$. Thus δ^i_α has the values.

(*i*) 0 if $i \neq \alpha, (\alpha, i = 1, 2, 3,)$

(*ii*) 1 if $i = \alpha$.

By counting the number of terms appearing in the sums. In general we have

$$\delta^i_\alpha = \frac{1}{n-1}\delta^{ij}_{\alpha j} \text{ and } \delta^{ij}_{ij} = n(n-1) \qquad \text{...(1)}$$

We can also deduce that

$$\delta^{i_1 i_2 \cdots i_r}_{j_1 j_2 \cdots j_r} = \frac{(n-k)!}{(n-r)!}\delta^{i_1 i_2 \cdots i_r i_{r-1} \cdots i_k}_{j_1 j_2 \cdots j_r j_{r-1} \cdots j_k} \qquad \text{...(2)}$$

and

$$\delta^{i_1 i_2 \cdots i_r}_{j_1 j_2 \cdots j_r} = n(n-1)(n-2)\cdots(n-r+1) = \frac{n!}{n-r!} \qquad \text{...(3)}$$

or

$$e^{i_1 i_2 \ldots i_n} e_{i_1 i_2 \ldots i_n} = n! \qquad \text{...(4)}$$

and from (2) we deduce the relation

$$e^{i_1 i_2 \cdots i_r i_{r+1} \cdots i_n} e_{j_1 j_2 \cdots j_r j_{r+1} \cdots i_n} = n! \qquad \text{...(5)}$$

EXERCISE

1. Expand for $n = 3$

   (*a*) $\delta^i_\alpha \delta^\alpha_j$ (*b*) $\delta^{12}_{ij} x^i x^j$ (*c*) $\delta^{\alpha\beta}_{ij} x^i y^j$ (*d*) δ^{ij}_{ij}

2. Expand for $n = 2$

   (*a*) $e^{ij} a^1_i a^2_j$ (*b*) $e^{ij} a^2_i a^1_j$ (*c*) $e^{ij} a^i_\alpha a^j_\beta = e^{ij}|a|$.

3. Show that $\delta^{ijk}_{ijk} = 3!$ if $i, j, k = 1, 2, 3$.

4. If a set of quantities $A_{i_1 i_2 \ldots i_k}$ is skew-symmetric in the subscripts (K in number) then

$$\delta^{i_1 \ldots i_k}_{j_1 \ldots jk} A_{i_1 \ldots i_k} = k!\ A_{j_1 \ldots j_k}$$

5. Prove that $\varepsilon_{ijk}\sqrt{g}$ is a covariant tensor of rank three where where ε_{ijk} is the usual permutation symbol.

CHAPTER – 7

GEOMETRY

7.1 LENGTH OF ARC

Consider the n-dimensional space R be covered by a coordinate system X and a curve C so that

$$C : x^i = x^i(t), \quad (i = 1,2...,n) \qquad ...(1)$$

which is one-dimensional subspace of R. Where t is a real parameter varying continuously in the interval $t_1 \le t \le t_2$. The one dimensional manifold C is called arc of a curve.

Let $F\left(x^1, x^2, ..., x^n, \frac{dx^1}{dt}, \frac{dx^2}{dt}, ..., \frac{dx^n}{dt}\right)$ be a continuous function in the interval $t_1 \le t \le t_2$. We assume that $F\left(x, \frac{dx}{dt}\right) > 0$, unless every $\frac{dx^i}{dt} = 0$ and that for every positive number k

$$F\left(x^1, x^2, ..., x^n, k\frac{dx^1}{dt}, k\frac{dx^2}{dt}, ..., k\frac{dx^n}{dt}\right) = kF\left(x^1, x^2, ..., x^n, \frac{dx^1}{dt}, \frac{dx^2}{dt}, ... \frac{dx^n}{dt}\right).$$

The integral

$$s = \int_{t_1}^{t_2} F\left(x, \frac{dx}{dt}\right) dt \qquad ...(2)$$

is called the length of C and the space R is said to be metrized by equation (2).

Different choices of functions $F\left(x, \frac{dx}{dt}\right)$ lead to different metric geometrices.

If one chooses to define the length of arc by the formula

$$s = \int_{t_1}^{t_2} \sqrt{g_{pq}(x) \frac{dx^p}{dt} \frac{dx^q}{dt}}\, dt, \quad (p,q = 1, 2, ..., n) \qquad ...(3)$$

where $g_{pq}(x) \frac{dx^p}{dt} \frac{dx^q}{dt}$ is a positive definite quadratic form in the variable $\frac{dx^p}{dt}$, then the resulting geometry is the Riemannian geometry and space R metrized in this way is the Riemannian n-dimensional space R_n.

Consider the coordinate transformation $T:\bar{x}^i = \bar{x}^i(x^1,...,x^n)$ such that the square of the element of arc ds,

$$ds^2 = g_{pq}dx^p dx^q \qquad ...(4)$$

can be reduced to the form

$$ds^2 = d\bar{x}^i d\bar{x}^i \qquad ...(5)$$

Then the Riemannian manifold R_n is said to reduce to an n-dimensional Euclidean manifold E_n.

The Y-coordinate system in which the element of arc of C in E_n is given by the equation (5) is called an orthogonal cartesian coordinate system. Obviously, E_n is a generalization of the Euclidean plane determined by the totality of pairs of real values $(\bar{x}^1, \bar{x}^2)$. If these values $(\bar{x}^1, \bar{x}^2)$ are associated with the points of the plane referred to a pair of orthogonal Cartesian axes then the square of the element of arc ds assumes the familiar form

$$ds^2 = (d\bar{x}^1)^2 + (d\bar{x}^2)^2.$$

THEOREM 7.1 *A function* $F\left(x, \frac{dx}{dt}\right)$ *satisfying the condition* $F\left(x, k\frac{dx}{dt}\right) = kF\left(x, \frac{dx}{dt}\right)$ *for every* $k > 0$. *This condition is both necessary and sufficient to ensure independence of the value of the integral* $s = \int_{t_1}^{t_2} F\left(x, \frac{dx}{dt}\right) dt$ *of a particular mode of parametrization of* C. *Thus if* t *in* $C: x^i = x^i(t)$ *is replaced by some function* $t = \phi(s)$ *and we denote* $x^i[\phi(s)]$ *by* $\xi^i(\lambda)$. *so that* $x^i(t)$ $\xi^i(s)$ *we have equality*

$$\int_{t_1}^{t_2} F\left(x, \frac{dx}{dt}\right) dt = \int_{s_1}^{s_2} F(\xi, \xi')\, ds$$

where $\xi'^i = \dfrac{ds^i}{ds}$ and $t_1 = \phi\,(s_1)$ and $t_2 = \phi\,(s_2)$.

Proof: Suppose that k is an arbitary positive number and put $t = ks$ so that $t_1 = ks_1$ and $t_2 = ks_2$. Then

$$C: x^i = x^i(t) \text{ becomes}$$

$$C: x^i(ks) = \xi^i(t)$$

and

$$\xi'^i(s) = \frac{dx^i(ks)}{ds} = k\frac{dx^i(ks)}{dt}$$

Substituting these values in $s = \int_{t_1}^{t_2} F\left(x, \frac{dx}{dt}\right) dt$ we get

$$s = \int_{s_1}^{s_2} F\left[x(ks), \frac{dx(ks)}{dt}\right] kds$$

or

$$s = \int_{s_1}^{s_2} F\,[\xi(s), \xi'(s)]ds$$

We must have the relation $F(\xi, \xi') = F\left(x, k\frac{dx}{dt}\right) = kF\left(x, \frac{dx}{dt}\right)$. Conversely, if this relation is true for every line element of C and each $k > 0$ then the equality of integrals is assumed for every choice of parameter $t = \phi(s), \phi'(s) > 0, s_1 \le s \le s_2$ with and $t_2 = \phi(s_2)$.

Note: (i) Here take those curves for which $x^i(t)$ and $\frac{dx^i}{dt}$ are continuous functions in $t_1 \le t \le t_2$.

(ii) A function $F\left(x, \frac{dx}{dt}\right)$ satisfying the condition $F\left(x, k\frac{dx}{dt}\right) = kF\left(x, \frac{dx}{dt}\right)$ for every $k > 0$ is called positively homogeneous of degree 1 in the $\frac{dx^i}{dt}$.

EXAMPLE 1

What is meant, consider a sphere S of radius a, immersed in a three-dimensional Euclidean manifold E_3, with centre at the origin (0, 0, 0) of the set of orthogonal cartesian axes $O - X^1X^2X^3$.

Solution

Let T be a plane tangent to S at $(0, 0, -a)$ and the points of this plane be referred to a set of orthogonal cartesian axes $O'-Y^1Y^2$ as shown in figure. If we draw from $O(0,0,0)$ a radial line OP, interesting the sphere S at $P(x^1, x^2, x^3)$ and plane T at $Q(\bar{x}^1, \bar{x}^2, -a)$ then the points P on the lower half of the sphere S are in one-to-one correspondence with points $(\bar{x}^1, \bar{x}^2)$ of the tangent plane T.

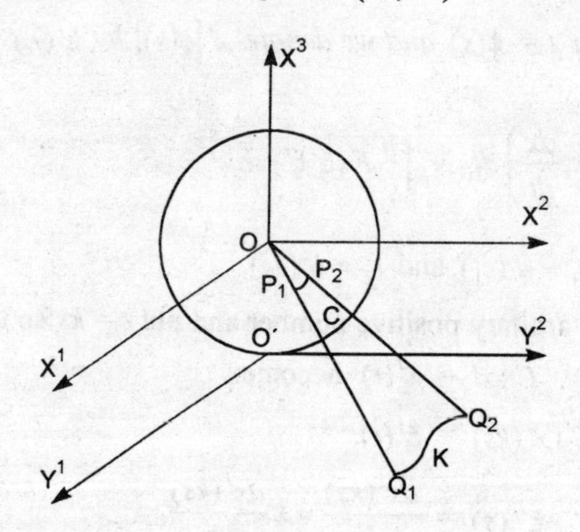

Fig. 7.1

If $P(x^1, x^2, x^3)$ is any point on the radial line OP, then symmetric equations of this line is

$$\frac{x^1 - 0}{\bar{x}^1 - 0} = \frac{x^2 - 0}{\bar{x}^2 - 0} = \frac{x^3 - 0}{-a - 0} = \lambda$$

or

$$x^1 = \lambda \bar{x}^1, \; x^2 = \lambda \bar{x}^2, \; x^3 = -\lambda a \qquad ...(6)$$

Since the images Q of points P lying on S, the variables x^i satisfy the equation of S,

$$(x^1)^2 + (x^2)^2 + (x^3)^2 = a^2$$

or

$$\lambda^2\left[\left(\bar{x}^1\right)^2 + \left(\bar{x}^2\right)^2 + a^2\right] = a^2$$

Solving for λ and substituting in equation (6), we get

$$x^1 = \frac{a\bar{x}^1}{\sqrt{(\bar{x}^1)^2 + (\bar{x}^2)^2 + a^2}}, \quad x^2 = \frac{a\bar{x}^2}{\sqrt{(\bar{x}^1)^2 + (\bar{x}^2)^2 + (\bar{x}^3)^2}}$$

and
$$x^3 = \frac{-a^2}{\sqrt{(\bar{x}^1)^2 + (\bar{x}^2)^2 + (\bar{x}^3)^2}} \qquad \text{...(7)}$$

These are the equations giving the analytical one-to-one correspondence of the points Q on T and points P on the portion of S under consideration.

Let $P_1(x^1, x^2, x^3)$ and $P_2(x^1 + dx^1, x^2 + dx^2, x^3 + dx^3)$ be two close points on some curve C lying on S. The Euclidean distance $\overline{P_1P_2}$ along C, is given by the formula

$$ds^2 = dx^i dx^i, \quad (i = 1, 2, 3) \qquad \text{...(8)}$$

Since

$$dx^i = \frac{\partial x^i}{\partial \bar{x}^p} d\bar{x}^p, \quad (P = 1, 2)$$

Thus equation (8) becomes

$$ds^2 = \frac{\partial x^i}{\partial \bar{x}^p} \frac{\partial x^i}{\partial \bar{x}^q} d\bar{x}^p d\bar{x}^q$$

$$= g_{pq}(\bar{x}) d\bar{x}^p d\bar{x}^q, \quad (p, q = 1, 2)$$

where $g_{pq}(\bar{x})$ are functions of $\bar{x}^i$ and $g_{pq} = \frac{\partial x^i}{\partial \bar{x}^p} \frac{\partial x^i}{\partial \bar{x}^q}$.

If the image K of C on T is given by the equations

$$K : \begin{cases} \bar{x}^1 = \bar{x}^1(t) \\ \bar{x}^2 = \bar{x}^2(t), \quad t_1 \le t \le t_2 \end{cases}$$

then the length of C can be computed from the integral

$$s = \int_{t_1}^{t_2} \sqrt{g_{pq} \frac{d\bar{x}^p}{dt} \frac{d\bar{x}^q}{dt}}\, dt$$

A straight forward calculation gives

$$ds^2 = \frac{(d\bar{x}^1)^2 + (d\bar{x}^2)^2 + \frac{1}{a^2}(\bar{x}^1 d\bar{x}^2 - \bar{x}^2 d\bar{x}^1)^2}{\left[1 + \frac{1}{a^2}\left\{(\bar{x}^1)^2 + (\bar{x}^2)^2\right\}\right]^2} \qquad \text{...(9)}$$

and

$$s = \int_{t_1}^{t_2} \frac{\sqrt{\left(\frac{d\overline{x}^1}{dt}\right)^2 + \left(\frac{d\overline{x}^2}{dt}\right)^2 + \frac{1}{a^2}\left(\overline{x}^1 \frac{d\overline{x}^2}{dt} - \overline{x}^2 \frac{d\overline{x}^1}{dt}\right)^2}}{1 + \frac{1}{a^2}\{(\overline{x}^1)^2 + (\overline{x}^2)^2\}} dt$$

So, the resulting formulas refer to a two-dimensional manifold determined by the variables $(\overline{x}^1, \overline{x}^2)$ in the cartesian plane T and that the geometry of the surface of the sphere imbeded in a three-dimensional Euclidean manifold can be visualized on a two-dimensional manifold R_2 with metric given by equation (9).

If the radius of S is very large then in equation (9) the terms involving $\frac{1}{a^2}$ can be neglected. Then equation (9) becomes

$$ds^2 = (d\overline{x}^1)^2 + (d\overline{x}^2)^2. \qquad ...(10)$$

Thus for large values of a, metric properties of the sphere S are indistinguishable from those of the Euclidean plane.

The chief point of this example is to indicate that the geometry of sphere imbedded in a Euclidean 3-space, with the element of arc in the form equation (8), is indistinguishable from the Riemannian geometry of a two-dimensional manifold R_2 with metric (9). the latter manifold, although referred to a cartesian coordinate system Y, is not Euclidean since equation (9) cannot be reduced by an admissible transformation to equation (10).

7.2 CURVILINEAR COORDINATES IN E_3

Let $P(\overline{x})$ be the point, in an Euclidean 3-space E_3, referred to a set of orthogonal Cartesian coordinates Y.

Consider a coordinate transformation

$$T : x^i = x^i(\overline{x}^1, \overline{x}^2, \overline{x}^3), \quad (i = 1, 2, 3)$$

Such that $J = \left|\frac{\partial x^i}{\partial \overline{x}^j}\right| \neq 0$ in some region R of E_3. The inverse coordinate transformation

$$T^{-1} : \overline{x}^i = \overline{x}^i(x^1, x^2, x^3), \quad (i = 1, 2, 3)$$

will be single values and the transformations T and T^{-1} establish one-to-one correspondence between the sets of values (x^1, x^2, x^3) and $(\overline{x}^1, \overline{x}^2, \overline{x}^3)$.

The triplets of numbers (x^1, x^2, x^3) is called curvilinear coordinates of the points P in R.

If one of the coordinates x^1, x^2, x^3 is held fixed and the other two allowed to vary then the point P traces out a surface, called coordinate surface.

If we set $x^1 = \text{constant}$ in T then

$$x^1(\overline{x}^1, \overline{x}^2, \overline{x}^3) = \text{constant} \qquad ...(1)$$

defines a surface. If constant is allowed to assume different values, we get a one-parameter family of surfaces. Similarly, $x^2(\overline{x}^1, \overline{x}^2, \overline{x}^3) = \text{constant}$ and $x^3(\overline{x}^1, \overline{x}^2, \overline{x}^3) = \text{constant}$ define two families of surfaces.

The surfaces

$$x^1 = c_1, \; x^2 = c_2, \; x^3 = c_3 \qquad \ldots (2)$$

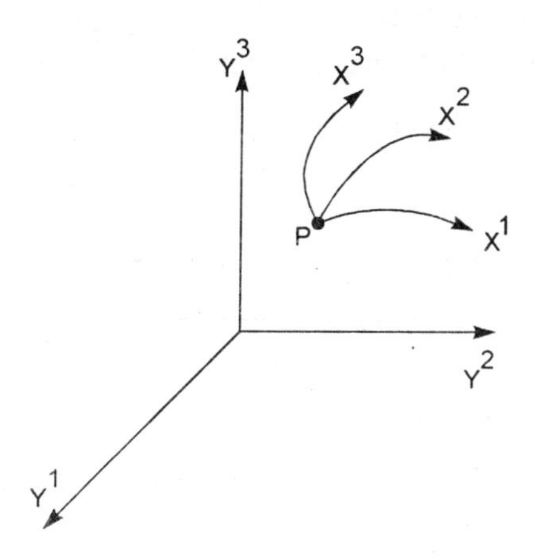

Fig. 7.2

intersect in one and only one point. The surfaces defined by equation (2) the coordinate surfaces and intersection of coordinate surface pair-by pair are the coordinate lines. Thus the line of intersection of $x^1 = c_1$ and $x^2 = c_2$ is the x^3 – coordinate line because along this the line the variable x^3 is the only one that is changing.

EXAMPLE 2

Consider a coordinate system defined by the transformation

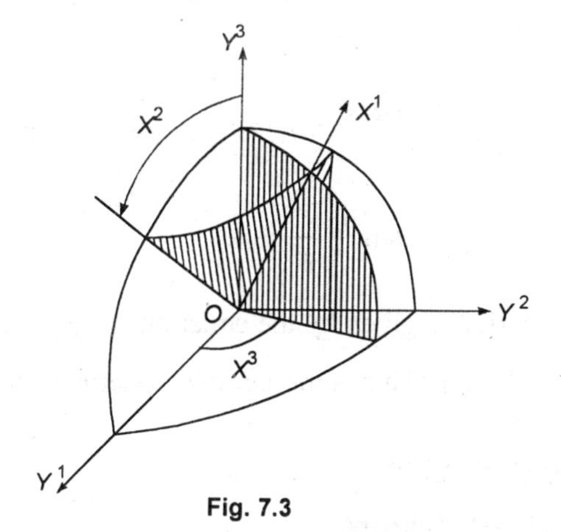

Fig. 7.3

$$\bar{x}^1 = x^1 \sin x^2 \cos x^3 \qquad \ldots(3)$$

$$\bar{x}^2 = x^1 \sin x^2 \sin x^3 \qquad \ldots(4)$$

$$\bar{x}^3 = x^1 \cos x^2 \qquad \ldots(5)$$

The surfaces $x^1 = \text{constant}$ are spheres, $x^2 = \text{constant}$ are circluar cones and x^3 = constant are planes passing through the Y^3-axis (Fig. 7.3).
The squaring and adding equations (3), (4) and (5) we get,

$$(\bar{x}^1)^2 + (\bar{x}^2)^2 + (\bar{x}^3)^2 = (x^1 \sin x^2 \cos x^3)^2 + (x^1 \sin x^2 \sin x^3)^2 + (x^1 \cos x^2)^2$$

On solving

$$(\bar{x}^1)^2 + (\bar{x}^2)^2 + (\bar{x}^3)^2 = (x^1)^2$$

$$x^1 = \sqrt{(\bar{x}^1)^2 + (\bar{x}^2)^2 + (\bar{x}^3)^2} \qquad ...(6)$$

Now, squaring and adding equations (3) and (4), we get

$$(\bar{x}^1)^2 + (\bar{x}^2)^2 = (x^1 \sin x^2 \sin x^3)^2 + (x^1 \sin x^2 \sin x^3)^2$$

$$(\bar{x}^1)^2 + (\bar{x}^2)^2 = (x^1)^2 (\sin x^2)^2$$

$$x^1 \sin x^2 = \sqrt{(\bar{x}^1)^2 + (\bar{x}^2)^2} \qquad ...(7)$$

Divide (7) and (5), we get

$$\tan x^2 = \frac{\sqrt{(\bar{x}^1)^2 + (\bar{x}^2)^2}}{\bar{x}^3}$$

or

$$x^2 = \tan^{-1}\left(\frac{\sqrt{(\bar{x}^1)^2 + (\bar{x}^2)^2}}{\bar{x}^3}\right) \qquad ...(8)$$

Divide (3) and (4), we get

$$\frac{\bar{x}^2}{\bar{x}^1} = \tan x^3$$

$$\Rightarrow \qquad x^3 = \tan^{-1}\left(\frac{\bar{x}^2}{\bar{x}^1}\right) \qquad ...(9)$$

So, the inverse transformation is given by the equations (6), (8) and (9).

If $x^1 > 0, 0 < x^2 < \pi, 0 \le x^3 < 2\pi$. This is the familiar spherical coordinate system.

7.3 RECIPROCAL BASE SYSTEMS

Covariant and Contravariant Vectors

Let a cartesian coordinate system be determined by a set of orthogonal base vectors $\vec{b}_1,\ \vec{b}_2, \vec{b}_3$ then the position vector $\vec{r}$ of any point $P(\bar{x}^1, \bar{x}^2, \bar{x}^3)$ can be expressed as

$$\vec{r} = \vec{b}_i \bar{x}^i \quad (i = 1, 2, 3) \qquad ...(1)$$

Since the base vectors $\vec{b}_i$ are independent of the position of the point $P(\bar{x}^1, \bar{x}^2, \bar{x}^3)$. Then from (1),

$$d\vec{r} = \vec{b}_i d\bar{x}^i \qquad ...(2)$$

If $P(\bar{x}^1, \bar{x}^2, \bar{x}^3)$ and $Q(\bar{x}^1 + d\bar{x}^1, \bar{x}^2 + d\bar{x}^2, \bar{x}^3 + d\bar{x}^3)$ be two closed point. The square of the element of arc ds between two points is

$$ds^2 = d\vec{r} \cdot d\vec{r}$$

from equation (2),

$$ds^2 = \vec{b}_i d\bar{x}^i \cdot \vec{b}_j d\bar{x}^j$$

$$= \vec{b}_i \cdot \vec{b}_j d\bar{x}^i d\bar{x}^j$$

$$ds^2 = \delta_{ij} d\bar{x}^i d\bar{x}^j; \text{ since } \vec{b}_i \cdot \vec{b}_j = \delta_{ij}$$

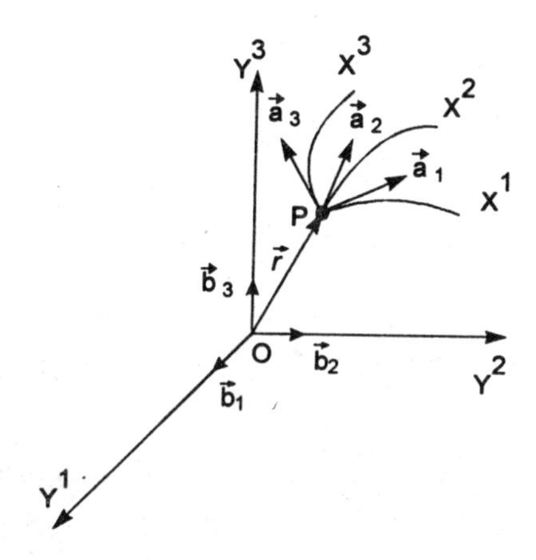

Fig. 7.4.

$(\vec{b}_1, \vec{b}_2, \vec{b}_3$ are orthogonal base vector *i.e.*, $\vec{b}_1 \cdot \vec{b}_1 = 1 \;\&\; \vec{b}_1 \cdot \vec{b}_2 = 0)$.

$$ds^2 = d\bar{x}^i d\bar{x}^i; \quad \text{as } \delta_{ij} = 1, \; i = j; \quad = 0, \; i \neq j$$

a familiar expression for the square of element of arc in orthogonal cartesian coordinates.
Consider the coordinate transformation

$$x^i = x^i(\bar{x}^1, \bar{x}^2, \bar{x}^3), \; (i = 1, 2, 3)$$

define a curvilinear coordinate system X. The position vector $\vec{r}$ is a function of coordinates x^i. *i.e.*,

$$\vec{r} = \vec{r}(x^i), \; (i = 1, 2, 3)$$

Then

$$d\vec{r} = \frac{\partial \vec{r}}{\partial x^i} dx^i \qquad ...(3)$$

and

$$ds^2 = d\vec{r} \cdot d\vec{r}$$

$$= \frac{\partial \vec{r}}{\partial x^i} \cdot \frac{\partial \vec{r}}{\partial x^j} dx^i dx^j$$

$$ds^2 = g_{ij} dx^i dx^j$$

where

$$g_{ij} = \frac{\partial \vec{r}}{\partial x^i} \cdot \frac{\partial \vec{r}}{\partial x^j} \qquad ...(4)$$

The vector $\frac{\partial \vec{r}}{\partial x^i}$ is a base vector directed tangentially to X^i-coordinate curve.

Put

$$\frac{\partial \vec{r}}{\partial x^i} = \vec{a}_i \qquad ...(5)$$

Then from (3) and (4)

$$d\vec{r} = \vec{a}_i dx^i \text{ and } g_{ij} = \vec{a}_i \cdot \vec{a}_j \qquad ...(6)$$

Now, from equations (2) and (6), we get

$$\vec{a}_j dx^j = \vec{b}_i d\bar{x}^i$$

$$\vec{a}_j dx^j = \vec{b}_i \frac{\partial \bar{x}^i}{\partial x^j} dx^j$$

$$\Rightarrow \qquad \vec{a}_j = \vec{b}_i \frac{\partial \bar{x}^i}{\partial x^j}, \text{ as } dx^j \text{ are arbitrary}$$

So, the base vectors $\vec{a}_j$ transform according to the law for transformation of components of covariant vectors.

The components of base vectors $\vec{a}_i$, when referred to X-coordinate system, are

$$\vec{a}_1 : (a_1, 0, 0), \quad \vec{a}_2 : (0, a_2, 0), \quad \vec{a}_3 : (0, 0, a_3). \qquad ...(7)$$

and they are not necessarily unit vectors.

In general,

$$g_{11} = \vec{a}_1 \cdot \vec{a}_1 \neq 1, \; g_{22} = \vec{a}_2 \cdot \vec{a}_2 \neq 1, \; g_{33} = \vec{a}_3 \cdot \vec{a}_3 \neq 1. \qquad ...(8)$$

If the curvilinear coordinate system X is orthogonal. Then

$$g_{ij} = \vec{a}_i \cdot \vec{a}_j = |\vec{a}_i| |\vec{a}_j| \cos\theta_{ij} = 0, \text{ if } i \neq j. \qquad ...(9)$$

Any vector $\vec{A}$ are can be written in the form $\vec{A} = k d\vec{r}$ where k is a scalar.

Since $d\vec{r} = \frac{\partial \vec{r}}{\partial x^i} dx^i$ we have

$$\vec{A} = \frac{\partial \vec{r}}{\partial x^i}(kdx^i)$$

$$\vec{A} = \vec{a}_i A^i$$

where $A^i = kdx^i$. The numbers A^i are the contravariant components of the vector $\vec{A}$
Consider three non-coplanar vectors

$$\vec{a}^1 = \frac{\vec{a}_2 \times \vec{a}_3}{[\vec{a}_1\vec{a}_2\vec{a}_3]},\ \vec{a}^2 = \frac{\vec{a}_3 \times \vec{a}_1}{[\vec{a}_1\vec{a}_2\vec{a}_3]},\ \vec{a}^3 = \frac{\vec{a}_1 \times \vec{a}_2}{[\vec{a}_1\vec{a}_2\vec{a}_3]} \quad ...(10)$$

where $\vec{a}_2 \times \vec{a}_3$, etc. denote the vector product of $\vec{a}_2$ and $\vec{a}_3$ and $[\vec{a}_1\vec{a}_2\vec{a}_3]$ is the triple scalar prduct $\vec{a}_1 \cdot \vec{a}_2 \times \vec{a}_3$.

Now,

$$\vec{a}^1 \cdot \vec{a}_1 = \frac{\vec{a}_2 \times \vec{a}_3 \cdot \vec{a}_1}{[\vec{a}_1\vec{a}_2\vec{a}_3]} = \frac{[\vec{a}_1\vec{a}_2\vec{a}_3]}{[\vec{a}_1\vec{a}_2\vec{a}_3]} = 1.$$

$$\vec{a}^1 \cdot \vec{a}_2 = \frac{\vec{a}_2 \times \vec{a}_3 \cdot \vec{a}_2}{[\vec{a}_1\vec{a}_2\vec{a}_3]} = \frac{[\vec{a}_2\vec{a}_2\vec{a}_3]}{[\vec{a}_1\vec{a}_2\vec{a}_3]} = 0.$$

Since $[\vec{a}_2\vec{a}_2\vec{a}_3] = 0$.
Similarly,

$$\vec{a}^1 \cdot \vec{a}_3 = \vec{a}^2 \cdot \vec{a}_1 = \cdots = 0$$

$$\vec{a}^2 \cdot \vec{a}_2 = 1,\ \vec{a}^3 \cdot \vec{a}_3 = 1$$

Then we can write

$$\vec{a}^i \cdot \vec{a}_j = \delta^i_j$$

EXAMPLE 3

To show that $[\vec{a}_1\vec{a}_2\vec{a}_3] = \sqrt{g}$ and $[\vec{a}^1\vec{a}^2\vec{a}^3] = \frac{1}{\sqrt{g}}$ where $g = |g_{ij}|$.

Solution

The components of base vectors a_i are

$$\vec{a}_1 : (a_1, 0, 0),\ \vec{a}_2 : (0, a_2, 0) \text{ and } \vec{a}_3 : (0, 0, a_3)$$

Then

$$[\vec{a}_1\vec{a}_2\vec{a}_3] = \begin{vmatrix} a_1 & 0 & 0 \\ 0 & a_2 & 0 \\ 0 & 0 & a_3 \end{vmatrix} = a_1a_2a_3 \quad ...(11)$$

and

$$g = |g_{ij}| = \begin{vmatrix} g_{11} & g_{12} & g_{13} \\ g_{21} & g_{22} & g_{23} \\ g_{31} & g_{32} & g_{33} \end{vmatrix}$$

from equations (8) and (9), we have

$$g_{11} = \vec{a}_1 \cdot \vec{a}_1 \Rightarrow a_1^2 = g_{11}$$

Similarly

$$a_2^2 = g_{22}, \; a_3^2 = g_{33}$$

and

$$g_{12} = \vec{a}_1 \cdot \vec{a}_2 = 0, \; g_{13} = \vec{a}_1 \cdot \vec{a}_3 = 0 \text{ etc.}$$

So,

$$g = \begin{vmatrix} a_1^2 & 0 & 0 \\ 0 & a_2^2 & 0 \\ 0 & 0 & a_3^2 \end{vmatrix} = a_1^2 a_2^2 a_3^2$$

$$\sqrt{g} = a_1 a_2 a_3 \qquad \text{...(12)}$$

from eqn. (11) and (12), we have

$$[\vec{a}_1 \vec{a}_2 \vec{a}_3] = \sqrt{g}$$

Since the triple products $[\vec{a}^1 \vec{a}^2 \vec{a}^3] = \dfrac{1}{\sqrt{g}}$. Moreover,

$$\vec{a}_1 = \frac{\vec{a}^2 \times \vec{a}^3}{[\vec{a}^1 \vec{a}^2 \vec{a}^3]}, \; \vec{a}_2 = \frac{\vec{a}^3 \times \vec{a}^1}{[\vec{a}^1 \vec{a}^2 \vec{a}^3]}, \; \vec{a}_3 = \frac{\vec{a}^1 \times \vec{a}^2}{[\vec{a}^1 \vec{a}^2 \vec{a}^3]}$$

The system of vectors $\vec{a}^1, \vec{a}^2, \vec{a}^3$ is called the *reciprocal base system.*

Hence if the vectors $\vec{a}^1, \vec{a}^2, \vec{a}^3$ are unit vectors associated with an orthogonal cartesian coordinates then the reciprocal system of vector defines the same system of coordinates. **Solved.**

The differential of a vector $\vec{r}$ in the reciprocal base system is $d\vec{r} = \vec{a}^i dx_i$.

where dx_i are the components of $d\vec{r}$. Then

$$ds^2 = d\vec{r} \cdot d\vec{r}$$

$$= (\vec{a}^i dx_i) \cdot (\vec{a}^j dx_j)$$

$$= \vec{a}^i \cdot \vec{a}^j dx_i dx_j$$

$$ds^2 = g^{ij} dx_i dx_j$$

where

$$g^{ij} = \vec{a}^i \cdot \vec{a}^j = g^{ji} \qquad \text{...(13)}$$

The system of base vectors determined by equation (10) can be used to represent an arbitrary vector A in the form $\vec{A} = \vec{a}^i A_i$, where A_i are the covariant components of $\vec{A}$.

Taking scalar product of vector $A_i\vec{a}^i$ with the base vector $\vec{a}_j$, we get

$$A_i\vec{a}^i \cdot \vec{a}_j = A_i\delta^i_j = A_j \quad \text{as } \vec{a}^i \cdot \vec{a}_j = \delta^i_j.$$

7.4 ON THE MEANING OF COVARIANT DERIVATIVES

THEOREM 7.2 *If $\vec{A}$ is a vector along the curve in E_3. Prove that $\dfrac{\partial \vec{A}}{\partial x^j} = A^\alpha_{,j}\vec{a}_\alpha$*

Also, prove that $A^i_{,j} = \dfrac{\partial A^i}{\partial x^j}$. Where A^i are component of $\vec{A}$.

Proof: A vector $\vec{A}$ can be expressed in the terms of base vectors $\vec{a}_i$ as

$$\vec{A} = A^i\vec{a}_i$$

where $\vec{a}_i = \dfrac{\partial \vec{r}}{\partial x^i}$ and A^i are components of $\vec{A}$.

The partial derivative of $\vec{A}$ with respect to x^j is

$$\frac{\partial \vec{A}}{\partial x^j} = \frac{\partial A^i}{\partial x^j}\vec{a}_i + A^i\frac{\partial \vec{a}_i}{\partial x^j} \qquad ...(1)$$

Since $g_{ij} = \vec{a}_i \cdot \vec{a}_j$.

Differentiating partially it w.r.t. x^k, we have

$$\frac{\partial g_{ij}}{\partial x^k} = \frac{\partial \vec{a}_i}{\partial x^k}\cdot\vec{a}_j + \frac{\partial \vec{a}_j}{\partial x^k}\cdot\vec{a}_i$$

Similarly,

$$\frac{\partial g_{jk}}{\partial x^i} = \frac{\partial \vec{a}_j}{\partial x^i}\cdot\vec{a}_k + \frac{\partial \vec{a}_k}{\partial x^i}\cdot\vec{a}_j$$

and

$$\frac{\partial g_{ik}}{\partial x^j} = \frac{\partial \vec{a}_i}{\partial x^j}\cdot\vec{a}_k + \frac{\partial \vec{a}_k}{\partial x^j}\cdot\vec{a}_i$$

Since $\vec{A}$ can be written as

$$\vec{A} = \vec{a}_i A^i = \vec{a}^i A_i$$

Taking scalar product with $\vec{a}_j$, we have

$$\vec{a}_i \cdot \vec{a}_j A^i = \vec{a}^i \cdot \vec{a}_j A_i$$

$$\Rightarrow \qquad g_{ij}A^i = \delta^j_i A_i = A_j$$

As $\vec{a}_i \cdot \vec{a}_j = g_{ij}, \vec{a}^i \cdot \vec{a}_j = \delta^i_j$ and $\delta^i_j A_i = A^j$.

We see that the vector obtained by lowering the index in A^i is precisely the covariant vector A_i.

The two sets of quantities A^i and A_i are represent the same vector $\vec{A}$ referred to two different base systems.

EXAMPLE 4

Show that $g_{i\alpha} g^{j\alpha} = \delta_i^j$.

Solution

Since we know that

$$g_{i\alpha} = \vec{a}_i \cdot \vec{a}_\alpha \text{ and } g^{j\alpha} = \vec{a}^j \cdot \vec{a}^\alpha$$

Then

$$g_{i\alpha} g^{j\alpha} = (\vec{a}_i \cdot \vec{a}_\alpha)(\vec{a}^j \cdot \vec{a}^\alpha)$$

$$= (\vec{a}_i \cdot \vec{a}^j)(\vec{a}_\alpha \cdot \vec{a}^\alpha)$$

$$= \delta_i^j \, \delta_\alpha^\alpha \text{ as } \vec{a}_i \cdot \vec{a}^j = \delta_i^j$$

$$g_{i\alpha} g^{j\alpha} = \delta_i^j \text{ as } \delta_\alpha^\alpha = 1.$$

But

$$\vec{a}_i = \frac{\partial \vec{r}}{\partial x^i}$$

$$\frac{\partial \vec{a}_i}{\partial x^j} = \frac{\partial^2 \vec{r}}{\partial x^j dx^i} = \frac{\partial^2 \vec{r}}{\partial x^i \partial x^j} = \frac{\partial \vec{a}_j}{\partial x^i}$$

So,

$$\frac{\partial \vec{a}_i}{\partial x^j} = \frac{\partial \vec{a}_j}{\partial x^i}$$

Now,

$$[ij,k] = \frac{1}{2}\left(\frac{\partial g_{ik}}{\partial x^j} + \frac{\partial g_{jk}}{\partial x^i} - \frac{\partial g_{ij}}{\partial x^k}\right), \text{ Christoffel's symbol}$$

Substituting the value of $\frac{\partial g_{ik}}{\partial x^j}, \frac{\partial g_{jk}}{\partial x^i}$ and $\frac{\partial g_{ij}}{\partial x^k}$, we get

$$[ij,k] = \frac{1}{2} \cdot 2 \frac{\partial \vec{a}_i}{\partial x^j} \cdot \vec{a}_k$$

$$[ij,k] = \frac{\partial \vec{a}_i}{\partial x^j} \cdot \vec{a}_k$$

or

$$\frac{\partial \vec{a}_i}{\partial x^j} = [ij,k]\vec{a}^k, \text{ as } \frac{1}{\vec{a}_k} = \vec{a}^k$$

Hence

$$\frac{\partial \vec{a}_i}{\partial x^j} \cdot \vec{a}^\alpha = [ij,k]\vec{a}^k \cdot \vec{a}^\alpha$$

$$= [ij,k]g^{k\alpha}, \text{ Since } g^{k\alpha} = \vec{a}^k \cdot \vec{a}^\alpha$$

$$\frac{\partial \vec{a}_i}{\partial x^j} . \vec{a}^\alpha = \begin{Bmatrix} \alpha \\ i\ j \end{Bmatrix}, \text{ as } [i\,j,k]g^{k\alpha} = \begin{Bmatrix} \alpha \\ i\ j \end{Bmatrix}$$

$$\frac{\partial \vec{a}_i}{\partial x^j} = \begin{Bmatrix} \alpha \\ i\ j \end{Bmatrix} \vec{a}_\alpha \quad \text{...(2)}$$

Substituting the values of $\frac{\partial \vec{a}_i}{\partial x^j}$ in equation (1), we get

$$\frac{\partial \vec{A}}{\partial x^j} = \frac{\partial A^i}{\partial x^j}\vec{a}_i + \begin{Bmatrix} \alpha \\ i\ j \end{Bmatrix} A^i \vec{a}_\alpha$$

$$= \frac{\partial A^\alpha}{\partial x^j}\vec{a}_\alpha + \begin{Bmatrix} \alpha \\ i\ j \end{Bmatrix} A^i \vec{a}_\alpha$$

$$\frac{\partial \vec{A}}{\partial x^j} = \left[\frac{\partial A^\alpha}{\partial x^j} + \begin{Bmatrix} \alpha \\ i\ j \end{Bmatrix} A^i \right] \vec{a}_\alpha$$

$$\frac{\partial \vec{A}}{\partial x^j} = A^\alpha_{,j}\, \vec{a}_\alpha \text{ since } A^\alpha_{,j} = \frac{\partial A^\alpha}{\partial x^j} + \begin{Bmatrix} \alpha \\ i\ j \end{Bmatrix} A^i$$

Thus, the covariant derivative $A^\alpha_{,j}$ of the vector A^α is a vector whose components are the components of $\frac{\partial \vec{A}}{\partial x^j}$ referred to the base system $\vec{a}_i$.

If the Christoffel symbols vanish identically *i.e.*, $\begin{Bmatrix} \alpha \\ i\ j \end{Bmatrix} = 0$ the $\frac{\partial \vec{a}_i}{\partial x^j} = 0$, from (2).

Substituting this value in equation (1), we get

$$\frac{\partial \vec{A}}{\partial x^j} = \frac{\partial A^i}{\partial x^j}\vec{a}_i$$

But
$$A^i_{,j} = \frac{\partial A^i}{\partial x^j} + \begin{Bmatrix} \cdot\, i \\ \beta\ j \end{Bmatrix} A^\beta$$

$$A^i_{,j} = \frac{\partial A^i}{\partial x^j} \text{ as } \begin{Bmatrix} i \\ \beta\ j \end{Bmatrix} = 0.$$

Proved.

THEOREM 7.3 *If $\vec{A}$ is a vector along the curve in E_3. Prove that $A_{j,k}\vec{a}^j$ where A_j are components of $\vec{A}$.*

Proof: If $\vec{A}$ can be expressed in the form

$$\vec{A} = A_i \vec{a}^i$$

where A_i are components of $\vec{A}$.

The partial derivative of $\vec{A}$ with respect to x^k is

$$\frac{\partial \vec{A}}{\partial x^k} = \frac{\partial A_i}{\partial x^k}\vec{a}^i + A_i\frac{\partial \vec{a}^i}{\partial x^k} \qquad ...(1)$$

Since $\vec{a}^i \cdot \vec{a}_j = \delta^i_j$, we have,

Differentiating it partially w.r. to x^k, we get

$$\frac{\partial \vec{a}^i}{\partial x^k}\cdot \vec{a}_j + \vec{a}^i \cdot \frac{\partial \vec{a}_j}{\partial x^k} = 0$$

$$\frac{\partial \vec{a}^i}{\partial x^k}\cdot \vec{a}_j = -\vec{a}^i \cdot \frac{\partial \vec{a}_j}{\partial x^k}$$

$$= -\vec{a}^i \cdot \vec{a}_\alpha \begin{Bmatrix} \alpha \\ j\ k \end{Bmatrix}, \quad \text{Since } \frac{\partial \vec{a}_j}{\partial x^k} = \begin{Bmatrix} \alpha \\ j\ k \end{Bmatrix}\vec{a}_\alpha$$

But $\vec{a}^i \cdot \vec{a}_\alpha = \delta^i_\alpha$. Then

$$\frac{\partial \vec{a}^i}{\partial x^k}\cdot \vec{a}_j = -\delta^i_\alpha \begin{Bmatrix} \alpha \\ j\ k \end{Bmatrix}$$

$$\frac{\partial \vec{a}^i}{\partial x^k}\cdot \vec{a}_j = -\begin{Bmatrix} i \\ j\ k \end{Bmatrix}$$

$$\frac{\partial \vec{a}^i}{\partial x^k} = -\begin{Bmatrix} i \\ j\ k \end{Bmatrix}\vec{a}^j, \quad \text{as } \frac{1}{\vec{a}_j} = \vec{a}^j$$

substituting the value of $\frac{\partial \vec{a}^i}{\partial x^k}$ in equation (1), we get

$$\frac{\partial \vec{A}}{\partial x^k} = \frac{\partial A_i}{\partial x^k}\vec{a}^i - A_i\begin{Bmatrix} i \\ j\ k \end{Bmatrix}\vec{a}^j$$

$$= \frac{\partial A_j}{\partial x^k}\vec{a}^j - A_i\begin{Bmatrix} i \\ j\ k \end{Bmatrix}\vec{a}^j$$

$$\frac{\partial \vec{A}}{\partial x^k} = \left[\frac{\partial A_j}{\partial x^k} - A_i\begin{Bmatrix} i \\ j\ k \end{Bmatrix}\right]\vec{a}^j$$

$$\frac{\partial \vec{A}}{\partial x^k} = A_{j,k}\vec{a}^j, \quad \text{Since } A_{j,k} = \frac{\partial A_j}{\partial x^k} - A_i\begin{Bmatrix} i \\ j\ k \end{Bmatrix}$$

Proved.

7.5 INTRINSIC DIFFERENTIATION

Let a vector field $\vec{A}(x)$ and

$$C : x^i = x^i(t), \quad t_1 \le t \le t_2$$

be a curve in some region of E_3. The vector $\vec{A}(x)$ depend on the parameter t and if $A(x)$ is a differentiable vector then

$$\frac{d\vec{A}}{dt} = \frac{\partial \vec{A}}{\partial x^j} \cdot \frac{dx^j}{dt}$$

$$\frac{d\vec{A}}{dt} = A^{\alpha}_{,j}\, \vec{a}_{\alpha} \frac{dx^j}{dt}$$

Since we know

$$\frac{\partial \vec{A}}{\partial x^j} = A^{\alpha}_{,j}\, \vec{a}_{\alpha} = \left[\frac{\partial A^{\alpha}}{\partial x^j} + \begin{Bmatrix} \alpha \\ i \; j \end{Bmatrix} \right] \vec{a}_{\alpha}$$ (See Pg. 127, Theo. 7.2)

So,

$$\frac{d\vec{A}}{dt} = \left[\frac{\partial A^{\alpha}}{\partial x^j} + \begin{Bmatrix} \alpha \\ i \; j \end{Bmatrix} A^i \right] \vec{a}_{\alpha} \frac{dx^j}{dt}$$

$$\frac{d\vec{A}}{dt} = \left[\frac{dA^{\alpha}}{dt} + \begin{Bmatrix} \alpha \\ i \; j \end{Bmatrix} A^i \frac{dx^j}{dt} \right] \vec{a}_{\alpha}$$

The formula $\frac{dA^{\alpha}}{dt} + \begin{Bmatrix} \alpha \\ i \; j \end{Bmatrix} A^i \frac{dx^j}{dt}$ is called the *absolute or Intrinsic* derivative of A^{α} with respect to parameter t and denoted by $\frac{\delta A^{\alpha}}{\delta t}$.

So, $\frac{\delta A^{\alpha}}{\delta t} = \frac{dA^{\alpha}}{dt} + \begin{Bmatrix} \alpha \\ i \; j \end{Bmatrix} A^i \frac{dx^j}{dt}$ is contravariant vector. If A is a scalar then, obviously, $\frac{\delta A}{\delta t} = \frac{\delta A}{\delta t}$.

Some Results

(*i*) If A_i be covariant vector

$$\frac{\delta A_i}{\delta t} = \frac{dA_i}{dt} - \begin{Bmatrix} \alpha \\ i \; \beta \end{Bmatrix} A_{\alpha} \frac{dx^{\beta}}{dt}$$

(*ii*) $$\frac{\delta A^{ij}}{\delta t} = \frac{\delta A^{ij}}{dt} + \begin{Bmatrix} i \\ \alpha \; \beta \end{Bmatrix} A^{\alpha j} \frac{dx^{\beta}}{dt} + \begin{Bmatrix} j \\ \alpha \; \beta \end{Bmatrix} A^{i\alpha} \frac{dx^{\beta}}{dt}$$

(*iii*) $$\frac{\delta A^i_j}{\delta t} = \frac{\delta A^i_j}{dt} + \begin{Bmatrix} i \\ \alpha \; \beta \end{Bmatrix} A^{\alpha}_j \frac{dx^{\beta}}{dt} - \begin{Bmatrix} \alpha \\ i \; \beta \end{Bmatrix} A^i_{\alpha} \frac{dx^{\beta}}{dt}$$

$$(iv)\quad \frac{\delta A^i_{jk}}{\delta t} = \frac{\delta A^i_{jk}}{dt} + \begin{Bmatrix} i \\ \alpha\ \beta \end{Bmatrix} A^\alpha_{jk} \frac{dx^\beta}{dt} - \begin{Bmatrix} \alpha \\ j\ \beta \end{Bmatrix} A^i_{\alpha k} \frac{dx^\beta}{dt} - \begin{Bmatrix} \alpha \\ k\ \beta \end{Bmatrix} A^i_{j\alpha} \frac{dx^\beta}{dt}$$

EXAMPLE 5

If g_{ij} be components of metric tensor, show that $\frac{\delta g_{ij}}{\delta t} = 0.$

Solution

The intrinsic derivative of g_{ij} is

$$\frac{\delta g_{ij}}{\delta t} = \frac{dg_{ij}}{dt} - \begin{Bmatrix} \alpha \\ i\ \beta \end{Bmatrix} g_{\alpha j} \frac{dx^\beta}{dt} - \begin{Bmatrix} \alpha \\ \beta\ j \end{Bmatrix} g_{i\alpha} \frac{dx^\beta}{dt}$$

$$= \frac{\partial g_{ij}}{\partial x^\beta} \frac{dx^\beta}{dt} - \begin{Bmatrix} \alpha \\ i\ \beta \end{Bmatrix} g_{\alpha j} \frac{dx^\beta}{dt} - \begin{Bmatrix} \alpha \\ \beta\ j \end{Bmatrix} g_{i\alpha} \frac{dx^\beta}{dt}$$

$$= \left[\frac{\partial g_{ij}}{\partial x^\beta} - \begin{Bmatrix} \alpha \\ i\ \beta \end{Bmatrix} g_{\alpha j} - \begin{Bmatrix} \alpha \\ \beta\ j \end{Bmatrix} g_{i\alpha} \right] \frac{dx^\beta}{dt}$$

$$\frac{\delta g_{ij}}{\delta t} = \left\{ \frac{\partial g_{ij}}{\partial x^\beta} - [i\beta, j] - [\beta j, i] \right\} \frac{dx^\beta}{dt}$$

as $\begin{Bmatrix} \alpha \\ i\ \beta \end{Bmatrix} g_{\alpha j} = [i\beta, j]$ and $\begin{Bmatrix} \alpha \\ \beta\ j \end{Bmatrix} g_{i\alpha} = [\beta j, i]$.

But $\frac{\partial g_{ij}}{\partial x^\beta} = [i\beta, j] + [j\beta, i].$

So, $\frac{\delta g_{ij}}{\delta t} = 0.$

EXAMPLE 6

Prove that

$$\frac{d(g_{ij} A^i A^j)}{dt} = 2 g_{ij} A^i \frac{\delta A^i}{\delta t}$$

Solution

Since $g_{ij} A^i a^j$ is scalar.

Then

$$\frac{d(g_{ij} A^i A^j)}{dt} = \frac{\delta(g_{ij} A^i A^j)}{\delta t}$$

$$= g_{ij}\frac{\delta(A^iA^j)}{\delta t}, \text{ since } g_{ij} \text{ is independent of } t.$$

$$= g_{ij}\left[\frac{\delta A^i}{\delta t}A^j + A^i\frac{\delta A^j}{\delta t}\right]$$

Interchange i and j in first term, we get

$$\frac{d(g_{ij}A^iA^j)}{dt} = g_{ij}\left[A^i\frac{\delta A^j}{\delta t} + A^i\frac{\delta A^j}{\delta t}\right] \text{ as } g_{ij} \text{ is symmetric.}$$

$$\frac{d(g_{ij}A^iA^j)}{dt} = 2g_{ij}A^i\frac{\delta A^j}{\delta t}$$

Proved.

EXAMPLE 7

Prove that if A is the magnitude of A^i then

$$A_{,j} = \frac{A_{i,j}A^i}{A}$$

Solution

Given that A is magnitude of A^i. Then
Since

$$g_{ik}A^iA^k = A^iA^i$$

$$g_{ik}A^iA^k = A^2$$

Taking covariant derivative w.r. to x^j, we get

$$g_{ik}A^i_{,j}A^k + g_{ik}A^iA^k_{,j} = 2AA_{,j}$$

Interchange the dummy index in first term, we get

$$g_{ki}A^k_{,j}A^i + g_{ik}A^iA^k_{,j} = 2AA_{,j}$$

$$2g_{ik}A^iA^k_{,j} = 2AA_{,j}$$

$$g_{ik}A^iA^k_{,j} = AA_{,j}$$

$$A^i\left(g_{ik}A^k_{,j}\right) = AA_{,j}$$

$$A^iA_{i,j} = AA_{,j} \quad \text{since } g_{ik}A^k_{,j} = A_{i,j}$$

$$A_{,j} = \frac{A_{i,j}A^i}{A}$$

Proved.

7.6 PARALLEL VECTOR FIELDS

Consider a curve

$$C : x^i = x^i(t), \quad t_1 \le t \le t_2, \quad (i = 1, 2, 3)$$

in some region of E_3 and a vector $\vec{A}$ localized at point P of C. If we construct at every point of C a vector equal to A in magnitude and parallel to it in direction, we obtain a parallel field of vector along the curve C.

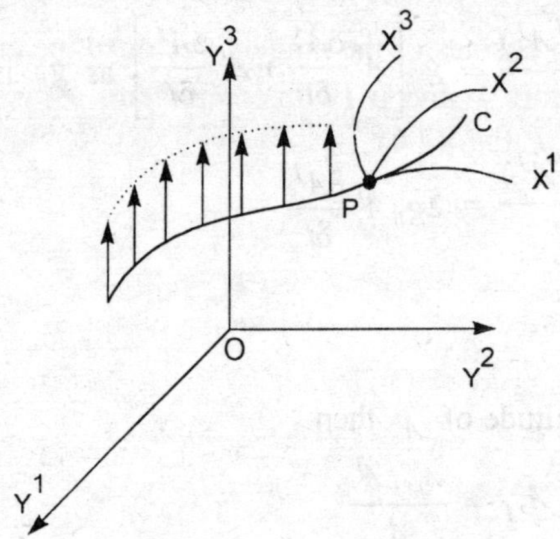

if $\vec{A}$ is a parallel field along C then the vector $\vec{A}$ do not change along the curve and we can write $\frac{d\vec{A}}{dt} = 0$. It follows that the components A^i of $\vec{A}$ satisfy a set of simultaneous differential equations

$\frac{\partial A^i}{\partial t} = 0$ or

$$\frac{dA^i}{dt} + \begin{Bmatrix} i \\ \alpha\ \beta \end{Bmatrix} A^\alpha \frac{dx^\beta}{dt} = 0$$

This is required condition for the vector field A^i is parallel.

7.7 GEOMETRY OF SPACE CURVES

Let the parametric equations of the curve C in E_3 be

$$C : x^i = x^i(t), \quad t_1 \le t \le t_2 \quad (i = 1, 2, 3).$$

The square of the length of an element of C is given by

$$ds^2 = g_{ij} dx^i dx^j \qquad ...(1)$$

and the length of arc s of C is defined by the integral

$$s = \int_{t_1}^{t_2} \sqrt{g_{ij} \frac{dx^i}{dt} \frac{dx^j}{dt}} dt \qquad ...(2)$$

from (1), we have

$$g_{ij} \frac{dx^i}{ds} \frac{dx^j}{ds} = 1 \qquad ...(3)$$

Put $\dfrac{dx^i}{ds} = \lambda^i$. Then equation (3) becomes

$$g_{ij}\lambda^i\lambda^j = 1 \qquad ...(4)$$

The vector $\vec{\lambda}$, with components λ^i, is a unit vector. Moreover, $\vec{\lambda}$ is tangent to C, since its components $\overline{\lambda}^i$, when the curve C is referred to a rectangular Cartesian coordinate Y, becomes $\overline{\lambda}^i = \dfrac{d\overline{x}^i}{ds}$. These are precisely the direction cosines of the tangent vector to the curve C.

Consider a pair of unit vectors $\vec{\lambda}$ and $\vec{\mu}$ (with components λ^i and μ^i respectively) at any point P of C. Let $\vec{\lambda}$ is tangent to C at P Fig. (7.6).

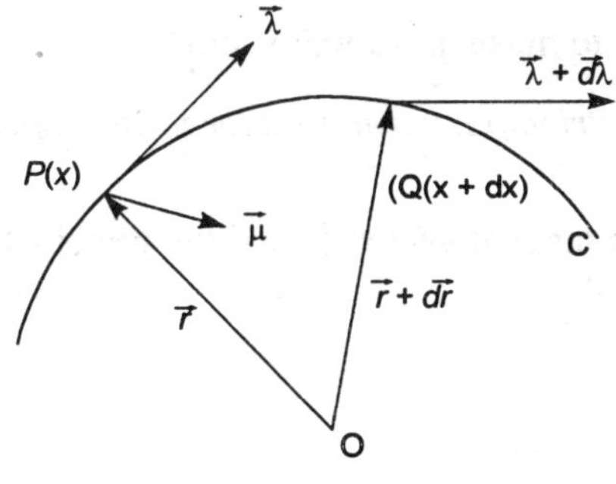

Fig. 7.6

The cosine of the angle θ between $\vec{\lambda}$ and $\vec{\mu}$ is given by the formula

$$\cos\theta = g_{ij}\lambda^i\lambda^j \qquad ...(5)$$

and if $\vec{\lambda}$ and $\vec{\mu}$ are orthogonal, then equation (5) becomes

$$g_{ij}\lambda^i\mu^j = 0 \qquad ...(6)$$

Any vector $\vec{\mu}$ satisfying equation (6) is said to be *normal* to C at P.

Now, differentiating intrinsically, with respect to the are parameter s, equation (4), we get

$$g_{ij}\frac{\delta\lambda^i}{\delta s}\lambda^j + g_{ij}\frac{\delta\lambda^j}{\delta s}\lambda^i = 0 \qquad ...(7)$$

as g_{ij} is constant with respect to s.

Interchange indices i and j in second term of equation (7) we get

$$g_{ij}\frac{\delta\lambda^i}{\delta s}\lambda^j + g_{ij}\frac{\delta\lambda^i}{\delta s}\lambda^j = 0$$

Since g_{ij} is symmetric. Then

$$2g_{ij}\lambda^i\frac{\delta\lambda^j}{\delta s} = 0$$

$$\Rightarrow \qquad g_{ij}\lambda^i \frac{\delta\lambda^j}{\delta s} = 0$$

we see that the vector $\frac{\delta\lambda^j}{\delta s}$ either vanishes or is normal to C and if does not vanish we denote the unit vector co-directional with $\frac{\delta\lambda^j}{\delta s}$ by μ^j and write

$$\mu^j = \frac{1}{K}\frac{\delta\lambda^j}{\delta s}, \quad K = \left|\frac{\delta\lambda^j}{\delta s}\right| \qquad \text{...(8)}$$

where $K > 0$ is so chosen as to make μ^j a unit vector.

The vector μ^j is called the *Principal normal vector* to the curve C at the point P and K is the *curvature* of C.

The plane determined by the tangent vector $\vec{\lambda}$ and the principal normal vector $\vec{\mu}$ is called the *osculating plane* to the curve C at P.

Since $\vec{\mu}$ is unit vector

$$g_{ij}\mu^i\mu^j = 1 \qquad \text{...(9)}$$

Also, differentiating intrinsically with respect to s to equation (6), we get

$$g_{ij}\frac{\delta\lambda^j}{\delta s}\mu^j + g_{ij}\lambda^i\frac{\delta\lambda^j}{\delta s} = 0$$

or

$$g_{ij}\lambda^i\frac{\delta\mu^j}{\delta s} = -g_{ij}\frac{\delta\lambda^i}{\delta s}\mu^j$$

$$= -Kg_{ij}\mu^i\mu^j \quad \text{Since } \frac{\delta\lambda^i}{\delta s} = K\mu^i$$

$$g_{ij}\lambda^i\frac{\delta\mu^j}{\delta s} = -K, \quad \text{since } g_{ij}\mu^i\mu^j = 1.$$

$$g_{ij}\lambda^i\frac{\delta\mu^j}{\delta s} + K = 0$$

$$g_{ij}\lambda^i\frac{\delta\mu^j}{\delta s} + g_{ij}\lambda^i\lambda^j K = 0 \text{ as } g_{ij}\lambda^i\lambda^j = 1$$

$$g_{ij}\lambda^i\left(\frac{\delta\mu^j}{\delta s} + K\lambda^j\right) = 0$$

This shows that the vector $\frac{\delta\mu^j}{\delta s} + K\lambda^j$ is orthogonal to λ^i.

Now, we define a unit vector $\vec{v}$, with components v^j, by the formula

$$v^i = \frac{1}{\tau}\left(\frac{\delta\mu^j}{\delta s} + K\lambda^j\right) \qquad ...(10)$$

where $\tau = \frac{\delta\mu^i}{\delta s} + K\lambda^i$ the vector $\vec{v}$ will be orthogonal to both $\vec{\lambda}$ and $\vec{\mu}$.

To choose the sign of τ in such a way that

$$\sqrt{g}e_{ijk}\lambda^i\mu^j v^k = 1 \qquad ...(11)$$

so that the triad of unit vectors $\vec{\lambda}$, $\vec{\mu}$ and $\vec{v}$ forms a right handed system of axes.

Since e_{ijk} is a relative tensor of weight -1 and $g = \left|\frac{\partial \bar{x}^i}{\partial x^j}\right|^2$ it follows that $\varepsilon_{ijk} = \sqrt{g}e_{ijk}$ is an *obsolute tensor* and hence left hand side of equation (11) is an invariant v^k in equation (11) is determined by the formula

$$v^k = \varepsilon^{ijk}\lambda_i\mu_j \qquad ...(12)$$

where λ_i and μ_j are the associated vectors $g_{i\alpha}\lambda^\alpha$ and $g_{i\alpha}\mu^\alpha$ and $\varepsilon^{ijk} = \frac{1}{\sqrt{g}}e^{ijk}$ is an absolute tensor.

The number τ appearing in equation (10) is called the *torsion* of *C* at *P* and the vector $\vec{v}$ is the *binormal.*

We have already proved that in Theorem 7.2, Pg. 127.

$$\frac{\partial \vec{A}}{\partial x^i} = A^\alpha_{,i}\vec{a}_\alpha$$

if the vector field $\vec{A}$ is defined along *C*, we can write

$$\frac{\partial \vec{A}}{\partial x^i}\frac{\partial x^i}{\partial s} = A^\alpha_{,i}\frac{\partial x^i}{\partial s}\vec{a}_\alpha \qquad ...(13)$$

Using definition of intrinsic derivative,

$$\frac{\delta A^\alpha}{\delta s} = A^\alpha_{,i}\frac{dx^i}{ds}$$

Then equation (13) becomes

$$\frac{d\vec{A}}{ds} = \frac{\delta A^\alpha}{\delta s}\vec{a}_\alpha \cdots; \text{ as } \frac{\partial \vec{A}}{\partial x^i}\frac{dx^i}{ds} = \frac{d\vec{A}}{ds} \qquad ...(14)$$

Let $\vec{r}$ be the position vector of the point *P* on *C* then the tangent vector $\vec{\lambda}$ is determined by

$$\frac{d\vec{r}}{ds} = \lambda^i\vec{a}_i = \vec{\lambda}$$

from equation (14), we get

$$\Rightarrow \quad \frac{d^2\vec{r}}{ds^2} = \frac{d\vec{\lambda}}{ds} = \frac{\delta\lambda^\alpha}{\delta s}\vec{a}_\alpha = \vec{c} \qquad ...(15)$$

where $\vec{c}$ is a vector perpendicular to $\vec{\lambda}$.

With each point P of C we can associate a constant K, such that $\vec{c}/K = \vec{\mu}$ is a unit vector. Since

$$\frac{\vec{c}}{K} = \vec{\mu}$$

$$\vec{\mu} = \frac{1}{K}\frac{\delta\lambda^\alpha}{\delta s}\vec{a}_\alpha, \text{ from (15)}$$

from equation (8), we get

$$\vec{\mu} = \mu^\alpha \vec{a}_\alpha, \text{ since } \mu^\alpha = \frac{1}{K}\frac{\delta\lambda^\alpha}{\delta s}$$

7.8 SERRET-FRENET FORMULA

The serret-frenet formulas are given by

(*i*) $\mu^i = \frac{1}{K}\frac{\delta\lambda^i}{\delta s}$ or $\frac{\delta\lambda^i}{\delta s} = K\mu^i, K > 0$ where $K = \left|\frac{\delta\lambda^i}{\delta s}\right|$

(*ii*) $\nu^i = \frac{1}{\tau}\left(\frac{\delta\mu^i}{\delta s} + K\lambda^i\right)$ or $\frac{\delta\mu^i}{\delta s} = \tau\nu^i - K\lambda^i$ where $\tau = \left|\frac{\delta\mu^i}{\delta s} + K\lambda^i\right|$

(*iii*) $\frac{\delta\nu^k}{\delta s} = -\tau\mu^k$

First two formulas have already been derived in article (7.7), equation (8) and (10).

Proof of (*iii*)

From equation (12), article (7.7), we have

$$\varepsilon^{ijk}\lambda_i\mu_j = \nu^k$$

where λ^i, μ^i, ν^k are mutually orthogonal.

Taking intrinsic derivative with respect to *s*, we get

$$\varepsilon^{ijk}\frac{\delta\lambda_i}{\delta s}\mu_j + \varepsilon^{ijk}\lambda_i\frac{\delta\mu_j}{\delta s} = \frac{\delta\nu^k}{\delta s}$$

From formulas (*i*) and (*ii*), we get

$$\varepsilon^{ijk}K\mu_i\mu_j + \varepsilon^{ijk}\lambda_i(\tau\nu_j - K\lambda_j) = \frac{\delta\nu^k}{\delta s}$$

$$\varepsilon^{ijk}\lambda_i(\tau\nu_j - K\lambda_j) = \frac{\delta\nu^k}{\delta s}, \quad \text{Since } \varepsilon^{ijk}\mu_i\mu_j = 0$$

$$\varepsilon^{ijk}\lambda_i\nu_j\tau - K\varepsilon^{ijk}\lambda_j\lambda_i = \frac{\delta\nu^k}{\delta s}$$

Since $$\varepsilon^{ijk}\lambda_i\lambda_j = 0$$

$$\varepsilon^{ijk}\lambda_i \nu_j \tau = \frac{\delta \nu^k}{\delta s}$$

Since $\varepsilon^{ijk}\lambda_i\nu_j = \mu^k$, but ε^{ijk} are skew-symmetric.

Then

$$\varepsilon^{ijk}\lambda_i\nu_j = -\mu^k$$

So,

$$-\mu^k\tau = \frac{\delta \nu^k}{\delta s}$$

or $$\frac{\delta \nu^k}{\delta s} = -\tau\mu^k$$

$\Rightarrow$ $$\frac{\delta \nu^i}{\delta s} = -\tau\mu^i$$

This is the proof of third Serret-Frenet Formula

Expanded form of *Serret-Frenet Formula.*

(*i*) $$\frac{d\lambda^i}{ds} + \begin{Bmatrix} i \\ j\ k \end{Bmatrix} \lambda^j \frac{dx^k}{ds} = K\mu^i \quad \text{or} \quad \frac{d^2x^i}{ds^2} + \begin{Bmatrix} i \\ j\ k \end{Bmatrix} \frac{dx^j}{ds}\frac{dx^k}{ds} = K\mu^i$$

(*ii*) $$\frac{d\mu^i}{ds} + \begin{Bmatrix} i \\ jk \end{Bmatrix} \mu^j \frac{dx^k}{ds} = \tau\nu^i - K\lambda^i$$

(*iii*) $$\frac{d\nu^i}{ds} + \begin{Bmatrix} i \\ jk \end{Bmatrix} \nu^j \frac{dx^k}{ds} = -\tau\mu^i$$

EXAMPLE 8

Consider a curve defined in cylindrical coordinates by equation

$$\begin{cases} x^1 = a \\ x^2 = \theta(s) \\ x^3 = 0 \end{cases}$$

This curve is a circle of radius a.

The square of the element of arc in cylindrical coordinates is

$$ds^2 = (dx^1)^2 + (x^1)^2(dx^2)^2 + (dx^3)^2$$

so that $g_{11} = 1, \quad g_{22} = (x^1)^2, \quad g_{33} = 1, \quad g_{ij} = 0, \quad i \neq j$

It is easy to verify that the non-vanishing Christoffel symbols are (see Example 3, Page 61)

$$\begin{Bmatrix} 1 \\ 22 \end{Bmatrix} = x^1, \quad \begin{Bmatrix} 2 \\ 12 \end{Bmatrix} = \begin{Bmatrix} 2 \\ 21 \end{Bmatrix} = \frac{1}{x^1}.$$

The components of the tangent vector λ to the circle C are $\lambda^i = \frac{dx^i}{ds}$ so that $\lambda^1 = 0,\ \lambda^2 = \frac{d\theta}{ds}$, $\lambda^3 = 0$.

Since λ is a unit vector, $g_{ij}\lambda^i\lambda^j = 1$ at all points of C and this requires that

$$\left(x^1\right)^2\left(\frac{d\theta}{ds}\right)^2 = a^2\left(\frac{d\theta}{ds}\right)^2 = 1$$

So, $\left(\frac{d\theta}{ds}\right)^2 = \frac{1}{a^2}$ and by Serret-Frenet first formula (expanded form), we get

$$K\mu^1 = \frac{d\lambda^1}{ds} + \begin{Bmatrix} 1 \\ j\ k \end{Bmatrix}\lambda^j \frac{dx^k}{ds} = \begin{Bmatrix} 1 \\ 2\ 2 \end{Bmatrix}\lambda^2 \frac{dx^2}{ds} = -\frac{1}{a}$$

$$K\mu^2 = \frac{d\lambda^2}{ds} + \begin{Bmatrix} 2 \\ j\ k \end{Bmatrix}\lambda^j \frac{dx^k}{ds} = \begin{Bmatrix} 2 \\ 2\ 1 \end{Bmatrix}\lambda^2 \frac{dx^1}{ds} = 0$$

$$K\mu^3 = \frac{d\lambda^3}{ds} + \begin{Bmatrix} 3 \\ j\ k \end{Bmatrix}\lambda^j \frac{dx^k}{ds} = 0$$

Since μ is unit vector, $g_{ij}\mu^i\mu^j = 1$ and it follows that $K = \frac{1}{a}$, $\mu^1 = -1$, $\mu^2 = 0$, $\mu^3 = 0$

Similarly we can shows that $\tau = 0$ and $\nu^1 = 0, \nu^2 = 0, \nu^3 = 1$.

7.9 EQUATIONS OF A STRAIGHT LINE

Let A^i be a vector field defined along a curve C in E_3 such that

$$C : x^i = x^i(s). \quad s_1 \le s \le s_2, \quad (i = 1, 2, 3).$$

s being the arc parameter.

If the vector field A^i is parallel then from article 7.6 we have

$$\frac{\delta A^i}{\delta s} = 0$$

or
$$\frac{dA^i}{ds} + \begin{Bmatrix} i \\ \alpha\ \beta \end{Bmatrix} A^\alpha \frac{dx^\beta}{ds} = 0 \tag{1}$$

We shall make use of equation (1) to obtain the equations of a straight line in general curvilinear coordinates. The characteristic property of straight lines is the tangent vector $\bar{\lambda}$ to a straight line is directed along the straight line. So that the totality of the tangent vectors $\bar{\lambda}$ forms a parallel vector field.

Thus the field of tangent vector $\lambda^i = \dfrac{dx^i}{ds}$ must satisfy equation (1), we have

$$\frac{\delta\lambda^i}{\delta s} = \frac{d^2x^i}{ds^2} + \begin{Bmatrix} i \\ \alpha\ \beta \end{Bmatrix} \frac{dx^\alpha}{ds}\frac{dx^\beta}{ds} = 0$$

The equation $\dfrac{d^2x^i}{ds^2} + \begin{Bmatrix} i \\ \alpha\ \beta \end{Bmatrix} \dfrac{dx^\alpha}{ds}\dfrac{dx^\beta}{ds} = 0$ is the differential *equation* of the *straight line.*

EXERCISE

1. Show that $\dfrac{d(g_{ij}A^iB^j)}{dt} = g_{ij}\dfrac{\delta A^i}{\delta t}B^j + g_{ij}A^i\dfrac{\delta B^i}{\delta t}$

2. Show that $A_{i,j} - A_{j,i} = \dfrac{\partial A_i}{\partial x^j} - \dfrac{\partial A_j}{\partial x^i}$

3. If $A_i = g_{ij}A^j$ show that $A_{i,k} = g_{i\alpha}A^\alpha_{,k}$

4. Show that $\dfrac{d(g_{ij}A^iB^j)}{dx^k} = A_{i,k}B^i + B_{i,k}A^i$

5. Show that

$$\frac{\delta^2\lambda^i}{ds^2} = \frac{dK}{ds}\mu^i - K(\tau\nu^i - K\lambda^i)$$

$$\frac{\delta^2\mu^i}{\delta s^2} = \frac{d\tau}{ds}\nu^i - (K^2+\tau^2)\mu^i - \frac{dK}{ds}\lambda^i$$

$$\frac{\delta^2\nu^i}{\delta s^2} = \tau(K\lambda^i - \tau\nu^i) - \frac{d\tau}{ds}\mu^i$$

6. Find the curvature and tension at any point of the circular helix C whose equations in cylindrical coordinates are

$$C : x^1 = a,\ \ x^2 = \theta,\ x^3 = \theta$$

Show that the tangent vector λ at every point of C makes a constant angle with the direction of X^3-axis. Consider C also in the form $y^1 = a\cos\theta$, $y^2 = a\sin\theta$, $y^3 = \theta$. Where the coordinates y^i are rectangular Cartesion.

CHAPTER – 8

ANALYTICAL MECHANICS

8.1 INTRODUCTION

Analytical mechanics is concerned with a mathematical description of motion of material bodies subjected to the action of forces. A material body is assumed to consist of a large number of minute bits of matter connected in some way with one another. The attention is first focused on a single particle, which is assumed to be free of constraints and its behaviour is analyzed when it is subjected to the action of external forces. The resulting body of knowledge constitutes the mechanics of a particle. To pass from mechanics of a single particle to mechanics of aggregates of particles composing a material body, one introduces the principle of superposition of effects and makes specific assumptions concerning the nature of constraining forces, depending on whether the body under consideration is rigid, elastic, plastic, fluid and so on.

8.2 NEWTONIAN LAWS

1. Every body continues in its state of rest or of uniform motion in a straight line, except in so far as it is compelled by impressed forces to change that state.
2. The change of motion is proportional to the impressed motive force and takes place in the direction of the straight line in which that force is impressed.
3. To every action there is always an equal and contrary reaction; or the mutual actions of two bodies are always and oppositely directed along the same straight line.

The first law depends for its meaning upon the dynamical concept of force and on the kinematical idea of uniform rectilinear motion.

The second law of motion intorduces the kinematical concept of motion and the dynamical idea of force. To understand its meaning it should be noted that Newton uses the term motion in the sense of momentum, *i.e.*, the product of mass by velocity, this, "change of motion" means the time of change of momentum.

In vector notation, the second law can be stated as

$$\vec{F} = \frac{d(m\vec{v})}{dt} \qquad \ldots (1)$$

If we postulate the invariance of mass then equation (1) can be written as

$$\vec{F} = m\vec{a} \qquad \text{... (2)}$$

from (1) if $\vec{F} = 0$ then $\dfrac{d(m\vec{v})}{dt} = 0.$

So that

$$m\vec{v} = \text{constant}.$$

hence $\vec{v}$ is constant vector.

Thus the first law is a consequence of the second.

The third law of motion states that accelerations always occur in pairs. In term of force we may say that if a force acts on a given body, the body itself exerts an equal and oppositely directed force on some other body. Newton called the two aspects of the force of action and reaction.

8.3 EQUATIONS OF MOTION OF A PARTICLE

THEOREM 8.1 *The work done in displacing a particle along its trajectory is equal to the change in the kinetic energy of particle.*

Proof: Let the equation of path C of the particle in E_3 be

$$C : x^i = x^i(t) \qquad \text{... (1)}$$

and the curve C the trajectory of the particle. Let at time t, particle is at $P\ \{x^i(t)\}$.

If v^i be the component of velocity of moving particle then

$$v^i = \frac{dx^i}{dt} \qquad \text{... (2)}$$

and if a^i be the component of acceleration of moving particle then

$$a^i = \frac{\delta v^i}{\delta t} = \frac{dv^i}{dt} + \begin{Bmatrix} i \\ j\,k \end{Bmatrix} v^j \frac{dx^k}{dt}$$

$$a^i = \frac{d^2x^i}{dt^2} + \begin{Bmatrix} i \\ j\,k \end{Bmatrix} \frac{dx^j}{dt}\frac{dx^k}{dt} \qquad \text{... (3)}$$

where $\dfrac{\delta v^i}{\delta t}$ is the intrinsic derivative and the $\begin{Bmatrix} i \\ j\,k \end{Bmatrix}$ are the Christoffel symbols calculated from the metric tensor g_{ij}

If m be the mass of particle. Then by Newton's second law of motion

$$F^i = m\frac{\delta v^i}{\delta t} = ma^i \qquad \text{... (4)}$$

We define the element of work done by the force $\vec{F}$ in producing a displacement $d\vec{r}$ by invariant $dW = \vec{F}.d\vec{r}$.

Since the components of $\vec{F}$ and $d\vec{r}$ are F^i and dx^i respectively.

Then

$$dW = g_{ij} F^i dx^j \qquad \dots (5)$$

$$= F_j dx^j \text{ where } F_j = g_{ij} F^i$$

The work done in displacing a particle along the trajectory C, joining a pair of points P_1 and P_2, is line integral

$$W = \int_{P_1}^{P_2} F_i \, dx^i \qquad \dots (6)$$

using equation (4) then equation (6) becomes

$$W = \int_{P_1}^{P_2} m\, g_{ij} \frac{\delta v^i}{\delta t} dx^j$$

$$= \int_{P_1}^{P_2} m\, g_{ij} \frac{\delta v^i}{\delta t} \frac{dx^j}{dt} dt$$

$$W = \int_{P_1}^{P_2} m\, g_{ij} \frac{\delta v^i}{\delta t} v^j dt \qquad \dots (7)$$

Since $g_{ij}\, v^i\, v^j$ is an invariant then

$$\frac{\delta (g_{ij} v^i v^j)}{\delta t} = \frac{d}{dt}(g_{ij} v^i v^j)$$

or

$$\frac{d}{dt}(g_{ij} v^i v^j) = 2\, g_{ij} \frac{\delta v^i}{\delta t} v^j$$

$\Rightarrow$

$$g_{ij} \frac{\delta v^i}{\delta t} v^j = \frac{1}{2} \frac{d}{dt}(g_{ij} v^i v^j)$$

using this result in equation (7), we get

$$W = \int_{P_1}^{P_2} \frac{m}{2} \frac{d}{dt}(g_{ij} v^i v^j)\, dt$$

$$W = \frac{m}{2} [g_{ij} v^i v^j]_{P_1}^{P_2}$$

Let T_2 and T_1 is kinetic energy at P_2 and P_1 respectively.

$$W = T_2 - T_1$$

where $T = \frac{m}{2} g_{ij} v^i v^j = \frac{m v^2}{2}$ is kinetic energy of particle.

We have the result that the work done by force F_i in displacing the particle from the point P_1 to the point P_2 is equal to the difference of the values of the quantity $T = \frac{1}{2} mv^2$ at the end and the beginning of the displacement.

8.4 CONSERVATIVE FORCE FIELD

The force field F_i is such that the integral $W = \int_{P_1}^{P_2} F_i \, dx^i$ is independent of the path.

Therefore the integrand $F_i\, dx^i$ is an exact differential

$$dW = F_i \, dx^i \qquad \dots (8)$$

of the work function W. The negative of the work function W is called the force potential or potential energy.

We conclude from equation (8) that

$$F_i = -\frac{\partial V}{\partial x^i} \qquad \dots (9)$$

where potential energy V is a function of coordinates x^i. Hence, the fields of force are called conservative if $F_i = -\dfrac{\partial V}{\partial x^i}$.

THEOREM 8.2 *A necessary and sufficient condition that a force field F_i, defined in a simply connected region, be conservative is that $F_{i,j} = F_{j,i}$.*

Proof: Suppose that F_i conservative. Then $F_i = -\dfrac{\partial V}{\partial x^i}$

Now,

$$F_{i,j} = \frac{\partial F_i}{\partial x^j} - \begin{Bmatrix} k \\ i\,j \end{Bmatrix} F_k$$

$$F_{i,j} = \frac{\partial\left(-\dfrac{\partial V}{\partial x^i}\right)}{\partial x^j} - \begin{Bmatrix} k \\ i\,j \end{Bmatrix} F_k$$

$$= -\frac{\partial^2 V}{\partial x^j \partial x^i} - \begin{Bmatrix} k \\ j\,i \end{Bmatrix} F_k \qquad \dots (1)$$

and

$$F_{j,i} = \frac{\partial F_j}{\partial x^i} - \begin{Bmatrix} k \\ j\,i \end{Bmatrix} F_k$$

Similarly,

$$F_{j,i} = -\frac{\partial^2 V}{\partial x^i \partial x^j} - \begin{Bmatrix} k \\ j\,i \end{Bmatrix} F_k \qquad \dots (2)$$

From equation (1) and (2), we get

$$F_{i,j} = F_{j,i}$$

conversely, suppose that $F_{i,j} = F_{j,i}$

Then

$$\frac{\partial F_i}{\partial x^j} - \begin{Bmatrix} k \\ i\,j \end{Bmatrix} F_k = \frac{\partial F_j}{\partial x^i} - \begin{Bmatrix} k \\ j\,i \end{Bmatrix} F_k$$

$\Rightarrow$

$$\frac{\partial F_i}{\partial x^j} = \frac{\partial F_j}{\partial x^i} \text{ as } \begin{Bmatrix} k \\ i\,j \end{Bmatrix} \text{ due to symmetry.}$$

Take $F_i = -\frac{\partial V}{\partial x^i}$

Then

$$\frac{\partial F_i}{\partial x^j} = -\frac{\partial^2 V}{\partial x^j \partial x^i}$$

$$= -\frac{\partial^2 V}{\partial x^i \partial x^j}$$

$$= \frac{\partial}{\partial x^i}\left(-\frac{\partial V}{\partial x^j}\right)$$

$$\frac{\partial F_i}{\partial x^j} = \frac{\partial F_j}{\partial x^i}$$

So, we can take $\qquad F_i = -\frac{\partial V}{\partial x^i}.$

Hence, F_i is conservative.

8.5 LAGRANGEAN EQUATIONS OF MOTION

Consider a particle moving on the curve

$$C : x^i = x^i(t)$$

At time t, let particle is at point $P(x^i)$.

The kinetic energy $T = \frac{1}{2}mv^2$ can be written as

$$T = \frac{1}{2} m\, g_{ij} \dot{x}^i \dot{x}^j$$

Since $\dot{x}^i = v^i$.

or

$$T = \frac{1}{2} m\, g_{jk}\, \dot{x}^j \dot{x}^k \qquad \ldots (1)$$

Differentiating it with respect to $\dot{x}^i$, we get

$$\frac{\partial T}{\partial \dot{x}^i} = \frac{1}{2} m\, g_{jk}\left(\frac{\partial \dot{x}^j}{\partial \dot{x}^i}\dot{x}^k + \dot{x}^j \frac{\partial \dot{x}^k}{\partial \dot{x}^i}\right)$$

$$= \frac{1}{2} m\, g_{jk}\left(\delta_i^j \dot{x}^k + \delta_i^k \dot{x}^j\right)$$

$$= \frac{1}{2} m\, g_{jk}\delta_i^j \dot{x}^k + \frac{1}{2} m\, g_{jk}\delta_i^k \dot{x}^j$$

$$= \frac{1}{2} m\left(g_{ik}\dot{x}^k + g_{ji}\dot{x}^j\right)$$

$$= \frac{1}{2} m(g_{ij}\dot{x}^j + g_{ij}\dot{x}^j) \quad \text{as} \quad g_{ij} = g_{ji}$$

$$\frac{\partial T}{\partial \dot{x}^i} = m g_{ij} \dot{x}^j$$

or
$$\frac{\partial T}{\partial \dot{x}^i} = m g_{ik} \dot{x}^k \qquad \text{...(2)}$$

Differentiating equation (2) with respect to t, we get

$$\frac{d}{dt}\left(\frac{\partial T}{\partial \dot{x}^i}\right) = m\frac{d}{dt}\left(g_{ik}\dot{x}^k\right)$$

$$= m\left[\frac{d}{dt} g_{ik}\dot{x}^k + g_{ik}\ddot{x}^k\right]$$

$$= m\left[\frac{\partial g_{ik}}{\partial x^j}\frac{dx^j}{dt}\dot{x}^k + g_{ik}\ddot{x}^k\right]$$

$$\frac{d}{dt}\frac{\partial T}{\partial \dot{x}^i} = m\frac{\partial g_{ik}}{\partial x^j}\dot{x}^j\dot{x}^k + m g_{ik}\ddot{x}^k \qquad \text{... (3)}$$

Since $T = \frac{1}{2} m g_{jk}\dot{x}^j\dot{x}^k$

Differentiating it with respect to x^i, we get

$$\frac{\partial T}{\partial x^i} = \frac{1}{2} m\frac{\partial g_{jk}}{\partial x^i}\dot{x}^j\dot{x}^k \qquad \text{... (4)}$$

Now,

$$\frac{d}{dt}\left(\frac{\partial T}{\partial \dot{x}^i}\right) - \frac{\partial T}{\partial x^i} = m\frac{\partial g_{ik}}{\partial x^j}\dot{x}^j\dot{x}^k + m g_{ik}\ddot{x}^k - \frac{1}{2}m\frac{\partial g_{jk}}{\partial x^i}\dot{x}^j\dot{x}^k$$

$$= m g_{ik}\ddot{x}^k + \frac{1}{2}m\frac{g_{ik}}{\partial x^j}\dot{x}^j\dot{x}^k + \frac{1}{2}m g_{ik}\dot{x}^j\dot{x}^k - \frac{1}{2}m\frac{\partial g_{jk}}{\partial x^i}\dot{x}^j\dot{x}^k$$

$$= m g_{ik}\ddot{x}^k + \frac{1}{2}m\frac{g_{ik}}{\partial x^j}\dot{x}^j\dot{x}^k + \frac{1}{2}m\frac{\partial g_{ij}}{\partial x^k}\dot{x}^k\dot{x}^j - \frac{1}{2}m\frac{\partial g_{jk}}{\partial x^i}\dot{x}^j\dot{x}^k$$

$$= m g_{ik}\ddot{x}^k + \frac{1}{2}m\left[\frac{\partial g_{ik}}{\partial x^j} + \frac{\partial g_{ij}}{\partial x^k} - \frac{\partial g_{jk}}{\partial x^i}\right]\dot{x}^j\dot{x}^k$$

$$= m g_{ik}\ddot{x}^k + m[jk,i]\dot{x}^j\dot{x}^k$$

$$= m g_{ik}\ddot{x}^k + m g^{il} g_{il}[jk,i]\dot{x}^j\dot{x}^k$$

$$= m g_{il}\ddot{x}^l + m g^{il} g_{il}[jk,i]\dot{x}^j\dot{x}^k$$

$$= m\, g_{il}\left[\ddot{x}^l + g^{il}[jk,i]\dot{x}^j\dot{x}^k\right]$$

$$= m\, g_{il}\left[\ddot{x}^l + \begin{Bmatrix} l \\ j\,k \end{Bmatrix}\dot{x}^j\dot{x}^k\right], \qquad \text{as } g^{il}[jk,i] = \begin{Bmatrix} l \\ j\,k \end{Bmatrix}$$

$$\frac{d}{dt}\left(\frac{\partial T}{\partial \dot{x}^i}\right) - \frac{\partial T}{\partial x^i} = m\, g_{il} a^l, \quad \text{Since } a^l = \ddot{x}^l + \begin{Bmatrix} l \\ j\,k \end{Bmatrix}\dot{x}^j\dot{x}^k$$

where a^l is component of acceleration

or
$$\frac{d}{dt}\left(\frac{\partial T}{\partial \dot{x}^i}\right) - \frac{\partial T}{\partial x^i} = m\, a_i$$

$$\frac{d}{dt}\left(\frac{\partial T}{\partial \dot{x}^i}\right) - \frac{\partial T}{\partial x^i} = F_i \qquad \text{... (5)}$$

where $F_i = m\, a_i$, component of force field. The equation (5) is *Lagrangean equation of Motion.*

For a conservative system, $F_i = -\dfrac{\partial V}{\partial x^i}$. Then equation (5) becomes

$$\frac{d}{dt}\left(\frac{\partial T}{\partial \dot{x}^i}\right) - \frac{\partial T}{\partial x^i} = -\frac{\partial V}{\partial x^i}$$

or
$$\frac{d}{dt}\left(\frac{\partial T}{\partial \dot{x}^i}\right) - \frac{\partial (T-V)}{\partial x^i} = 0 \qquad \text{... (6)}$$

Since the potential V is a function of the coordinates x^i alone. If we introduce the Lagrangean function

$$L = T - V$$

Then equation (6) becomes

$$\frac{d}{dt}\left(\frac{\partial L}{\partial \dot{x}^i}\right) - \frac{\partial L}{\partial x^i} = 0 \qquad \text{... (7)}$$

EXAMPLE 1

Show that the covariant components of the acceleration vector in a spherical coordinate system with

$$ds^2 = (dx^1)^2 + (x^1 dx^2)^2 + (x^1)^2 \sin^2 x^2 (dx^3)^2 \quad \text{are}$$

$$a_1 = \ddot{x}^1 - x^1(\dot{x}^2)^2 - x^1(\dot{x}^3 \sin x^2)^2$$

$$a_2 = \frac{d}{dt}\left[(x^1)^2 \dot{x}^2\right] - (x^1)^2 \sin x^2 \cos x^2 (\dot{x}^3)^2$$

and
$$a_3 = \frac{d}{dt}\left[(x^1 \sin x^2)^2 \dot{x}^3\right]$$

Solution

In spherical coordinate system, the metric is given by

$$ds^2 = (dx^1)^2 + (x^1 dx^2)^2 + (x^1)^2 \sin^2 x^2 (dx^3)^2$$

If v is velocity of the paiticle then

$$v^2 = \left(\frac{ds}{dt}\right)^2 = \left(\frac{dx^1}{dt}\right)^2 + (x^1)^2\left(\frac{dx^2}{dt}\right)^2 + (x^1 \sin x^2)^2\left(\frac{dx^3}{dt}\right)^2$$

$$v^2 = (\dot{x}^1)^2 + (x^1)^2(\dot{x}^2)^2 + (x^1 \sin x^2)^2(\dot{x}^3)^2$$

If T be kinetic energy then

$$T = \frac{1}{2}mv^2$$

$$T = \frac{1}{2}m\left[(\dot{x}^1)^2 + (x^1)^2(\dot{x}^2)^2 + (x^1 \sin x^2)^2(\dot{x}^3)^2\right] \quad \dots (1)$$

By Lagrangean equation of motion

$$\frac{d}{dt}\left(\frac{\partial T}{\partial \dot{x}^i}\right) - \frac{\partial T}{\partial x^i} = F_i \text{ and } m\, a_i = F_i$$

where F_i and a_i are covariant component of force field and acceleration vector respectively.

So,

$$\frac{d}{dt}\left(\frac{\partial T}{\partial \dot{x}^i}\right) - \frac{\partial T}{\partial x^i} = m\, a_i \quad \dots (2)$$

Take $i = 1$,

$$m\, a_1 = \frac{d}{dt}\left(\frac{\partial T}{\partial \dot{x}^1}\right) - \frac{\partial T}{\partial x^1}$$

from (1), we get

$$m\, a_1 = \frac{1}{2}m\frac{d}{dt}(2\dot{x}^1) - \frac{m}{2}\left[2x^1(\dot{x}^2)^2 + 2x^1(\sin x^2)^2(\dot{x}^3)^2\right]$$

$$a_1 = \frac{d\dot{x}^1}{dt} - \left[x^1(\dot{x}^2)^2 + x^1(\sin x^2)^2(\dot{x}^3)^2\right]$$

$$a_1 = \ddot{x}^1 - x^1(\dot{x}^2)^2 - x^1(\dot{x}^3 \sin x^2)^2$$

Take $i = 2$,

$$m\, a_2 = \frac{d}{dt}\left(\frac{\partial T}{\partial \dot{x}^2}\right) - \frac{\partial T}{\partial x^2}$$

$$m\, a_2 = \frac{1}{2}m\frac{d}{dt}\left[(x^1)^2 2\dot{x}^2\right] - \frac{1}{2}m\, 2(x^1)^2 \sin x^2 \cos x^2 (\dot{x}^3)^2$$

$$a_2 = \frac{d}{dt}\left[(x^1)^2 \dot{x}^2\right] - (x^1)^2 \sin x^2 \cos x^2 (\dot{x}^3)^2$$

Take $i = 3$

$$m\, a_3 = \frac{d}{dt}\left(\frac{\partial T}{\partial \dot{x}^3}\right) - \frac{\partial T}{\partial x^3}$$

$$m\, a_3 = \frac{1}{2} m \frac{d}{dt}\left[2(\dot{x}^3)(x^1 \sin x^2)^2\right] - 0$$

$$a_3 = \frac{d}{dt}\left[\dot{x}^3 (x^1)^2 (\sin x^2)^2\right]$$

EXAMPLE 2

Use Lagrangean equations to show that, if a particle is not subjected to the action of forces then its trajectory is given by $y^i = a^i t + b^i$ where a^i and b^i are constants and the y^i are orthogonal cartesian coordinates.

Solution

If v is the velocity of particle. Then we know that,

$$v^2 = g_{ij} \dot{y}^i \dot{y}^j$$

where y^i are orthogonal cartesian coordinates.

Since

$$g_{ij} = 0, \; i \neq j$$
$$g_{ij} = 1, \; i = j$$

So,

$$v^2 = (\dot{y}^i)^2$$

But,

$$T = \frac{1}{2} m v^2, \quad T \text{ is kinetic energy.}$$

$$T = \frac{1}{2} m (\dot{y}^i)^2$$

The Lagrangean equation of motion is

$$\frac{d}{dt}\left(\frac{\partial T}{\partial \dot{y}^i}\right) - \frac{\partial T}{\partial y^i} = F_i$$

Since particle is not subjected to the action of forces.

So, $F_i = 0$

Then $$\frac{d}{dt}\left(\frac{1}{2} m 2 \dot{y}^i\right) - 0 = 0$$

$$m \frac{d\dot{y}^i}{dt} = 0$$

or $$\frac{d\dot{y}^i}{dt} = 0$$

$\Rightarrow$ $$\dot{y}^i = a^i$$

$\Rightarrow$ $$y^i = a^i t + b^i$$

where a^i and b^i are constant.

EXAMPLE 3

Prove that if a particle moves so that its velocity is constant in magnitude then its acceleration vector is either orthogonal to the velocity or it is zero.

Solution

If v^i be the component of velocity of moving particle then

$$v^i = \frac{dx^i}{dt} \quad \text{or} \quad v^i = \dot{x}^i$$

given $|v|$ = constant.

Since

$$g_{ij}v^i v^j = |v|^2 = \text{constant}$$

Taking intrinsic derivative with respect to t, we get

$$\frac{\delta}{\delta t}(g_{ij}v^i v^j) = 0$$

$$g_{ij}\left(\frac{\delta v^i}{\delta t}v^j + v^i\frac{\delta v^j}{\delta t}\right) = 0$$

$$g_{ij}\frac{\delta v^i}{\delta t}v^j + g_{ij}v^i\frac{\delta v^j}{\delta t} = 0$$

$$g_{ij}\frac{\delta v^i}{\delta t}v^j + g_{ji}v^j\frac{\delta v^i}{\delta t} = 0, \text{ (Interchange dummy index } i \text{ and } j \text{ in second term)}$$

$$2g_{ij}\frac{\delta v^i}{\delta t}v^j = 0 \text{ as } g_{ij} = g_{ji}$$

$$g_{ij}\frac{\delta v^i}{\delta t}v^j = 0$$

This shows that acceleration vector $\frac{\delta v^i}{\delta t}$ is either orthogonal to v^i or zero *i.e.*, $\frac{\delta v^i}{\delta t} = 0$.

8.6 APPLICATIONS OF LAGRANGEAN EQUATIONS

(i) *Free-Moving Particle*

If a particle is not subjected to the action of forces, the right hand side of equation (5), 148, vanishes. Then we have

$$\frac{d}{dt}\left(\frac{\partial T}{\partial \dot{x}^i}\right) - \frac{\partial T}{\partial x^i} = 0 \qquad \text{... (1)}$$

If x^i be rectangular coordinate system, then $T = \frac{1}{2} m \dot{y}^i \dot{y}^i$.

Hence, the equation (1) becomes $m \ \ddot{y}^i = 0$. Integrating it we get $y^i = a^i t + b^i$, which represents a straight line.

(ii) *Simple Pendulum*

Let a pendulum bob of mass m be supported by an extensible string. In spherical coordinates, the metric is given by

$$ds^2 = dr^2 + r^2 d\phi^2 + r^2 \sin^2 \phi \, d\theta^2$$

If T be the kinetic energy, then

$$T = \frac{1}{2} m v^2 = \frac{1}{2} m (\dot{r}^2 + r^2 \dot{\phi}^2 + r^2 \sin^2 \phi \, \dot{\theta}^2) \qquad \text{... (1)}$$

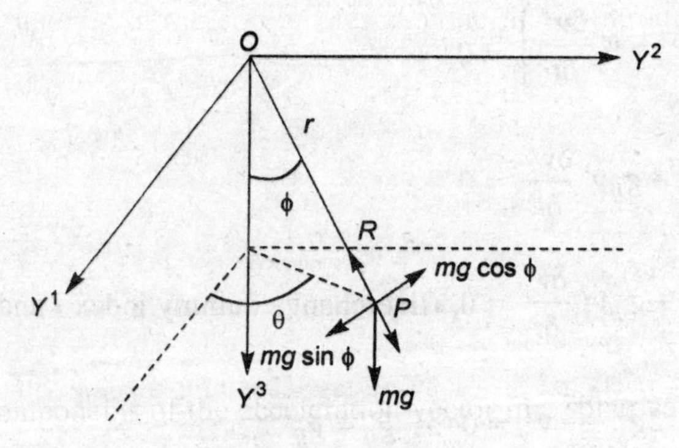

Fig. 8.1.

from Lagrangean equation of motion

$$\frac{d}{dt}\left(\frac{\partial T}{\partial \dot{x}^i}\right) - \frac{\partial T}{\partial x^i} = F_i \qquad i = 1, 2, 3$$

$$x^1 = r, x^2 = \phi, x^3 = \theta.$$

So, take $\quad x^1 = r$

$$\frac{d}{dt}\left(\frac{\partial T}{\partial \dot{r}}\right)-\frac{\partial T}{\partial r} = mg \cos \phi - R$$

from (1), we have

$$\ddot{r}-r\dot{\phi}^2 - r\sin^2\phi\dot{\theta}^2 = g \cos \phi - \frac{R}{m} \quad \dots (2)$$

Take $x^2 = \phi$, we have

$$r\ddot{\phi}+2\dot{r}\dot{\phi}-r\sin\phi\cos\phi\dot{\theta}^2 = -g \sin \phi \quad \dots (3)$$

and take $x^3 = \theta$, we have

$$\frac{d}{dt}(r^2\dot{\theta}\sin^2\phi) = 0 \quad \dots (4)$$

If the motion is in one plane, we obtain from equations (2), (3), and (4), by taking $\dot{\theta}=0$.

$$\ddot{r}-r\dot{\phi}^2 = g \cos \phi - \frac{R}{m}$$

$$r\ddot{\phi}+2\dot{r}\dot{\phi} = -g \sin \phi$$

If $\dot{r}$ = 0, we get, $\ddot{\phi}=-\left(\frac{g}{r}\right)$ sinϕ which is equation of simple pendulum supported by an inextensible string. For small angles of oscillation the vibration is simple harmonic. For large vibration the solution is given in the term of elliptic functions.

8.7 HAMILTON'S PRINCIPLE

If a particle is at the point P_1 at the time t_1 and at the point P_2 at the time t_2, then the motion of the particle takes place in such a away that

$$\int_{t_1}^{t_2}(\delta T + F_i\delta x^i)\,dt = 0$$

where $x^i = x^i(t)$ are the coordinates of the particle along the trajectory and $x^i + \delta x^i$ are the coordinates along a varied path beginning at P_1 at time t_1 and ending at P_2 at time t_2.

Proof: Consider a particle moving on the curve

$$C: x^i = x^i(t), \qquad t_1 \le t \le t_2$$

At time t, let particle is at $P(x^i)$. If T is kinetic energy. Then

$$T = \frac{1}{2}m g_{ij}\dot{x}^i\dot{x}^j$$

or $T = T(x^i, \dot{x}^i)$ *i.e*, T is a function of x^i and $\dot{x}^i$. Let C' be another curve, joining t_1 and t_2 close to be C is

$$C': \bar{x}^i(\varepsilon, t) = x^i(t)+\delta x^i(t)$$

At t_1 and t_2

$$x^i = \bar{x}^i = x^i + \varepsilon \delta x^i$$

$$\Rightarrow \qquad \delta x^i(t_1) = 0 \text{ and } \delta x^i(t_2) = 0$$

But $T = T(x^i, \dot{x}^i)$.

If δT be small variation in T. Then

$$\delta T = \frac{\partial T}{\partial \dot{x}^i}\delta \dot{x}^i + \frac{\partial T}{\partial x^i}\delta x^i$$

Now,

$$\int_{t_1}^{t_2}\left\{(\delta T + F_i)\delta x^i\right\}dt = \int_{t_1}^{t_2}\left(\frac{\partial T}{\partial x^i}\delta x^i + \frac{\partial T}{\partial \dot{x}^i}\delta \dot{x}^i + F_i \delta x^i\right)dt$$

$$= \int_{t_1}^{t_2}\frac{\partial T}{\partial x^i}\delta x^i dt + \int_{t_1}^{t_2}\frac{\partial T}{\partial \dot{x}^i}\delta \dot{x}^i dt + \int_{t_1}^{t_2} F_i \delta x^i\, dt$$

Integrating second term by taking $\dfrac{\partial T}{\partial \dot{x}^i}$ as 1st term

$$= \int_{t_1}^{t_2}\frac{\partial T}{\partial x^i}\delta x^i dt + \left[\frac{\partial T}{\partial \dot{x}^i}\delta x^i\right]_{t_1}^{t_2} - \int_{t_1}^{t_2}\frac{d}{dt}\left(\frac{\partial T}{\partial \dot{x}^i}\right)\delta x^i dt + \int_{t_1}^{t_2} F_i\ \delta x^i dt$$

Since $\delta x^i(t_1) = 0, \delta x^i(t_2) = 0$.

then $\left(\dfrac{\partial T}{\partial \dot{x}^i}\delta x^i\right)_{t_1}^{t_2} = 0.$

So,

$$\int_{t_1}^{t_2}\left(\delta T + F_i \delta x^i\right)dt = \int_{t_1}^{t_2}\frac{\partial T}{\partial x^i}\delta x^i dt - \int_{t_1}^{t_2}\frac{d}{dt}\left(\frac{\partial T}{\partial \dot{x}^i}\right)\delta x^i dt$$

$$+ \int_{t_1}^{t_2} F_i \delta x^i dt$$

$$\int_{t_1}^{t_2}\left(\delta T + F_i \delta x^i\right)dt = \int_{t_1}^{t_2}\left[\frac{\partial T}{\partial x^i} - \frac{d}{dt}\left(\frac{\partial T}{\partial \dot{x}^i}\right) + F_i\right]\delta x^i dt$$

since particle satisfies the Lagrangean equation of motion. Then

$$\frac{d}{dt}\left(\frac{\partial T}{\partial \dot{x}^i}\right) - \frac{\partial T}{\partial x^i} = F_i$$

or

$$\frac{\partial T}{\partial x^i} - \frac{d}{dt}\left(\frac{\partial T}{\partial \dot{x}^i}\right) = -F_i$$

So,

$$\int_{t_1}^{t_2}(\delta T + F_i\delta x^i)\,dt = \int_{t_1}^{t_2}[-F_i + F_i]\delta x^i dt$$

$$\int_{t_1}^{t_2}(\delta T + F_i\delta x^i)\,dt = 0$$

Proved.

8.8 INTEGRAL OF ENERGY

THEOREM 8.3 *The motion of a particle in a conservative field of force is such that the sum of its kinetic and potential energies is a constant.*

Proof: Consider a particle moving on the curve

$$C : x^i = x^i(t), \qquad t_1 \le t \le t_2$$

At time t, let particle is at $P\,(x^i)$. If T is kinetic energy. Then

$$T = \frac{1}{2} m\, g_{ij}\dot{x}^i\dot{x}^j$$

or

$$T = \frac{1}{2} m\, g_{ij}v^i v^j$$

As T is invariant. Then

Taking intrinsic derivative with respect to t, we get

$$\frac{dT}{dt} = \frac{\delta T}{\delta t}$$

$$= \frac{\delta}{\delta t}\left(\frac{1}{2} m\, g_{ij}v^i v^j\right)$$

$$= \frac{1}{2} m\, g_{ij}\left(\frac{\delta v^i}{\delta t}v^j + v^i\frac{\delta v^j}{\delta t}\right)$$

$$= \frac{1}{2} m\left(g_{ij}\frac{\delta v^i}{\delta t}v^j + g_{ij}\frac{\delta v^j}{\delta t}v^i\right)$$

$$= \frac{1}{2} m\left(g_{ij}\frac{\delta v^i}{\delta t}v^j + g_{ji}\frac{\delta v^i}{\delta t}v^j\right), \text{ Since } i \text{ and } j \text{ are dummy indices.}$$

$$= \frac{1}{2} m\,2g_{ij}\frac{\delta v^i}{\delta t}v^j \quad \text{as } g_{ij} = g_{ji}$$

$$\frac{dT}{dt} = m\, g_{ij}\frac{\delta v^i}{\delta t}v^j$$

or

$$\frac{dT}{dt} = m\, g_{ij}\frac{\delta v^j}{\delta t}v^i$$

$$= m\, g_{ij} a^j v^i \text{ as } \frac{\delta v^j}{\delta t} = a^j$$

$$\frac{dT}{dt} = m\, a_i\, v^i, \quad \text{since } g_{ij}\, a^j = a_i$$

$$\frac{dT}{dt} = F_i\, v^i,$$

Since $F_i = m\, a_i$ is a covariant component of force field.

But given F_i is conservative, then

$$F_i = -\frac{\partial V}{\partial x^i}, \text{ where } V \text{ is potential energy.}$$

So,

$$\frac{dT}{dt} = -\frac{\partial V}{\partial x^i} v^i$$

$$= -\frac{\partial V}{\partial x^i}\frac{dx^i}{dt}$$

$$\frac{dT}{dt} = -\frac{dV}{dt}$$

$$\frac{dT}{dt} + \frac{dV}{dt} = 0 \Rightarrow \frac{d}{dt}(T+V) = 0$$

$\Rightarrow$ $T + V = h$, where h is constant.

Proved.

8.9 PRINCIPLE OF LEAST ACTION

Let us consider the integral

$$A = \int_{P_1}^{P_2} mv.ds \tag{1}$$

evaluated over the path

$$C : x^i = x^i(t), \quad t_1 \le t \le t_2$$

where C is the trajectory of the particle of mass m moving in a conservative field of force. In the three dimensional space with curvilinear coordinates, the integral (1) can be written as

$$A = \int_{P_1}^{P_2} m\, g_{ij} \frac{dx^i}{dt} dx^j$$

$$= \int_{t(P_1)}^{t(P_2)} m\, g_{ij} \frac{dx^i}{dt}\frac{dx^j}{dt} dt$$

Since $T = \frac{1}{2} m g_{ij} \frac{dx^i}{dt} \frac{dx^j}{dt}$, we have

$$A = \int_{t(P_1)}^{t(P_2)} 2T \, dt$$

This integral has a physical meaning only when evaluated over the trajectory C, but its value can be computed along any varied path joining the points P_1 and P_2.

Let us consider a particular set of admissible paths C' along which the function $T + V$, for each value of parameter t, has the same constant value h. The integral A is called the action integral.

The principle of least action stated as "of all curves C' passing through P_1 and P_2 in the neighbourhood of the trajectory C, which are traversed at a rate such that, for each C', for every value of t, $T + V = h$, that one for which the action integral A is stationary is the trajectory of the particle."

8.10 GENERALIZED COORDINATES

In the solution of most of the mechanical problems it is more convenient to use some other set of coordinates instead of cartesian coordinates. For example, in the case of a particle moving on the surface of a sphere, the correct coordinates are spherical coordinates r, θ, ϕ where θ and ϕ are only two variable quantities.

Let there be a particle or system of n particles moving under possible constraints. For example, a point mass of the simple pendulum or a rigid body moving along an inclined plane. Then there will be a minimum number of independent coordinates required to specify the motion of particle or system of particles. The set of independent coordinates sufficient in number to specify unambiguously the system configuration is called generalized coordinates and are denoted by $q^1, q^2, ... q^n$ where n is the total number of generalized coordinates or degree of freedom.

Let there be N particles composing a system and let $x^i_{(\alpha)}, (i = 1,2,3), (\alpha = 1,2,...N)$ be the positional coordinates of these particles referred to some convenient reference frame in E_3. The system of N free particles is described by $3N$ parameters. If the particles are constrained in some way, there will be certain relations among the coordinates $x^i_{(\alpha)}$ and suppose that there are r such independent relations,

$$f^i(x^1_{(1)}, x^2_{(1)}, x^3_{(1)}; x^1_{(2)}, x^2_{(2)}, x^3_{(2)}; ... x^1_{(N)} x^2_{(N)} x^3_{(N)}) = 0, (i = 1, 2, ..., r) \quad ... (1)$$

By using these r equations of constraints (1), we can solve for some r coordinates in terms of the remaining $3N - r$ coordinates and regard the latter as the independent generalized coordinates q^i. It is more convenient to assume that each of the $3N$ coordinates is expressed in terms of $3N - r = n$ independent variables q^i and write $3N$ equations.

$$x^i_{(\alpha)} = x^i_{(\alpha)} \, (q^1, q^2, ..., q^n, t) \quad ... (2)$$

where we introduced the time parameter t which may enter in the problem explicitly if one deals with moving constraints. If t does not enter explicitly in equation (2), the dynamical system is called a natural system.

The velocity of the particles are given by differentiating equations (2) with respect to time. Thus

$$\dot{x}^i_{(\alpha)} = \frac{\partial x^i_{(\alpha)}}{\partial q^j} \dot{q}^j + \frac{\partial x^i_{(\alpha)}}{\partial t} \quad ... (3)$$

The time derivatives $\dot{q}^i$ of generalized coordinates q^i the generalized velocities.

For symmetry reasons, it is desirable to introduced a number of superfluous coordinates q^i and describe the system with the aid of $k > n$ coordinates $q^1, q^2, ..., q^k$. In this event there will exist certain relations of the form

$$f^j(q^1 ...., q^k, t) = 0 \qquad ... (4)$$

Differentiating it we get

$$\frac{\partial f^j}{\partial q^i}\dot{q}^i + \frac{\partial f^j}{\partial t} = 0 \qquad ... (5)$$

It is clear that they are integrable, so that one can deduce from them equations (4) and use them to eliminate the superfluous coordinates .

In some problems, functional relations of the type

$$F^j(q^1, q^2, ..., q^k; \dot{q}^1, ..., \dot{q}^k, t) = 0, \ (j = 1, 2, 3, ..., m) \qquad ... (6)$$

arise which are non-integrable. If non-integrable relations (6) occurs in the problems we shall say that the given system has $k - m$ degrees of freedom, where m is the number or independent non-integrable relations (6) and k is the number of independent coordinates. The dynamical systems involving non-integrable relations (6) are called non-holonomic to distinguish them from holonomic systems in which the number of degrees of freedom is equal to the number of independent generalized coordinates.

In other words, a holonomic system is one in which there are no non-integrable relations involving the generalized velocities.

8.11 LAGRANGEAN EQUATIONS IN GENERALIZED COORDINATES

Let there be a system of particle which requires n independent generalized coordinates or degree of freedom to specify the states of its particle.

The position vectors x^r are expressed as the function of generalized coordinates $q^i, (i = 1, 2, ..., n)$ and the time t *i.e.*,

$$x^r = x^r(q^1, q^2, ..., q^n, t); \qquad (r = 1, 2, 3,)$$

The velocity $\dot{x}^r$ of any point of the body is given by

$$\dot{x}^r = \frac{\partial x^r}{\partial q^j}\frac{dq^j}{dt} + \frac{\partial x^r}{\partial t}$$

$$= \frac{\partial x^r}{\partial q^j}\dot{q}^j + \frac{\partial x^r}{\partial t}, \ (j = 1, 2, ..., n)$$

where $\dot{q}^j$ are the generalized velocities.

Consider the relation, with n degree of freedom,

$$x^r = x^r(q^1, q^2, ..., q^n) \qquad ... (1)$$

involve n independent parameters q^i. The velocities $\dot{x}^r$ in this case are given by

$$\dot{x}^r = \frac{\partial x^r}{\partial q^j}\dot{q}^j, \quad (r = 1, 2, 3; \ j = 1, 2, ..., n) \qquad ...(2)$$

where $\dot{q}^j$ transform under any admissible transformation,

$$\bar{q}^k = \bar{q}^k(q^1,...,q^n), \qquad (k = 1, 2, ..., n) \qquad ... (3)$$

in accordance with the contravariant law.

The kinetic energy of the system is given by the expression of the form

$$T = \frac{1}{2}\Sigma m g_{rs}\, \dot{x}^r \dot{x}^s, \quad (r,s = 1,2,3,) \qquad ... (4)$$

where m is the mass of the particle located at the point x^r. The g_{rs} are the components of the metric tensor.

Substituting the value of $\dot{x}^r$ from equation (2), then equation (4) becomes

$$T = \frac{1}{2}\Sigma m g_{rs} \frac{\partial x^r}{\partial q^i}\frac{\partial x^s}{\partial q^j}\dot{q}^i \dot{q}^j$$

$$T = \frac{1}{2}a_{ij}\dot{q}^i\dot{q}^j \qquad ... (5)$$

where $$a_{ij} = \Sigma m g_{rs} \frac{\partial x^r}{\partial q^i}\frac{\partial r^s}{\partial q^j}, \quad (r, s = 1, 2, 3), (i, j = 1, ..., n)$$

Since $T = \frac{1}{2}a_{ij}\dot{q}^i\dot{q}^j$ is an invariant and the quantities a_{ij} are symmetric, we conclude that the a_{ij} are components of a covariant tensor of rank two with respect to the transformations (3) of generalized coordinates.

Since the kinetic energy T is a positive definite form in the velocities $\dot{q}^i, |a_{ij}| > 0$. Then we construct the reciprocal tensor a^{ij}.

Now, from art. 8.5, *Pg.* 146, by using the expression for the kinetic energy in the form (5), we obtain the formula,

$$\frac{d}{dt}\left(\frac{\partial T}{\partial \dot{q}^i}\right) - \frac{\partial T}{\partial q^i} = a_{il}\left(\ddot{q}^l + \begin{Bmatrix} l \\ j\,k \end{Bmatrix}\dot{q}^j\dot{q}^k\right) \qquad (6)$$

where the Christoffel symbol $\begin{Bmatrix} l \\ j\,k \end{Bmatrix}$ are constructed from the tensor a_{kl}.

Put

$$\ddot{q}^l + \begin{Bmatrix} l \\ j\,k \end{Bmatrix}\dot{q}^j\dot{q}^k = Q^l$$

so, the equation (6), becomes

$$\frac{d}{dt}\left(\frac{\partial T}{\partial \dot{q}^i}\right) - \frac{\partial T}{\partial q^i} = a_{il}\, Q^l$$

$$= Q_i \; (i = 1, 2, ..., n) \qquad ... (7)$$

Now, from the realtions $\frac{\partial \dot{x}^r}{\partial \dot{q}^j} = \frac{\partial x^r}{\partial q^j}, \frac{\partial \dot{x}^r}{\partial q^i} = \frac{\partial^2 x^r}{\partial x^i \partial q^j}\dot{q}^j$ and $\frac{\partial \dot{x}^r}{\partial q^i} = \frac{d}{dt}\left(\frac{\partial x^r}{\partial q^i}\right)$ and using equations (2) and (4).

Then by straightforward calculation, left hand member of equation (7) becomes

$$\frac{d}{dt}\left(\frac{\partial T}{\partial \dot{q}^i}\right) - \frac{\partial T}{\partial q^i} = \sum m\, a_r \frac{\partial x^r}{\partial q^i} \qquad \text{... (8)}$$

in which $a_j = g_{ij}\, a^i$ is acceleration of the point P.

Also, Newton's second law gives

$$m\, a_r = F_r \qquad \text{... (9)}$$

where $F_r^{'s}$ are the components of force F acting on the particle located at the point P.

From the equation (9), we have

$$\sum m\, a_r \frac{\partial x^r}{\partial q^i} = \sum F_r \frac{\partial x^r}{\partial q^i}$$

and equation (8) can be written as

$$\frac{d}{dt}\left(\frac{\partial T}{\partial \dot{q}^i}\right) - \frac{\partial T}{\partial q^i} = \sum F_r \frac{\partial x^r}{\partial q^i} \qquad \text{... (10)}$$

comparing (7) with (8), we conclude that

$$Q_i = \sum F_r \frac{\partial x^r}{\partial q^i}$$

where vector Q_i is called generalized force.

The equations

$$\frac{d}{dt}\left(\frac{\partial T}{\partial \dot{q}^i}\right) - \frac{\partial T}{\partial q^i} = Q_i \qquad \text{... (11)}$$

are known as Lagrangean equations in generalized coordinates.

They give a system of n second order ordinary differential equations for the generalized coordinates q^i. The solutions of these equations in the form

$$C : q^i = q^i\ (\text{t})$$

Represent the dynamical trajectory of the system.

If there exists a functions $V(q^1, q^2, ..., q^n)$ such that the system is said to be conservative and for such systems, equation (11) assume the form

$$\frac{d}{dt}\left(\frac{\partial L}{\partial \dot{q}^i}\right) - \frac{\partial L}{\partial q^i} = 0 \qquad \text{... (12)}$$

where $L = T - V$ is the kinetic potential.

Since $L\,(q,\ \dot{q})$ is a function of both the generalized coordinates and velocities.

$$\frac{dL}{dt} = \frac{\partial L}{\partial \dot{q}^i}\ddot{q}^i + \frac{\partial L}{\partial q^i}\dot{q}^i \qquad \ldots (13)$$

from (12), we have $\dfrac{\partial L}{\partial q^i} = \dfrac{d}{dt}\left(\dfrac{\partial L}{\partial \dot{q}^i}\right)$.

Then equation (13), becomes

$$\frac{dL}{dt} = \frac{\partial L}{\partial \dot{q}^i}\ddot{q}^i + \frac{d}{dt}\left(\frac{\partial L}{\partial \dot{q}^i}\right)\dot{q}^i$$

$$= \frac{d}{dt}\left(\frac{\partial L}{\partial \dot{q}^i}\dot{q}^i\right) \qquad \ldots (14)$$

since $L = T - V$ but the potential energy V is not a function of the $\dot{q}^i$

$$\frac{\partial L}{\partial \dot{q}^i}\dot{q}^i = \frac{\partial T}{\partial \dot{q}^i}\dot{q}^i = 2T$$

since $$T = \frac{1}{2}a_{ij}\dot{q}^i\,\dot{q}^j.$$

Thus, the equation (14) can be written in the form

$$\frac{d(L-2T)}{dt} = \frac{d(T+V)}{dt} = 0$$

which implies that $T + V = h$ (constant).

Thus, along the dynamical trajectory, the sum of the kinetic and potential energies is a constant.

8.12 DIVERGENCE THEOREM, GREEN'S THEOREM, LAPLACIAN OPERATOR AND STOKE'S THEOREM IN TENSOR NOTATION

(i) Divergence Theorem

Let $\vec{F}$ be a vector point function in a closed region V bounded by the regular surface S. Then

$$\int_V \operatorname{div}\vec{F} = \int_S \vec{F}\cdot\hat{n}\,ds \qquad \ldots (1)$$

where $\hat{n}$ is outward unit normal to S.

Briefly the theorem states that the integral with subscript V is evaluated over the volume V while the integral in the right hand side of (1) measures the flux of the vector quantity $\vec{F}$ over the surface S.

In orthogonal cartesian coordinates, the divergence of $\vec{F}$ is given by the formula

$$\operatorname{div}\,\vec{F} = \frac{\partial F^1}{\partial x^1} + \frac{\partial F^2}{\partial x^2} + \frac{\partial F^3}{\partial x^3} \qquad \ldots (2)$$

If the components of $\vec{F}$ relative to an arbitrary curvilinear coordinate system X are denoted by F^i then the covariant derivative of F^i is

$$F^i_{,j} = \frac{\partial F^i}{\partial x^j} + \begin{Bmatrix} i \\ k\ j \end{Bmatrix} F^k$$

The invariant $F^i_{,j}$ in cartesian coordinates represents the divergence of the vector field $\vec{F}$.

Also,

$$\vec{F}\cdot\hat{n} = g_{ij}F^i n^j = F^i n_i \quad \text{since } g_{ij}n^j = n_i$$

Hence we can rewrite equation (1) in the form

$$\int_V F^i_{,i}dV = \int_S F^i n_i\, dS \qquad \text{... (3)}$$

(ii) Symmetrical form of Green's Theorem

Let $\phi\,(x^1, x^2, x^3)$ and $\psi\,(x^1, x^2, x^3)$ be two scalar function in V. Let ϕ_i and ψ_i be the gradients of ϕ and ψ respectively, so that

$$\nabla\phi = \phi_i = \frac{\partial\phi}{\partial x^i} \quad \text{and} \quad \nabla\psi = \psi_i = \frac{\partial\psi}{\partial x^i}$$

Put $F_i = \phi\psi_i$ and from the divergence of we get

$$F^j_{,i} = g^{ij}F_{i,j} = g^{ij}(\phi\psi_{i,j} + \psi_i\phi_j)$$

Substituting this in equation (3), we get

$$\int_V g^{ij}(\phi\psi_{i,j} + \psi_i\phi_j)dV = \int_S \phi\psi_i n^i\, dS \qquad \text{... (4)}$$

Since $\nabla\psi = \psi_i$, then

$$g^{ij}\psi_{i,j} = \nabla^2\psi \qquad \text{... (5)}$$

Also, the inner product $g^{ij}\psi_i\phi_j$ can be written as

$$g^{ij}\psi_i\phi_j = \nabla\phi.\nabla\psi$$

where ∇ denote the gradient and ∇^2 denote the Laplacian operator.

Hence the formula (4) can be written in the form

$$\int_V (g^{ij}\phi\psi_{i,j} + g^{ij}\psi_i.\phi_j)dV = \int_S \phi\hat{n}\cdot\nabla\psi\, dS$$

$$\int_V (\phi\nabla^2\psi + \nabla\phi\cdot\nabla\psi)dV = \int_S \phi\hat{n}\cdot\nabla\psi\, dS$$

$$\int_V \phi\nabla^2\psi\, dV = \int_S \phi\hat{n}\cdot\nabla\psi - \int_V \nabla\phi\cdot\nabla\psi\, dV \qquad \text{... (6)}$$

where $\hat{n}\cdot\nabla\psi = \psi_i n^i = \dfrac{\partial\psi}{\partial n}$.

Interchanging ϕ and ψ in equation (5), we get

$$\int_V \psi \nabla^2 \phi \, dV = \int_S \psi \hat{n} \cdot \nabla \phi - \int_V \nabla \psi \cdot \nabla \phi \, dV \qquad \dots (7)$$

Subtracting equation (5) from equation (6), we get

$$\int_V (\phi \nabla^2 \psi - \psi \nabla^2 \phi) \, dV = \int_S \left(\phi \frac{\partial \psi}{\partial n} - \psi \frac{\partial \phi}{\partial n} \right) dS \qquad \dots (8)$$

This result is called a *symmetric form of Green's theorem.*

(iii) Expansion form of the Laplacian Operator

The Laplacian of ψ is given by

$$\nabla^2 \psi = g^{ij} \psi_{i,j} \quad \text{from (5)}$$

when written in the terms of the christoffel symbols associated with the curvilinear coordinates x^i covering E_3,

$$\nabla^2 \psi = g^{ij} \left(\frac{\partial^2 \psi}{\partial x^i \partial x^j} - \begin{Bmatrix} k \\ i\,j \end{Bmatrix} \frac{\partial \psi}{\partial x^k} \right) \qquad \dots (9)$$

and the divergence of the vector F^i is

$$F^i_{,i} = \frac{\partial F^i}{\partial x^i} + \begin{Bmatrix} i \\ j\,i \end{Bmatrix} F^j \qquad \dots (10)$$

But we know that $\qquad \begin{Bmatrix} i \\ j\,i \end{Bmatrix} = \frac{\partial}{\partial x^j} \log \sqrt{g}$

The equation (10) becomes

$$F^i_{,i} = \frac{\partial F^i}{\partial x^i} + \left(\frac{\partial}{\partial x^j} \log \sqrt{g} \right) F^j$$

or

$$F^i_{,i} = \frac{1}{\sqrt{g}} \frac{\partial (\sqrt{g} F^i)}{\partial x^i} \qquad \dots (11)$$

If putting $F^i = g^{ij} \dfrac{\partial \psi}{\partial x^j} = g^{ij} \psi_j$ in equation (11), we get

$$g^{ij} \psi_{j,i} = \frac{1}{\sqrt{g}} \frac{\partial \left(\sqrt{g} g^{ij} \dfrac{\partial \psi}{\partial x^j} \right)}{\partial x^i} \qquad \dots (12)$$

But from equation (5), we know that

$$\nabla^2 \psi = g^{ij} \psi_{j,i}$$

Hence equation (12) becomes

$$\nabla^2\psi = g^{ij}\psi_{j,i} = \frac{1}{\sqrt{g}}\frac{\partial\left(\sqrt{g}g^{ij}\frac{\partial\psi}{\partial x^j}\right)}{\partial x^i}$$

It is expansion form of Laplacian operator.

(iv) Stoke's Theorem

Let a portion of regular surface S be bounded by a closed regular curve C and let $\vec{F}$ be any vector point function defined on S and on C. The theorem of Stokes states that

$$\int_S \hat{n}.\text{curl}\,\vec{F}\,ds = \int_C \vec{F}.\hat{\lambda}\,ds \qquad \dots (13)$$

where λ is the unit tangent vector to C and curl $\vec{F}$ is the vector whose components in orthogonal cartesian coordinates are determined from

$$\text{curl } \vec{F} = \begin{vmatrix} e_1 & e_2 & e_3 \\ \frac{\partial}{\partial \bar{x}^1} & \frac{\partial}{\partial \bar{x}^2} & \frac{\partial}{\partial \bar{x}^3} \\ F^1 & F^2 & F^3 \end{vmatrix} = \nabla\times\vec{F} \qquad \dots (14)$$

where e_i being the unit base vectors in a cartesian frame.

We consider the covariant derivative $F_{i,j}$ of the vector F_i and form a contravariant vector

$$G^i = -\varepsilon^{ijk}F_{j,k} \qquad \dots (15)$$

we define the vector G to be the curl of $\vec{F}$.

Since $\hat{n}$. curl $\vec{F} = n_iG^i = -\varepsilon^{ijk}F_{j,k}n_i$ and the components of the unit tangent vector λ and $\frac{dx^i}{ds}$.

Then equation (13) may be written as

$$-\int_S \varepsilon^{ijk}F_{j,k}n_i\,ds = \int_C F_i\frac{dx^i}{ds}ds \qquad \dots (16)$$

The integral $\int_c F_i dx^i$ is called the circulation of $\vec{F}$ along the contour C.

8.13 GAUSS'S THEOREM

The integral of the normal component of the gravitational flux computed over a regular surface S containing gravitating masses within it is equal to $4\pi m$ where m is the total mass enclosed by S.

Proof: According to Newton's Law of gravitation, a particle P of mass m exerts on a particle Q of unit mass located at a distancer r from P. Then a force of magnitude $F = \frac{m}{r^2}$.

Consider a closed regular surface S drawn around the point P and let θ be the angle between the unit outward normal to $\hat{n}$ to S and the axis of a cone with its vertex at P. This cone subtends an element of surface dS.

The flux of the gravitational field produced by m is

$$\int_S \vec{F}.\hat{n}\, dS = \int_S \frac{m\cos\theta}{r^2}\frac{r^2 dw}{\cos\theta}$$

where $dS = \dfrac{r^2 dw}{\cos\theta}$ and dw is the solid angle subtended by dS.

Thus, we have,

$$\int_S \vec{F}.\hat{n}\, dS = \int_S m\, dw = 4\pi m \qquad \dots (1)$$

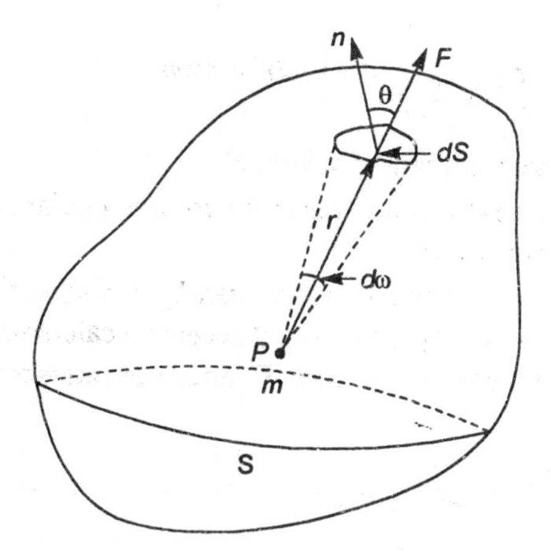

Fig. 8.2.

If there are n discrete particles of masses m_i located within S, then

$$\vec{F}.\hat{n} = \sum_{i=1}^{n} \frac{m_i \cos\theta_i}{r_i^2}$$

and total flux is

$$\int_S \vec{F}.\hat{n}\, dS = 4\pi \sum_{i=1}^{n} m_i \qquad \dots (2)$$

The result (2) can be easily generalized to continuous distributions of matter whenever such distribution no where melt the surface S.

The contribution to the flux integration from the mass element $\rho\, dV$ contained within V, is

$$\int_S \vec{F}.\hat{n}\, dS = \int_S \frac{\cos\theta \rho\, dV}{r^2} dS$$

and the contribution from all masses contained easily within S is

$$\int_S \vec{F}.\hat{n}\, dS = \int_S \left(\int_V \frac{\cos\theta\, \rho\, dV}{r^2} \right) dS \qquad \dots (3)$$

where $\int_V$ denotes the volume integral over all bodies interior to S. Since all masses are assumed to be interior to S, r never vanishes. So that the integrand in equation (3) is continuous and one can interchange to order of integration to obtain

$$\int_S \vec{F}.\hat{n}\,dS = \int_V \rho\left(\int_s \frac{\cos\theta\,dS}{r^2}\right)dV \quad \text{... (4)}$$

But $\int_S \frac{\cos\theta\,dS}{r^2} = 4\pi$. Since it represents the flux due to a unit mass contained within S.

Hence

$$\int_S \vec{F}.\hat{n}\,dS = 4\pi\int_V \rho\,dV = 4\pi m \quad \text{... (5)}$$

where m denotes the total mass contained within S. **Proved.**

Gauss's theorem may be extended to cases where the regular surface S cuts the masses, provided that the density S is piecewise continuous.

Let S cut some masses. Let S' and S'' be two nearby surfaces, the first of which lies wholly within S and the other envelopes S. Now apply Gauss's theorem to calculate the total flux over S'' produced by the distribution of masses enclosed by S since S'' does not intersect them.

We have

$$\int_{S''} (\vec{F}.\hat{n})_i\,dS = 4\pi m$$

where the subscript i on $\vec{F}\cdot\hat{n}$ refers to the flux due to the masses located inside S *and* m *is* the total mass within S. On the other hand, the net flux over S' due to the masses outside S, by Gauss's theorem is

$$\int_{S'} (\vec{F}\cdot\hat{n})_o\,ds = 0$$

where the subscript o on $\vec{F}\cdot\hat{n}$ refers to the flux due to the masses located outside S.

Now if we S' and S'' approach S, we obtain the same formula (5) because the contribution to the total flux from the integral $\int_{S'} (\vec{F}\cdot\hat{n})_o\,dS$ is zero.

8.14 POISSON'S EQUATION

By divergence theorem, we have

$$\int_S \vec{F}.\hat{n}\,ds = \int_V \text{div}\,\vec{F}\,dV$$

and by Gauss's Theorem,

$$\int_S \vec{F}.\hat{n}\,ds = 4\pi\int_V \rho\,dV$$

from these, we have

$$\int_V (\text{div}\,\vec{F} - 4\pi\rho)\,dV = 0$$

Since this relation is true for an arbitrary V and the integrand is piecewise continuous, then

$$\text{div}\ \vec{F} = 4\pi\rho$$

By the definition of potential function V, we have

$$\vec{F} = -\nabla V$$

and $$\text{div}\,\nabla V = \nabla^2 V$$

So,

$$\text{div}\ \vec{F} = 4\pi\rho$$

$$\text{div}\,(-\nabla V) = 4\pi\rho$$

$$\nabla^2 V = -\,4\pi\rho$$

which is *equation of poisson.*

If the point P is not occupied by the mass, then $\rho = 0$. Hence at all points of space free of matter the potential function V satisfies *Laplace's equation*

$$\nabla^2 V = 0$$

8.15 SOLUTION OF POISSON'S EQUATION

We find the solution of Poisson's Equation by using Green's symmetrical formula. We know that Green's symmetrical formula

$$\int_V (\phi\nabla^2\psi - \psi\nabla^2\phi)\,dV = \int_S \left(\phi\frac{\partial\psi}{\partial n} - \psi\frac{\partial\phi}{\partial n}\right)dS \qquad \ldots (1)$$

where V is volume enclosed by S and ϕ and ψ are scalar point functions.

Put $\phi = \frac{1}{r}$ where r is the distance between the points $P\,(x^1, x^2, x^3)$ and $Q\,(y^1, y^2, y^3)$ and V is the gravitational potential.

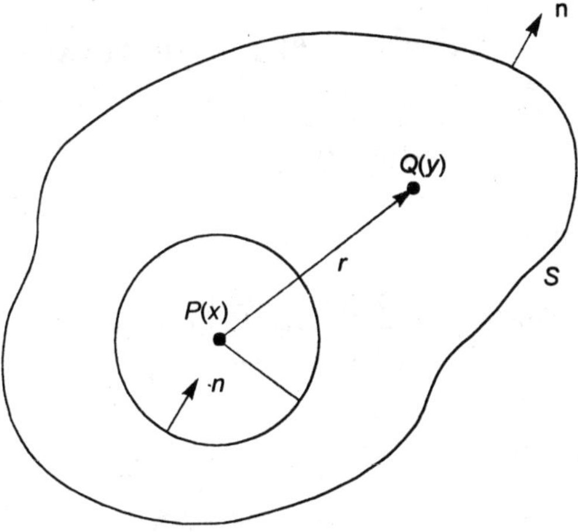

Fig. 8.3.

Since $\frac{1}{r}$ has a discontinuity at $x^i = y^i$, delete the point $P(x)$ from region of integration by surrounding it with a sphere of radius ε and volume V'. Apply Green's symmetrical formula to the region $V - V'$ within which $\frac{1}{r}$ and V possess the desired properties of continuity.

In region $V - V', \nabla^2\phi = \nabla^2 \frac{1}{r} = 0.$

Then equation (1) becomes

$$\int_{V-V'} \frac{1}{r}\nabla^2\psi\, dV = \int_S \left(\frac{1}{r}\frac{\partial\psi}{\partial n} - \psi\frac{\partial\frac{1}{r}}{\partial n}\right) dS + \int_{S'} \left(\frac{1}{r}\frac{\partial\psi}{\partial n} - \psi\frac{\partial\frac{1}{r}}{\partial n}\right) ds \quad \ldots (2)$$

where $\hat{n}$ is the unit outward normal to the surface $S + S'$ bounding $V - V'$. S' being the surface of the sphere of radius ε and $\frac{\partial}{\partial n} = -\frac{\partial}{\partial r}$.

Now

$$\int_{S'} \left(\frac{1}{r}\frac{\partial\psi}{\partial n} - \psi\frac{\partial\frac{1}{r}}{\partial n}\right) dS = \int_{S'} \left(-\frac{1}{r}\frac{\partial\psi}{\partial r} - \psi\frac{\partial\frac{1}{r}}{\partial r}\right) dS$$

$$= \int_{S'} \left(-\frac{1}{r}\frac{\partial\psi}{\partial r} - \frac{\psi}{r^2}\right) r^2\, dw$$

$$= -\int_{S'} \left(r\frac{\partial\psi}{\partial r} + \psi\right) dw$$

$$\int_{S'} \left(\frac{1}{r}\frac{\partial\psi}{\partial n} - \psi\frac{\partial\left(\frac{1}{r}\right)}{\partial n}\right) = -\varepsilon \int_{S'} \left(\frac{\partial\psi}{\partial r}\right)_{r=\varepsilon} dw - 4\pi\overline{\psi} \quad \ldots (3)$$

where $\overline{\psi}$ is the mean value of V over the sphere S' and w denote the solid angle.

Let $\psi(x^1, x^2, x^3) = \psi(P)$ as $r \to 0$ then as $\varepsilon \to 0$ from (3), we have

$$\int_{S'} \left(\frac{1}{r}\frac{\partial\psi}{\partial n} - \psi\frac{\partial\frac{1}{r}}{\partial n}\right) = -4\pi\psi(P)$$

Then equation (2) becomes

$$\int_V \frac{1}{r}\nabla^2\psi\, dV = \int_S \left(\frac{1}{r}\frac{\partial\psi}{\partial n} - \psi\frac{\partial\frac{1}{r}}{\partial n}\right) dS - 4\pi\psi(P)$$

Since $\varepsilon \to 0$ then $\int_{V'} \frac{1}{r}\nabla^2\psi\, dV = 0$.

$$\psi(P) = -\frac{1}{4\pi}\int_V \frac{1}{r}\nabla^2\psi\, dV + \frac{1}{4\pi}\int_S \frac{1}{r}\frac{\partial\psi}{\partial n} dS - \frac{1}{4\pi}\int_S \psi \frac{\partial \frac{1}{r}}{\partial n} dS \qquad \ldots (4)$$

This gives the *solution of Poisson's equation* at the origin.

If ψ is regular at infinity, *i.e.*, for sufficiently large value of r, ψ is such that

$$(\psi) \leq \frac{m}{r} \text{ and } \frac{\partial\psi}{\partial r} \leq \frac{m}{r^2} \qquad \ldots (5)$$

where m is constant.

If integration in equation (4) is extended over all space, so that $r \to \infty$. Then, using equation (5), equation (4) becomes

$$\psi(P) = -\frac{1}{4\pi}\int_\infty \frac{\nabla^2\psi}{r} dV \qquad \ldots (6)$$

But ψ is a potential function satisfying the Poisson's equation *i.e.* $\nabla^2\psi = -4\pi\rho$.

Hence, from (6), we get

$$\psi(P) = \int_\infty \rho \frac{dV}{r}$$

This solution is Unique.

EXERCISES

1. Find, with aid of Lagrangian equations, the trajectory of a particle moving in a uniform gravitational field.
2. A particle is constrained to move under gravity along the line $y^i = c^i s$ $(i = 1, 2, 3)$. Discuss the motion.
3. Deduce from Newtonian equations the equation of energy $T + V = h$, where h is constant.
4. Prove that

$$\int_S \psi_{,i} n^i dS = \int_V \nabla^2\psi\, dV$$

where $\psi_{,i} = \dfrac{\partial\psi}{\partial x^i}$.

5. Prove that the curl of a gradient vector vanishes identically.

CHAPTER – 9

CURVATURE OF CURVE, GEODESIC

9.1 CURVATURE OF CURVE: PRINCIPAL NORMAL

Let C be a curve in a given V_n and let the coordinates x^i of the current point on the curve expressed as functions of the arc length s. Then the unit tangent t to the curve the contravariant components

$$t^i = \frac{dx^i}{ds} \quad ...(1)$$

The intrinsic derivative (or desired vector) of t^i along the curve C is called the *first curvature vector of curve C* relative to V_n and is denoted by $\vec{p}$. The magnitude of curvature vector $\vec{p}$ is called *first curvature* of C relative V_n and is denoted by K.

So,

$$K = \sqrt{g_{ij}\,p^i p^j}$$

where p^i are contravariant components of $\vec{p}$ so that

$$p^i = t^i{}_{,j}\frac{dx^j}{ds}$$

$$= \left(\frac{\partial t^i}{\partial x^j} + t^\alpha \begin{Bmatrix} i \\ \alpha \;\; j \end{Bmatrix}\right)\frac{dx^i}{ds}$$

$$= \frac{\partial t^i}{\partial x^j}\frac{dx^j}{ds} + t^\alpha \frac{dx^j}{ds}\begin{Bmatrix} i \\ \alpha \;\; j \end{Bmatrix}$$

$$= \frac{dt^i}{ds} + \frac{dx^\alpha}{ds}\frac{dx^j}{ds}\begin{Bmatrix} i \\ \alpha \;\; j \end{Bmatrix}$$

$$= \frac{d^2x^i}{ds^2} + \frac{dx^j}{ds}\frac{dx^k}{ds}\begin{Bmatrix} i \\ k \;\; j \end{Bmatrix}, \text{ Replacing dummy index } \alpha \text{ by } k$$

$$p^i = \frac{d^2x^i}{ds^2} + \frac{dx^j}{ds}\frac{dx^k}{ds}\begin{Bmatrix} i \\ j\ k \end{Bmatrix} \quad as \quad \begin{Bmatrix} i \\ j\ k \end{Bmatrix} = \begin{Bmatrix} i \\ k\ j \end{Bmatrix}$$

If $\hat{n}$ is a unit vector in the direction of $\vec{p}$, then we have

$$\vec{p} = k\,\hat{n}$$

The vector $\hat{n}$ is called the *Unit principal normal.*

9.2 GEODESICS

Geodesics on a surface in Euclidean three dimensional space may be defined as the curve along which lies the shortest distance measured along the surface between any two points in its plane.
But when the problem of find the shortest distance between any two given points on a surface is treated properly, it becomes very complicated and therefore we define the geodesics in V_3 as follows:

(*i*) Geodesic in a surface is defined as the curve of stationary length on a surface between any two points in its plane.

(*ii*) In V_3 geodesic is also defined as the curve whose curvature relative to the surface is everywhere zero.

By generalising these definitions we can define geodesic in Riemannian V_n as

(*i*) Geodesic in a Riemannian V_n is defined as the curve of minimum (or maximum) length joining two points on it.

(*ii*) Geodesic is the curve whose first curvature relative to V_n is zero at all points.

9.3 EULER'S CONDITION

THEOREM 9.1 *The Euler condition for the integral*

$$\int_{t_0}^{t_1} f(x^i, \dot{x}^i)\,dt$$

to be staionary are

$$\frac{\partial f}{\partial x^i} - \frac{d}{dt}\left(\frac{\partial f}{\partial \dot{x}^i}\right) = 0$$

where $$\dot{x}^i = \frac{dx^i}{dt} \quad i = 1, 2, 3, \ldots$$

Proof: Let C be a curve in a V_n and A, B two fixed points on it. The coordinates x^i of the current point P on C are functions of a single parameter t. Let t_0 and t_1 be the values of the parameter for the points A and B respectively.

To find the condition for the integral

$$\int_{t_0}^{t_1} f(x^i, \dot{x}^i)\,dt \qquad \ldots(1)$$

to be stationary.

Let the curve suffer an infinitesimal deformation to C', the points A and B remaining fixed while the current points $P(x^i)$ is displaced to $P'\,(x^i + \eta^i)$ such that $\eta^i = 0$ at A and B both.

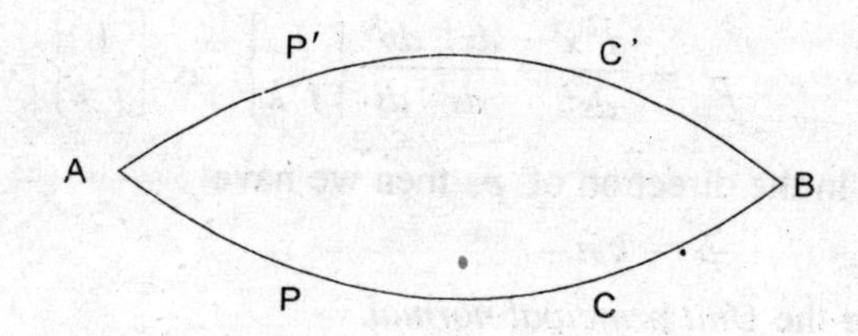

Fig. 9.1

In this case the value of integral (1) becomes I'

So,

$$I' = \int_{t_0}^{t_1} F\left(x^i + \eta^i, \dot{x}^i + \dot{\eta}^i\right) dt$$

By Taylor's theorem

$$F(x+h, y+k) = f(x,y) + \left(h\frac{\partial f}{\partial x} + k\frac{\partial f}{\partial y}\right) + \cdots$$

Then

$$I' = \int_{t_0}^{t_1}\left[F(x^i, \dot{x}^i) + \left(\frac{\partial F}{\partial x^i}\eta^i + \frac{\partial F}{\partial \dot{x}^i}\dot{\eta}^i\right) + \cdots\right] dt$$

$$I' = \int_{t_0}^{t_1} F(x^i, \dot{x}^i)\, dt + \int_{t_0}^{t_1}\left(\frac{\partial F}{\partial x^i}\eta^i + \frac{\partial F}{\partial \dot{x}^i}\dot{\eta}^i\right) dt$$

(Neglecting higher order terms in small quantities η^i)

$$I' = I + \int_{t_0}^{t_1}\left(\frac{\partial F}{\partial x^i}\eta^i + \frac{\partial f}{\partial \dot{x}^i}\dot{\eta}^i\right) dt$$

$$\delta I = I' - I = \int_{t_0}^{t_1}\left(\frac{\partial F}{\partial x^i}\eta^i + \frac{\partial F}{\partial \dot{x}^i}\dot{\eta}^i\right) dt \qquad \text{...(2)}$$

where $\dot{\eta}^i = \dfrac{\partial z^i}{\partial x^j}\dot{x}^j$

Now,

$$\int_{t_0}^{t_1}\frac{\partial F}{\partial \dot{x}^i}\dot{\eta}^i dt = \left[\frac{\partial F}{\partial \dot{x}^i}\eta^i\right]_{t_0}^{t_1} - \int_{t_0}^{t_1}\frac{d}{dt}\left(\frac{\partial F}{\partial \dot{x}^i}\right)\eta^i dt$$

$$= -\int_{t_0}^{t_1}\frac{d}{dt}\left(\frac{\partial F}{\partial \dot{x}^i}\right)\eta^i dt \qquad \text{...(3)}$$

$$\left[\text{since}\left[\frac{\partial F}{\partial \dot{x}^i}\eta^i(t)\right]_{t_0}^{t_1} = 0, \eta^i(t_1) = \eta^i(t_0) = 0\right]$$

Then equation (2) becomes

$$\delta I = \int_{t_0}^{t_1}\left[\frac{\partial F}{\partial x^i} - \frac{d}{dt}\left(\frac{\partial F}{\partial \dot{x}^i}\right)\right]\eta^i dt \qquad \text{...(4)}$$

The integral I is stationary if $\delta I = 0$.

i.e., if $\int_{t_0}^{t_1}\left[\frac{\partial F}{\partial x^i} - \frac{d}{dt}\left(\frac{\partial F}{\partial \dot{x}^i}\right)\right]\eta^i dt = 0$

Since η^i are arbitrary and hence the integrand of the last integral vanishes, so that

$$\frac{\partial F}{\partial x^i} - \frac{d}{dt}\left(\frac{\partial F}{\partial \dot{x}^i}\right) = 0, \quad (i = 1, 2, ..., n) \qquad \text{... (5)}$$

Hence the necessary and sufficient condition for the integral (1) to be stationary are

$$\frac{\partial F}{\partial x^i} - \frac{d}{dt}\left(\frac{\partial F}{dx^i}\right) = 0, \qquad (i = 1, 2, \ldots, n)$$

These are called *Euler's conditions* for the integral I to be stationary.

9.4 DIFFERENTIAL EQUATIONS OF GEODESICS

To obtain the differential equations of a geodesic in a V_n, using the property that it is a path of minimum (or maximum) length joining two points A and B on it.

Proof: Consider a curve C in V_n joining two fixed points A and B on it and $x^i(t)$ be the coordinates of point P on it.

The length of curve C is

$$s = \int_A^B \sqrt{g_{ij}\frac{dx^i}{dt}\frac{dx^j}{dt}}dt \qquad \text{...(1)}$$

$$\frac{ds}{dt} = \sqrt{g_{ij}\frac{dx^i}{dt}\frac{dx^j}{dt}}$$

Put

$$\frac{ds}{dt} = \sqrt{g_{ij}\frac{dx^i}{dt}\frac{dx^j}{dt}} = F \text{ (say)} \qquad \text{...(2)}$$

or

$$\dot{s} = \sqrt{g_{ij}\dot{x}^i\dot{x}^j} = F$$

Then equation (1) becomes

$$s = \int_A^B F\, dt \qquad \text{...(3)}$$

Since curve C is geodesic, then the integral (3) should be stationary, we have from Euler's condition

$$\frac{\partial F}{\partial x^i} - \frac{d}{dt}\left(\frac{\partial F}{\partial \dot{x}^i}\right) = 0 \qquad ...(4)$$

Differentiating equation (2) with respect to x^k and $\dot{x}^k$ we get,

$$\frac{\partial F}{\partial x^k} = \frac{1}{2\dot{s}} \frac{\partial g_{ij}}{\partial x^k} \dot{x}^i \dot{x}^j$$

and

$$\frac{\partial F}{\partial \dot{x}^k} = 2\left(\frac{1}{2\dot{s}} g_{ik} \dot{x}^i\right) = \frac{1}{\dot{s}} g_{ik} \dot{x}^i$$

$$\frac{d}{dt}\left(\frac{\partial F}{\partial \dot{x}^k}\right) = -\frac{1}{\dot{s}^2} \ddot{s}\, g_{ik}\, \dot{x}^i + \frac{1}{\dot{s}} \frac{\partial g_{ik}}{\partial x^j} \dot{x}^j \dot{x}^i + \frac{1}{\dot{s}} g_{ik} \ddot{x}^i$$

Putting these values in equation (4), we get

$$\frac{1}{2\dot{s}} \frac{\partial g_{ij}}{\partial x^k} \dot{x}^i \dot{x}^j - \left[-\frac{1}{\dot{s}^2} \ddot{s} g_{ik} \dot{x}^i + \frac{1}{\dot{s}} \frac{\partial g_{ik}}{\partial x^j} \dot{x}^j \dot{x}^i + \frac{1}{\dot{s}} g_{ik} \ddot{x}^i\right] = 0$$

$$g_{ik} \ddot{x}^i - \frac{\ddot{s}}{\dot{s}} g_{ik} \dot{x}^i + \left(\frac{\partial g_{ik}}{\partial x^j} - \frac{1}{2} \frac{\partial g_{ij}}{\partial x^k}\right) \dot{x}^i \dot{x}^j = 0$$

$$g_{ik} \ddot{x}^i - \frac{\ddot{s}}{\dot{s}} g_{ik} \dot{x}^i + [k, ij] \dot{x}^i \dot{x}^j = 0$$

multiplying it by g^{km}, we get

$$g^{km} g_{ik} \ddot{x}^i - \frac{\ddot{s}}{\dot{s}} g^{km} g_{ik} \dot{x}^i + g^{km} [k, ij] \dot{x}^i \dot{x}^j = 0$$

But

$$g^{km} g_{ik} = \delta_i^m \text{ and } g^{km}[k, ij] = \begin{Bmatrix} m \\ i \; j \end{Bmatrix}$$

$$\ddot{x}^m - \frac{\ddot{s}}{\dot{s}} \dot{x}^m + \begin{Bmatrix} m \\ i \; j \end{Bmatrix} \dot{x}^i \dot{x}^j = 0$$

$$\frac{d^2 x^m}{dt^2} - \frac{\ddot{s}}{\dot{s}} \frac{dx^m}{dt} + \begin{Bmatrix} m \\ j \; k \end{Bmatrix} \frac{dx^j}{dt} \frac{dx^k}{dt} = 0 \quad \left(\text{Replacing dummy index } i \text{ by } k\right) \qquad ...(5)$$

This is the *differential equation for the geodesic* in parameter t.

Taking $s = t, \dot{s} = 1, \ddot{s} = 0$. Then equation (5) becomes

$$\frac{d^2 x^m}{ds^2} + \begin{Bmatrix} m \\ j \; k \end{Bmatrix} \frac{dx^j}{ds} \frac{dx^k}{ds} = 0 \quad ...(6)$$

which may also written as

$$\frac{dx^k}{ds}\left(\frac{dx^m}{ds}\right)_{,k} = 0$$

Then the intrinsic derivative (or derived vector) of the unit tangent to a geodesic in the direction of the curve is everywhere zero. In otherwords, a geodesic of V_n is a line whose first curvature relative to V_n is identically zero.

THEOREM 9.2 *To prove that one and only one geodesic passes through two specified points lying in a neighbourhood of a point O of a V_n.*

OR

To prove that one and only one geodesic passes through a specified point O of V_n in a prescribed direction.

Proof: The differential equations of a geodesic curve in a V_n are

$$\frac{d^2x^m}{ds^2} + \begin{Bmatrix} m \\ j\ k \end{Bmatrix} \frac{dx^j}{ds}\frac{dx^k}{ds} = 0$$

These equations are n differential equations of the second order. Their complete integral involves $2n$ arbitrary constants. These may be determined by the n coordinates of a point P on the curve and the n components of the unit vector in the direction of the curve at P. Thus, in general, one and only one geodesic passes through a given point in a given direction.

9.5 GEODESIC COORDINATES

A cartesian coordinate system is one relative to which the coefficients of the fundamental form are constants. Coordinates of this nature do not exists for an arbitrary Riemannian V_n. It is, however, possible to choose a coordinate system relative to which the quantities g_{ij} are locally constant in the neighbourhood of an arbitrary point P_0 of V_n. Such a cartesian coordinate system is known as *geodesic coordinate system* with the pole at P_0.

The quantities g_{ij} are said to be locally constants in the neighbourhood of a point P_0 if

$$\frac{\partial g_{ij}}{\partial x^k} = 0 \text{ at } P_0$$

and

$$\frac{\partial g_{ij}}{\partial x^k} \neq 0 \text{ elsewhere}$$

This shows that $[ij,k] = 0, \begin{Bmatrix} k \\ i\ j \end{Bmatrix} = 0$ at P_0.

Since the covariant derivative of A_{ij} with respect to x^k is written as

$$A_{ij,k} = \frac{\partial A_{ij}}{\partial x^k} - \begin{Bmatrix} h \\ j\ k \end{Bmatrix} A_{ih} - \begin{Bmatrix} h \\ i\ k \end{Bmatrix} A_{hj}, \text{ see pg. 71}$$

The covariant derivative of A_{ij} at P_0 with respect to x^k reduces to the corresponding ordinary derivatives. Hence

$$A_{ij,k} = \frac{\partial A_{ij}}{\partial x^k} \text{ at } P_0$$

THEOREM 9.3 *The necessary and sufficient condition that a system of coordintes be geodesic with pole at P_0 are that their second covariant derivatives with respect to the metric of the space all vanish at P_0.*

Proof: We know that (equation 8, Pg. 65)

$$\frac{\partial^2 x^s}{\partial \bar{x}^j \partial \bar{x}^j} = \frac{\partial x^s}{\partial \bar{x}^k}\overline{\begin{Bmatrix} k \\ i\ j \end{Bmatrix}} - \frac{\partial x^p}{\partial \bar{x}^i}\frac{\partial x^q}{\partial \bar{x}^j}\begin{Bmatrix} s \\ p\ q \end{Bmatrix}$$

or
$$\frac{\partial}{\partial \bar{x}^j}\left(\frac{\partial x^s}{\partial \bar{x}^i}\right) = \frac{\partial x^s}{\partial \bar{x}^k}\overline{\begin{Bmatrix} k \\ i\ j \end{Bmatrix}} - \frac{\partial x^p}{\partial \bar{x}^i}\frac{\partial x^q}{\partial \bar{x}^j}\begin{Bmatrix} s \\ p\ q \end{Bmatrix} \quad ...(1)$$

Interchanging the coordinate system x^i and $\bar{x}^i$ in equation (1), we get

$$\frac{\partial}{\partial x^j}\left(\frac{\partial \bar{x}^s}{\partial x^i}\right) = \frac{\partial \bar{x}^s}{\partial x^k}\begin{Bmatrix} k \\ i\ j \end{Bmatrix} - \frac{\partial \bar{x}^p}{\partial x^i}\frac{\partial \bar{x}^q}{\partial x^j}\overline{\begin{Bmatrix} s \\ p\ q \end{Bmatrix}}$$

$$-\frac{\partial \bar{x}^p}{\partial x^i}\frac{\partial \bar{x}^q}{\partial x^j}\overline{\begin{Bmatrix} s \\ p\ q \end{Bmatrix}} = \frac{\partial}{\partial x^j}\left(\frac{\partial \bar{x}^s}{\partial x^i}\right) - \frac{\partial \bar{x}^s}{\partial x^k}\begin{Bmatrix} k \\ i\ j \end{Bmatrix}$$

$$= \frac{\partial}{\partial x^j}(\bar{x}^s_{,i}) - \bar{x}^s_{,k}\begin{Bmatrix} k \\ i\ j \end{Bmatrix} \text{ since } \frac{\partial \bar{x}^s}{\partial x^k} = \bar{x}^s_{,k} \text{ at } P_0$$

$$= (\bar{x}^s_{,i})_{,j} \quad \text{since } \begin{Bmatrix} k \\ i\ j \end{Bmatrix} = 0 \text{ at } P_0$$

Thus,

$$\bar{x}^s_{,ij} = -\frac{\partial \bar{x}^p}{\partial x^i}\frac{\partial \bar{x}^q}{\partial x^j}\overline{\begin{Bmatrix} s \\ p\ q \end{Bmatrix}} \quad ...(2)$$

Necessary Condition

Let $\bar{x}^s$ be a geodesic coordinate system with the pole at P_0 so that

$$\overline{\begin{Bmatrix} s \\ p\ q \end{Bmatrix}} = 0 \text{ at } P_0$$

Hence from (2), we have

$$\bar{x}^s_{,ij} = 0 \text{ at } P_0$$

Sufficient Condition

Conversely suppose that $\bar{x}^s_{,ij} = 0$ at P_0.

Then equation (2) becomes

$$\begin{Bmatrix} \bar{s} \\ p\ q \end{Bmatrix} \frac{\partial \bar{x}^p}{\partial x^i} \frac{\partial \bar{x}^q}{\partial x^j} = 0$$

$$\Rightarrow \qquad \begin{Bmatrix} \bar{s} \\ p\ q \end{Bmatrix} = 0 \text{ at } P_0, \text{ as } \frac{\partial \bar{x}^p}{\partial x^i} \neq 0 \text{ and } \frac{\partial \bar{x}^q}{\partial x^j} \neq 0 \text{ at } P_0$$

So, $\bar{x}^s$ is a geodesic coordinate system with the pole at P_0.

9.6 RIEMANNIAN COORDINATES

A particular type of geodesic coordinates introduced by Riemann and known as Riemannian coordinates. Let C be any geodesic through a given point P_0, s the length of the curve measured from P_0 and ξ^i the quantities defined by

$$\xi^i = \left(\frac{dx^i}{ds}\right)_0 \qquad ...(1)$$

the subscript zero indicating as usual that the function is to be evaluated at P_0. The quantities ξ^i represents that only one geodesic will pass through P_0 in the direction of ξ^i in V_n. Let y^i be the coordinates of a point P on the geodesic C such that

$$y^i = s\xi^i \qquad ...(2)$$

where s is the arc length of the curve from P_0 to P. The coordinates y^i are called *Riemannian coordinates.*

The differential equation of geodesic C in terms of coordinates y^i relative to V_n is given by

$$\frac{d^2y^i}{ds^2} + \begin{Bmatrix} i \\ j\ k \end{Bmatrix} \frac{dy^i}{ds} \frac{dy^j}{ds} = 0 \qquad ...(3)$$

where $\begin{Bmatrix} i \\ j\ k \end{Bmatrix}$ is a christoffel symbol relative to the coordinates y^i.

The differential equation (3) will be satisfied by (2), we have,

$$0 + \begin{Bmatrix} i \\ j\ k \end{Bmatrix} \xi^i \xi^j = 0 \text{ since } \frac{dy^i}{ds} = \xi^i$$

or $$\begin{Bmatrix} i \\ j\ k \end{Bmatrix} \xi^i \xi^j = 0 \qquad ...(4)$$

using equation (2), equation (4) becomes

$$\begin{Bmatrix} i \\ j\ k \end{Bmatrix} \frac{y^i}{s} \frac{y^j}{s} = 0 \text{ as } \frac{y^i}{s} = \xi^i$$

or $$\begin{Bmatrix} i \\ j\ k \end{Bmatrix} y^i y^j = 0 \qquad ...(5)$$

The equation (5) hold throughout the Riemannian V_n.

Since $y^i \neq 0, y^j \neq 0,$ from (5) we get

$$\left\{ \begin{matrix} i \\ j \ k \end{matrix} \right\} = 0 \text{ at } P_0$$

Hence the Riemannian coordinates are geodesic coordinate with the pole at P_0.

THEOREM 9.4 *The necessary and sufficient condition that the coordinates y^i be Riemannian coordinates is that $\left\{ \begin{matrix} i \\ j \ k \end{matrix} \right\} y^i y^j = 0$ hold throughout the Riemannian V_n.*

Proof: If y^i are Riemannian coordinates then the condition $\left\{ \begin{matrix} i \\ j \ k \end{matrix} \right\} y^i y^j = 0$ (from equation 5) throughout the Riemannian V_n.

Conversely if $\left\{ \begin{matrix} i \\ j \ k \end{matrix} \right\} y^i y^j = 0$ hold then $\frac{d^2 y}{ds^2} + \left\{ \begin{matrix} i \\ j \ k \end{matrix} \right\} \frac{dy^i}{ds} \frac{dy^j}{ds} = 0$ are saitsfied by $y^i = s\xi^i$.

Hence y^i are Riemannian coordinates.

9.7 GEODESIC FORM OF A LINE ELEMENT

Let ϕ be a scalar invariant whose gradiant is not zero. Let the hypersurface $\phi = 0$ be taken as coordinates hypersurface $x^1 = 0$ and the geodesics which cut this hypersurface orthogonally as the coordinate lines of parameter x^1, this parameter measuring the length of arc along a geodesic from the hypersurface $x^1 = 0$.

Since dx^1 is the length of the vector μ^i is given by

$$u^2 = g_{ij}\, u^i\, u^j$$

i.e.,

$$\left(dx^1\right)^2 = g_{11} dx^1 dx^1$$

$$\Rightarrow \qquad g_{11} = 1 \qquad \text{...(1)}$$

Now, if v^i is the tangent vector to the hypersurface $x^1 = 0$ then we have

$$v^i = (0, dx^2, dx^3, ..., dx^n)$$

since the vectors u^i and v^i are orthogonal vectors.

Then,

$$g_{ij} u^i v^j = 0$$

$$\Rightarrow \qquad g_{1j} u^1 v^j = 0, \quad [u^i = 0, i = 2, 3, ..., n]$$

$$\Rightarrow \qquad g_{1j} v^j = 0, \text{ as } u^1 \neq 0.$$

$$\Rightarrow \qquad g_{1j} = 0, \text{ for } j = 2, 3, \ldots, n. \qquad \text{...(2)}$$

Again the coordinate curves of parameter x^1 are geodesics. Then $s = x^1$.

If t^i is unit tangent vector to a geodesic at any point then

$$t^1 = 1 \text{ and } t^i = 0, \text{ for } i = 2, 3, \ldots, n.$$

Now,

$$t^i = \frac{dx^i}{ds} = \frac{dx^i}{dx^1}$$

$$\Rightarrow \qquad \frac{dx^1}{dx^1} = 1 \text{ and } \frac{dx^i}{dx^1} = 0 \text{ for } i \neq 1$$

and
$$\frac{d^2x^i}{ds^2} = \frac{d^2x^i}{dx^{12}} = 0 \quad \text{for } i = 1, 2, ..., n$$

Also, the differential equation of geodesic is

$$\frac{d^2x^i}{ds^2} + \begin{Bmatrix} i \\ j\ k \end{Bmatrix} \frac{dx^j}{ds}\frac{dx^k}{ds} = 0$$

using above results, we have

$$\begin{Bmatrix} i \\ 1\,1 \end{Bmatrix} \frac{dx^1}{ds}\frac{dx^1}{ds} = 0$$

$$\Rightarrow \qquad \begin{Bmatrix} i \\ 1\,1 \end{Bmatrix} = 0$$

$$\Rightarrow \qquad g^{ij}[1\,1, j] = 0$$

$$\Rightarrow \qquad [1\,1, j] = 0 \text{ as } \left|g^{ij}\right| \neq 0$$

$$\frac{1}{2}\left(\frac{2\partial g_{1j}}{\partial x^i} - \frac{\partial g_{11}}{\partial x^j}\right) = 0, \text{ since } g_{11} = i \Rightarrow \frac{\partial g_{11}}{\partial x^j} = 0$$

So,

$$\frac{\partial g_{1j}}{\partial x^1} = 0 \text{ for } j \neq 1 \qquad ...(3)$$

from equations (1), (2), and (3), we have

$$g_{11} = 1,\ g_{1j} = 0;\ (j = 2, 3, ..., n),\quad \frac{\partial g_{1j}}{\partial x^1} = 0,\ (j = 2, 3, ..., n)$$

The line element is given by

$$ds^2 = g_{ij}dx^i dx^j$$

$$ds^2 = g_{11}dx^1dx^1 + g_{jk}dx^j dx^k$$

$$ds^2 = (dx^1)^2 + g_{jk}dx^j dx^k;\quad (j = 2, 3, ..., n,\ k = 2, 3, ..., n) \qquad ...(4)$$

The line element (4) is called *geodesic form of the line element.*

Note 1: We note that the coordinate curves with parameter x^1 are orthogonal to the coordinate curve $x^i = c^i$ $(i = 1, 2, ..., n)$ at all points and hence to the hypersurfaces $x^1 = c$ at each point.

Note 2: The existence of geodesic form of the line element proves that the hypersurfaces $\phi = x^1$ = constant form a system of parallels *i.e.*, the hypersurfaces $\phi = x^1$ = constant are geodesically parallel hypersurfaces.

THEOREM 9.5 *The necessary and sufficient condition that the hypersurfaces $\phi =$ constant form a system of parallel is that* $(\nabla\phi)^2 = 1$.

Proof: Necessary Condition

Suppose that hypersurface ϕ = constant form a system of parallels then prove that $(\nabla\phi^2) = 1$.

Let us take the hypersurface $\phi = 0$ as the coordinate hypersurface $x^1 = 0$. Let the geodesics cutting this hypersurface orthogonally, be taken as coordinate lines of parameter x^1. Then the parameters x^1 measures are length along these geodesics from the hypersurface $x^1 = 0$. This implies the existence of geodesic form of the line element namely

$$ds^2 = (dx^1)^2 + g_{ij}dx^i dx^j \qquad ...(1)$$

where $i, j = 2, 3, ..., n$.

From (1), we have

$$g_{11} = 1, \; g_{1i} = 0 \;\; \text{for } i \neq 1.$$

from these values, it follows that

$$g^{11} = 1, \; g^{1i} = 0, \;\; \text{for } i \neq 1$$

Now,

$$(\nabla\phi)^2 = \nabla\phi \cdot \nabla\phi = g^{ij}\frac{\partial\phi}{\partial x^i}\frac{\partial\phi}{dx^j}$$

$$= g^{ij}\frac{\partial x^1}{\partial x^i}\frac{\partial x^1}{\partial x^j} = g^{ij}\delta^1_i\delta^1_j$$

so

$$(\nabla\phi)^2 = g^{11} = 1$$

$$(\nabla\phi)^2 = 1$$

Sufficient Condition

Suppose that $(\nabla\phi)^2 = 1$ then prove that the hypersurface ϕ = constant from a system of parallels.

Let us taken $\phi = x^1$ and orthogonal trajectories of the hypersurfaces $\phi = x^1$ constant as the coordinate lines of parameter x^1. Then the hypersurfaces

$$x^1 = \text{constant}$$

x^i = constant $(i \neq 1)$ are orthogonal to each other. The condition for this $g^{1i} = 0$ for $i \neq 1$.

Now, given that $(\nabla\phi)^2 = 1$

$$\Rightarrow \qquad g^{ij}\frac{\partial\phi}{\partial x^i}\frac{\partial\phi}{\partial x^j} = 1$$

$$\Rightarrow \qquad g^{ij}\frac{\partial x^1}{\partial x^i}\frac{\partial x^1}{\partial x^j} = 1$$

$$\Rightarrow \qquad g^{ij}\delta^1_i\delta^1_j = 1$$

$$\Rightarrow \qquad g^{11} = 1$$

Thus

$$g^{11} = 1 \text{ and } g^{1i} = 0 \quad \text{for } i \neq 1.$$

Consequently

$$g_{11} = 1, \quad g_{1i} = 0, \text{ for } i \neq 1.$$

Therefore, the line element

$$ds^2 = g_{ij} dx^i dx^j$$

is given by

$$ds^2 = (dx^1)^2 + g_{ik} dx^i dx^k; \ (i, k = 2, 3, \ldots, n)$$

which is geodesic form of the line element. It means that the hypersurfaces $\phi = x^1 =$ constant form a system of parallels.

9.8 GEODESICS IN EUCLIDEAN SPACE

Consider an Euclidean space S_n for n-dimensions. Let y^i be the Euclidean coordinates. The differential equation of geodesics in Euclidean space is given by

$$\frac{d^2 y^i}{ds^2} + \begin{Bmatrix} i \\ j \ k \end{Bmatrix} \frac{dy^j}{ds} \frac{dy^k}{ds} = 0 \quad \ldots(1)$$

In case of Euclidean coordinates the fundamental tensor g_{ij} is denoted by a_{ij} and

$$g_{ij} = a_{ij} = \delta^i_j = \begin{cases} 1, & \text{if } i = j \\ 0, & \text{if } i \neq j \end{cases}$$

$$\frac{\partial g_{ij}}{\partial x^k} = \frac{\partial a_{ij}}{\partial x^k} = 0$$

This implies that $\begin{Bmatrix} k \\ i \ j \end{Bmatrix} = 0, \ [ij, k] = 0$ relative to S_n.

Then equation (1) becomes

$$\frac{d^2 y^i}{ds^2} = 0$$

Integrating it, we get

$$\frac{dy^i}{ds} = a^i, \text{ where } a^i \text{ is constant of integration.}$$

Again Integrating, we get

$$y^i = a^i s + b^i, \text{ where } b^i \text{ is constant of integration} \quad \ldots(2)$$

The equation (2) is of the form $y = mx + c$.

Hence equation (2) represents a straight line. Since equation (2) is a solution of equation (1) and therefore the geodesic relative to S_n are given by equation (2). Hence geodesic curves in Euclidean space S_n are straight lines.

THEOREM 9.6 *Prove that the distance l between two points $P(y^i)$ and $Q(y'^i)$ in S_n is given by*

$$l = \sqrt{\sum_{i=1}^{n}\left(y'^i - y^i\right)^2}$$

Proof: We know that geodesics in S_n are straight line. Then equation of straight line in S_n may be taken as

$$y^i = a^i s + b^i \quad \text{...(1)}$$

Let $P(y^i)$ and $Q(y'^i)$ lie on equation (1). Then

$$y^i = a^i s + b^i, \quad y'^i = a^i s' + b^i$$

$$y'^i - y^i = a^i (s' - s) \quad \text{...(2)}$$

Then equation (2) becomes

$$y'^i - y^i = a^i l$$

$$l^2 \sum_{i=1}^{n} (a^i)^2 = \sum_{i=1}^{n} (y'^i - y^i)^2$$

But a^i is the unit tangent vector to the geodesics. Then

$$\sum_{i=1}^{n} (a^i)^2 = 1$$

So,

$$l^2 = \sum_{i=1}^{n} (y'^i - y^i)^2$$

$$l = \sqrt{\sum_{i=1}^{n} (y'^i - y^i)^2}$$

EXAMPLE 1

Prove that Pythagoras theorem holds in S_n.

Solution

Consider a triangle ABC right angled at A *i.e.*, $\angle BAC = 90°$.

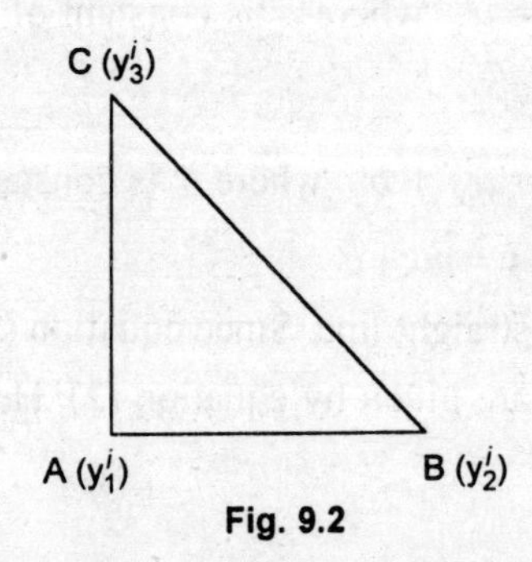

Fig. 9.2

Then the lines AB and AC are orthogonal to each other. So,

$$\overrightarrow{AB} \cdot \overrightarrow{AC} = 0$$

or $$a_{ij}(y_2^i - y_1^i)\,(y_3^i - y_1^i) = 0$$

or $$\sum_{i=1}^{n}(y_2^i - y_1^i)\,(y_3^i - y_1^i) = 0 \qquad ...(1)$$

By distance formula, we have

$$(AB)^2 = \sum_{i=1}^{n}(y_2^i - y_1^i)^2 \qquad ...(2)$$

$$(AC)^2 = \sum_{i=1}^{n}(y_3^i - y_1^i)^2 \qquad ...(3)$$

$$(BC)^2 = \sum_{i=1}^{n}(y_3^i - y_2^i)^2 \qquad ...(4)$$

Now, equation (4) can be written as

$$(BC)^2 = \sum_{i=1}^{n}[(y_3^i - y_1^i) + (y_1^i - y_2^i)]^2$$

$$= \sum_{i=1}^{n}[(y_3^i - y_1^i)^2 + (y_1^i - y_2^i)^2 + 2(y_3^i - y_1^i)(y_1^i - y_2^i)]$$

$$= \sum_{i=1}^{n}(y_3^i - y_1^i)^2 + \sum_{i=1}^{n}(y_1^i - y_2^i)^2 + 2\times 0, \text{[from (1)]}$$

$$= \sum_{i=1}^{n}(y_3^i - y_1^i)^2 + \sum_{i=1}^{n}(y_1^i - y_2^i)^2$$

$$(BC)^2 = (AC)^2 + (AB)^2$$

Hence Pythagoras theorem holds in S_n.

EXAMPLE 2

Prove that if θ is any solution of the differential equation $(\nabla\theta)^2 = f(\theta)$ then the hypersurfaces $\theta =$ constant constitute a system of parallels.

Solution

Given that

$$(\nabla\theta)^2 = f(\theta) \qquad ...(1)$$

Then prove that the hypersurfaces θ = constant form a system of parallel.

Suppose

$$\phi = \int \frac{d\theta}{\sqrt{f(\theta)}}, \quad \text{Then, } d\phi = \frac{d\theta}{\sqrt{f(\theta)}}$$

or

$$\frac{d\phi}{d\theta} = \frac{1}{\sqrt{f(\theta)}}$$

Now,

$$\nabla\phi = \frac{\partial\phi}{\partial x^i} = \frac{\partial\phi}{\partial\theta}\cdot\frac{\partial\theta}{\partial x^i} = \frac{1}{\sqrt{f(\theta)}}\nabla\theta$$

$$(\nabla\phi)^2 = \left(\frac{1}{\sqrt{f(\theta)}}\nabla\theta\right)^2$$

$$= \frac{1}{f(\theta)}(\nabla\theta)^2 = \frac{1}{f(\theta)}f(\theta); \text{ from (1)}$$

$$(\nabla\phi)^2 = 1$$

This proves that the hypersurfaces ϕ = *constant* form a system of parallels and therefore the hypersurfaces θ = constant.

EXAMPLE 3

Show that it is always possible to choose a geodesic coordinates system for any V_n with an arbitrary pole P_0.

Solution

Let P_0 be an arbitrary pole in a V_n. Let us consider general coordinate system x^i. suppose the value of x^i and P_0 are denoted by x_0^i. Now consider a new coordinate system $\bar{x}^j$ defined by the equation.

$$\bar{x}^j = a_m^j(x^m - x_0^m) + \frac{1}{2}a_h^j\begin{Bmatrix} h \\ l\ m \end{Bmatrix}(x^l - x_0^l)(x^m - x_0^m) \qquad ...(1)$$

The coefficients a_m^j being constants and as such that their determinant do not vanish.

Now we shall prove that this new system of coordinated $\bar{x}^j$ defined by equation (1) is a geodesic coordinate system with pole at P_0 *i.e.*, second covariant derivative of $\bar{x}^j$ vanishes at P.

Differentiating equation (1) with respect to x^m, we get

$$\frac{\partial\bar{x}^j}{\partial x^m} = a_m^j + \frac{1}{2}a_h^j\begin{Bmatrix} h \\ l\ m \end{Bmatrix}2(x^l - x_0^l) \qquad ...(2)$$

$$\left(\frac{\partial\bar{x}^j}{\partial x^m}\right)_0 = a_m^j \text{ at } P_0 \qquad ...(3)$$

Now, the Jacobian determinant

$$\left|\left(\frac{\partial \bar{x}^j}{\partial x^m}\right)_0\right| = \left|a_m^j\right| \neq 0$$

and therefore the transformation given by equation (1) is permissible in the neighbourhood of P_0.

Differentiating equation (2) with respect to $\bar{x}^j$, we get

$$\left(\frac{\partial^2 \bar{x}^j}{\partial x^j \partial x^m}\right)_0 = a_h^j \begin{Bmatrix} h \\ l\ m \end{Bmatrix}_0 \qquad ...(4)$$

But we know that

$$\left(x_{,lm}^j\right)_0 = \left(\frac{\partial^2 \bar{x}^j}{\partial x^l \partial x^m}\right)_0 - \begin{Bmatrix} h \\ l\ m \end{Bmatrix}_0 \left(\frac{\partial \bar{x}^j}{\partial x^h}\right)_0$$

$$= a_h^j \begin{Bmatrix} h \\ l\ m \end{Bmatrix}_0 - \begin{Bmatrix} h \\ lm \end{Bmatrix} a_h^j, \quad \text{(from (3) and (4))}$$

$$\left(x_{,lm}^j\right)_0 = 0$$

Hence equation (1) is a geodesic coordinate system with pole at P_0.

EXAMPLE 4

If the coordinates x^i of points on a geodesic are functions of arc lengths s and ϕ is any scalar function of the $x's$ show that

$$\frac{d^p \phi}{ds^p} = \phi_{,ij...l} \frac{dx^i}{ds} \frac{dx^j}{ds} \cdots \frac{dx^l}{ds} \qquad ...(1)$$

Solution

Since the coordinates x^i lie on a geodesic. Then

$$\frac{d^2 x^i}{ds^2} + \begin{Bmatrix} i \\ j\ k \end{Bmatrix} \frac{dx^j}{ds} \frac{dx^k}{ds} = 0 \qquad ...(2)$$

Here the number of suffices $i\,j...l$ is p.

We shall prove the theorem by mathematical induction method.

Since $x's$ are functions of s and ϕ is a scalar function of $x's$, we have

$$\frac{d\phi}{ds} = \frac{\partial \phi}{\partial x^i} \frac{dx^i}{ds} \quad \text{or} \quad \frac{d\phi}{ds} = \phi_{,i} \frac{dx^i}{ds} \qquad ...(3)$$

$$\frac{d^2 \phi}{ds^2} = \frac{\partial \phi_{,i}}{\partial x^j} \frac{dx^j}{ds} + \phi_{,i} \frac{d^2 x^i}{ds^2}$$

$$= \frac{\partial \phi_{,i}}{\partial x^j} \frac{dx^i}{ds} \frac{dx^j}{ds} - \phi_{,i} \begin{Bmatrix} i \\ j\ k \end{Bmatrix} \frac{dx^j}{ds} \frac{dx^k}{ds}, \text{ from (2)}$$

$$= \frac{\partial \phi_{,i}}{\partial x^i}\frac{dx^i}{ds}\frac{dx^j}{ds} - \phi_{,m}\begin{Bmatrix} m \\ j \quad k \end{Bmatrix}\frac{dx^j}{ds}\frac{dx^k}{ds}$$

$$= \frac{\partial \phi_{,i}}{\partial x^j}\frac{dx^i}{ds}\frac{dx^j}{ds} - \phi_{,m}\begin{Bmatrix} m \\ i \; j \end{Bmatrix}\frac{dx^i}{ds}\frac{dx^j}{ds} \text{ (adjusting the dummy index.)}$$

$$= \left(\frac{\partial \phi_{,i}}{\partial x^j} - \phi_{,m}\begin{Bmatrix} m \\ i \; j \end{Bmatrix}\right)\frac{dx^i}{ds}\frac{dx^j}{ds} \qquad \text{...(4)}$$

Equations (3) and (4) imply that the equation (1) holds for $p = 1$ and $p = 2$.

Suppose that the equation (1) holds for p indices $r_1, r_2, ..., r_p$ so that

$$\frac{d^p\phi}{ds^p} = \phi_{,r_1, r_2, ..., r_p}\frac{dx^{r_1}}{ds}\frac{dx^{r_1}}{ds}\cdots\frac{dx^{r_p}}{ds} \qquad \text{...(5)}$$

Differentiating the equation (5) with respect to s, we get

$$\frac{d^{p+1}\phi}{ds^{p+1}} = \frac{\partial \phi_{,r_1 r_2 ... r_p}}{\partial x^{r_{p+1}}}\frac{dx^{r_1}}{ds}\cdots\frac{dx^{r_p}}{ds}\frac{dx^{r_{p+1}}}{ds} + \phi_{,r_1 r_2 ... r_p}\frac{d^2x^{r_1}}{ds^2}\frac{dx^{r_2}}{ds}\cdots\frac{dx^{r_p}}{ds} + \cdots$$

$$+ \cdots \phi_{,r_1 r_2 ... r_p}\frac{dx^{r_1}}{ds}\cdots\frac{d^2x^{r_p}}{ds^2} \qquad \text{...(6)}$$

substituting value of $\frac{d^2x^{r_1}}{ds^2}$ etc. from (2) in (6) and adjusting dummy indices, we have

$$\frac{d^{p+1}\phi}{ds^{p+1}} = \left(\frac{\partial \phi_{,r_1 r_2 ... r_p}}{\partial x^{r_{p+1}}} - \phi_{,m r_2 ... r_p}\begin{Bmatrix} m \\ r_1 r_{p+1} \end{Bmatrix}\cdots\phi_{,r_1 r_2 ... m}\begin{Bmatrix} m \\ r_p r_{p+1} \end{Bmatrix}\right)$$

$$\frac{dx^{r_1}}{ds}\frac{dx^{r_2}}{ds}\cdots\frac{dx^{r_{p+1}}}{ds}$$

$$= \phi_{,r_1 r_2 .. r_p r_{p+1}}\frac{dx^{r_1}}{ds}\frac{dx^{r_2}}{ds}\cdots\frac{dx^{r_{p+1}}}{ds}$$

This shows that the equation (1) holds for next values of p. But equation (1) holds for $p = 1, 2, ...$ Hence equation (1) holds for all values of p.

EXERCISES

1. Prove that at the pole of a geodesic coordinate system, the components of first covariant derivatives are ordinary derivatives.
2. If $\bar{x}^i$ are geodesic coordinates in the neighbourhood of a point if they are subjected to the transformation

$$x^i = \bar{x}^i + \frac{1}{6}c^i_{jkl}\,\bar{x}^j\,\bar{x}^k\,\bar{x}^l$$

where C_s'' are constants then show that x^i are geodesic coordinates in the neighbourhood of O.

3. Show that the principal normal vector vanishes identically when the given curve is geodesic.
4. Show that the coordinate system $\bar{x}^i$ defined by

$$\bar{x}^i = x^i + \frac{1}{2}\begin{Bmatrix} i \\ j \; k \end{Bmatrix} x^j x^k$$

is geodesic coordinate system with the pole at the origin.

5. Obtain the equations of geodesics for the metric

$$ds^2 = e^{-2kt}(dx^2 + dy^2 + dz^2) + dt^2$$

6. Obtain the differential equations of geodesics for the metric

$$ds^2 = f(x)dx^2 + dy^2 + dz^2 + \frac{1}{f(x)}dt^2$$

$$\left[\textbf{Ans}: \frac{d^2x}{ds^2} - \frac{1}{2}\frac{d}{dx}(\log f)\left(\frac{dx}{ds}\right)^2 + \frac{1}{2f^2}\frac{d}{dx}(\log f)\frac{dt}{ds} = 0; \frac{d^2y}{ds^2} = 0; \frac{d^2z}{ds^2} = 0; \frac{d^2t}{ds^2} - \frac{d(\log f)}{dx}\frac{dx}{ds}\frac{dt}{ds} = 0 \right]$$

7. Find the differential equations for the geodesics in a cylindrical and spherical coordinates.
8. Find the rate of divergence of a given curve C from the geodesic which touches it at a given point.

CHAPTER – 10

PARALLELISM OF VECTORS

10.1 PARALLELISM OF A VECTOR OF CONSTANT MAGNITUDE (LEVI-CIVITA'S CONCEPT)

Consider a vector field whose direction at any point is that of the Unit Vector t^i. In ordinary space, the field is said to be parallel if the derivative of t^i vanishes for all directions u^i (say) and at every point of the field *i.e.*,

$$\frac{\partial t^i}{\partial x^j} u^j = 0$$

Similarly in a Riemannian V_n the field is said to be parallel if the derived vector of t^i vanishes at each point for every direction u^i at each point of V_n. *i.e.*,

$$t^i_{,j} = u^j = 0$$

It can be shown that it is not possible for an arbitrary V_n. Consequently we define parallelism of vectors with respect to a given curve C in a V_n.

A vector u^i of constant magnitude is parallel with respect to V_n along the curve C if its derived vector in the direction of the curve is zero at all points of C *i.e.*,

$$\frac{dx^j}{ds} u^i_{,j} = 0 \qquad \text{... (1)}$$

where s is arc-length of curve C.

The equation (1) can be written in expansion form as

$$\left[\frac{\partial u^i}{\partial x^j} + u^m \begin{Bmatrix} i \\ m\ j \end{Bmatrix}\right] \frac{dx^j}{ds} = 0$$

$$\frac{\partial u^i}{\partial x^j}\frac{dx^j}{ds} + u^m \begin{Bmatrix} i \\ m\ j \end{Bmatrix} \frac{dx^j}{ds} = 0$$

$$\frac{du^i}{ds} + u^m \begin{Bmatrix} i \\ m\ j \end{Bmatrix} \frac{dx^j}{ds} = 0 \qquad \text{... (2)}$$

This concept of parallelism is due to Levi-Civita. The vector u^i is satisfying the equation (1) is said to a *parallel displacement along the curve*

Now, multiplying equation (1) by g_{il}, we get

$$g_{il}\left(u^i_{,j}\frac{dx^j}{ds}\right) = 0$$

$$(g_{il}u^i_{,j})\frac{dx^j}{ds} = 0$$

$$(g_{il}u^i)_{,j}\frac{dx^j}{ds} = 0$$

$$u_{l,j}\frac{dx^j}{ds} = 0$$

or
$$u_{i,j}\frac{dx^j}{ds} = 0$$

$$\left[\frac{\partial u_i}{\partial x^j} - u_m\begin{Bmatrix} m \\ i\ j\end{Bmatrix}\right]\frac{dx^j}{ds} = 0$$

$$\frac{du_i}{ds} - u_m\begin{Bmatrix} m \\ i\ j\end{Bmatrix}\frac{dx^j}{ds} = 0 \qquad \ldots (3)$$

The equation (2) and (3) can be also written as

$$du^i = -u^m\begin{Bmatrix} i \\ m\ j\end{Bmatrix}dx^j \qquad \ldots (4)$$

and
$$du_i = u_m\begin{Bmatrix} m \\ i\ j\end{Bmatrix}dx^j \qquad \ldots (5)$$

The equation (4) and (5) give the increment in the components u^i and u_i respectively due to displacement dx^j along C.

THEOREM 10.1 *If two vectors of constant magnitudes undergo parallel displacements along a given curve then they are inclined at a constant angle.*

Proof: Let the vectors u^i and v^i be of constant magnitudes and undergo parallel displacement along a curve C, we have (from equation (1), Pg. 188.)

$$\left.\begin{aligned} u^i_{,j}\frac{dx^j}{ds} &= 0 \\ v^i_{,j}\frac{dx^j}{ds} &= 0 \end{aligned}\right\} \qquad \ldots (1)$$

at each point of C.

Multiplying (1) by g_{il}, we get

$$(g_{il}u^i_{,j})\frac{dx^j}{ds} = 0$$

$$u_{l,j}\frac{dx^j}{ds}=0$$

or $$u_{i,j}\frac{dx^j}{ds}=0 \qquad \ldots (2)$$

Similarly,

$$v_{i,j}\frac{dx^j}{ds}=0 \qquad \ldots (3)$$

Let ϕ be the angle between u^i and v^i then

$$u^i.v_i = uv\cos\theta$$

Differentiating it with respect to arc length s, we get

$$\frac{d}{ds}(uv\cos\theta)=\frac{d(u^iv_i)}{ds}$$

$$=(u^iv_i)_{,j}\frac{dx^j}{ds}$$

$$-uv\sin\theta\frac{d\theta}{ds}=u^i_{,j}\frac{dx^j}{ds}v_i+u^iv_{i,j}\frac{dx^j}{ds} \qquad \ldots (4)$$

Using equation (1) and (3), then equation (4) becomes

$$-uv\sin\theta\frac{d\theta}{ds}=0$$

$$\Rightarrow \qquad \sin\theta\frac{d\theta}{ds}=0, \quad \text{as} \quad u\neq 0, v\neq 0$$

$$\Rightarrow \qquad \text{Either } \sin\theta=0 \quad \text{or} \quad \frac{d\theta}{ds}=0$$

$$\Rightarrow \qquad \text{Either } \theta=0 \quad \text{or} \quad \theta=\text{constant.}$$

$\Rightarrow$ θ is constant. Since 0 is also a constant.

THEOREM 10.2 *A geodesic is an auto-parallel curve.*

Proof: The differential equation of the geodesic is given by (See Pg. 174, eqn. 6)

$$\frac{d^2x^m}{ds^2}+\begin{Bmatrix} m \\ j\,k \end{Bmatrix}\frac{dx^j}{ds}\frac{dx^k}{ds}=0$$

$$\frac{d}{ds}\left(\frac{dx^m}{ds}\right)+\begin{Bmatrix} m \\ j\,k \end{Bmatrix}\frac{dx^j}{ds}\frac{dx^k}{ds}=0$$

$$\frac{\partial}{\partial x^j}\left(\frac{dx^m}{ds}\right)\frac{dx^j}{ds}+\begin{Bmatrix} m \\ j\,k \end{Bmatrix}\frac{dx^j}{ds}\frac{dx^k}{ds}=0$$

$$\left[\frac{\partial}{\partial x^j}\left(\frac{dx^m}{ds}\right) + \begin{Bmatrix} m \\ j\,k \end{Bmatrix} \frac{dx^k}{ds}\right]\frac{dx^j}{ds} = 0$$

$$\left(\frac{dx^m}{ds}\right)_{,j} \frac{dx^j}{ds} = 0 \quad \text{or} \quad t^m_{,j}\frac{dx^j}{ds} = 0$$

This shows that the unit tangent vector $\frac{dx^m}{ds}$ suffer a parallel displacement along a geodesic curve.This confirms that geodesic is an auto-parallel curve. **Proved.**

10.2 PARALLELISM OF A VECTOR OF VARIABLE MAGNITUDE

Two vectors at a point are said to be parallel or to have the same direction if their corresponding components are proportional. Consequently the vector v^i will be parallel to u^i at each point of curve C provided

$$v^i = \phi u^i \qquad \ldots (1)$$

where ϕ is a function of arc length s.

If u^i is parallel with respect to Riemannian V_n along the curve C. Then,

$$u^i_{,j}\frac{dx^j}{ds} = 0 \qquad \ldots (2)$$

The equation (1) shows that v^i is of variable constant and parallel with respect to Riemannian V_n so that

$$v^i_{,j}\frac{dx^j}{ds} = (\phi u^i)_{,j}\frac{dx^j}{ds}$$

$$= (\phi_{,j}u^i + \phi u^i_{,j})\frac{dx^j}{ds}$$

$$= \phi_{,j}\frac{dx^j}{ds}u^i + \phi u^i_{,j}\frac{dx^j}{ds}$$

$$= \phi_{,j}\frac{dx^j}{ds}u^i \quad \text{Since} \quad u^i_{,j}\frac{dx^j}{ds} = 0$$

$$= \frac{\partial \phi}{\partial x^j}\frac{dx^j}{ds}u^i$$

$$v^i_{,j}\frac{dx^j}{ds} = \frac{d\phi}{ds}u^i$$

$$= \frac{d\phi}{ds}\frac{v^i}{\phi} \quad \text{from (1)}$$

$$= v^i \frac{d(\log \phi)}{ds}$$

$$v^i_{,j} \frac{dx^j}{ds} = v^i f(s) \quad \text{where} \quad f(s) = \frac{d(\log \phi)}{ds} \qquad \ldots (3)$$

Hence a vector v^i of variable magnitude will be parallel with respect to V_n if equation (3) is satisfied.

Conversely suppose that a vector v^i of variable magnitude such that

$$v^i_{,j} \frac{dx^j}{ds} = v^i f(s)$$

to show that v^i is parallel, with respect to V_n.

Take

$$u^i = v^i \psi(s) \qquad \ldots (5)$$

Then

$$u^i_{,j} \frac{dx^j}{ds} = (v^i \psi)_{,j} \frac{dx^j}{ds}$$

$$= v^i_{,j} \frac{dx^j}{ds} \psi + \psi_{,j} \frac{dx^j}{ds} v^i$$

$$= v^i f(s) \psi + \frac{\partial \psi}{\partial x^j} \frac{dx^j}{ds} v^i$$

$$u^i_{,j} \frac{dx^j}{ds} = v^i \left[\psi f(s) + \frac{d\psi}{ds} \right] \qquad \ldots (6)$$

Select ψ such that $\psi f(s) + \frac{d\psi}{ds} = 0.$

Then equation (6) becomes

$$u^i_{,j} \frac{dx^j}{ds} = 0$$

This equation shows that the vector u^i is of constant magnitude and suffers a parallel displacement along curve C. The equation (5) shows that v^i is parallel along C.

Hence necessary and sufficient condition that a vector v^i of variable magnitude suffers a parallel displacement along a curve C is that

$$v^i_{,j} \frac{dx^j}{ds} = v^i f(s).$$

EXAMPLE 1

Show that the vector v^i of variable magnitude suffers a parallel displacement along a curve C if and only if

$$(v^i v^i_{,k} - v^i v^l_{,k}) \frac{dx^k}{ds} = 0, \quad i = 1, 2, \ldots, n.$$

Solution

From equation (4), we have

$$v^i_{,j} \frac{dx^j}{ds} = v^i f(s) \qquad \ldots (1)$$

Multiplying by v^l, we get

$$v^l v^i_{,j} \frac{dx^j}{ds} = v^l v^i f(s)$$

Interchange the indices l and i, we get

$$v^i v^l_{,j} \frac{dx^j}{ds} = v^i v^l f(s) \quad \dots (2)$$

Subtract (1) and (2), we get

$$(v^l u^i_{,k} - v^i v^l_{,k})\frac{dx^k}{ds} = 0 \text{ by interchanging dummy indices } j \text{ and } k.$$

10.3 SUBSPACES OF A RIEMANNIAN MANIFOLD

Let V_n be Riemannian space of n dimensions referred to coordinates x^i and having the metric $ds^2 = g_{ij}\, dx^i\, dx^j$. Let V_m be Riemannian space of m dimensions referred to coordinates y^α and having the metric $ds^2 = a_{x\beta}\, dy^\alpha\, dy^\beta$, where $m > n$. Let Greek letters α, β, γ take the values 1, 2, ..., m and Latin indices i, j, k ... take the values 1, 2, ... n.

If the n independent variables x^i are such that the coordinates (y^α) of points in V_m are expressed as a function of x^i then V_n is immersed in V_m *i.e.* V_n is a subspace of V_m. Also V_m is called enveloping space of V_n.

Since the length ds of the element of arc connecting the two points is the same with respect to V_n or V_m, it follows that

$$g_{ij}\, dx^i\, dx^j = a_{\alpha\beta} dy^\alpha dy^\beta$$

$$g_{ij}\, dx^i\, dx^j = a_{\alpha\beta} \frac{\partial y^\alpha}{\partial x^i}\frac{\partial y^\beta}{\partial x^j} dx^i dx^j$$

$$\Rightarrow \qquad g_{ij} = a_{\alpha\beta} \frac{\partial y^\alpha}{\partial x^i}\frac{\partial y^\beta}{\partial x^j} \quad \dots (1)$$

As dx^i and dx^j are arbitrary.

This gives relation between g_{ij} and $a_{\alpha\beta}$.

THEOREM 10.3 *To show that the angle between any two vectors is the same whether it is calculated with respect to V_m or V_n.*

Proof: Consider two vectors dx^i and δx^j defined at any point of V_n and suppose that the same vectors in V_m are represented by dy^α and δy^α respectively. If θ is the angle between dx^i and δx^i then

$$\cos\theta = \frac{g_{ij} dx^i \delta x^j}{\sqrt{g_{ij} dx^i dx^j}\sqrt{g_{ij}\delta x^i \delta x^j}} \quad \dots (1)$$

If ϕ is the angle between the vectors dy^α and δy^α then

$$\cos\phi = \frac{a_{\alpha\beta} dy^\alpha \delta y^\beta}{\sqrt{a_{\alpha\beta} dy^\alpha dy^\beta}\sqrt{a_{\alpha\beta}\delta y^\alpha \delta y^\beta}}$$

$$= \frac{a_{\alpha\beta}\frac{\partial y^\alpha}{\partial x^i}dx^i\frac{\partial y^\beta}{\partial x^j}dx^j}{\sqrt{a_{\alpha\beta}\frac{\partial y^\alpha}{\partial x^i}dx^i\frac{\partial y^\beta}{\partial x^j}dx^j}\sqrt{a_{\alpha\beta}\frac{\partial y^\alpha}{\partial x^i}\delta x^i\frac{\partial y^\beta}{\partial x^j}\delta x^j}}.$$

$$= \frac{a_{\alpha\beta}\frac{\partial y^\alpha}{\partial x^i}\frac{\partial y^\beta}{\partial x^j}dx^i dx^j}{\sqrt{a_{\alpha\beta}\frac{\partial y^\alpha}{\partial x^i}\frac{\partial y^\beta}{\partial x^j}dx^i dx^j}\sqrt{a_{\alpha\beta}\frac{\partial y^\alpha}{\partial x^i}\frac{\partial y^\beta}{\partial x^j}\delta x^i \delta x^j}}.$$

$$\cos\phi = \frac{g_{ij}dx^i dx^j}{\sqrt{g_{ij}dx^i dx^j}\sqrt{g_{ij}\delta x^i \delta x^j}} \qquad \ldots (2)$$

Since $g_{ij} = a_{\alpha\beta}\frac{\partial y^\alpha}{\partial x^i}\frac{\partial y^\beta}{\partial x^j}$ (from equation (1), art. 10.3)

from (1) & (2)

$$\cos\theta = \cos\phi \Rightarrow \theta = \phi.$$

Proved.

THEOREM 10.4 *If U^α and u^i denote the components of the same vector in Riemannian V_m and V_n respectively then to show that*

$$U^\alpha = u^i\frac{\partial y^\alpha}{\partial x^i}$$

Proof: Let the given vector be unit vector at any point P of V_n. Let the component of the same vector x's and y's be a^i and A^α respectively. Let C be curve passing through s in the direction of the given vector then

$$A^\alpha = \frac{\partial y^\alpha}{\partial s} = \frac{\partial y^\alpha}{\partial x^i}\frac{dx^i}{ds} = \frac{\partial y^\alpha}{\partial x^i}a^i \qquad \ldots (1)$$

But the components of a vector of magnitude a are a times the corresponding components of the Unit vector in the same direction. Then

$$U^\alpha = aA^\alpha, \quad u^i = aa^i \qquad \ldots (2)$$

Multiplying (1) by a, we get

$$aA^\alpha = \frac{\partial y^\alpha}{\partial x^i}(aa^i)$$

Or

$$U^\alpha = \frac{\partial y^\alpha}{\partial x^i}u^i, \quad \text{using (2)}$$

Proved.

THEOREM 10.5 *To show that there are m-n linearly vector fields normal to a surface V_n immersed in a Riemannian V_m.*

Proof: Since V_n is immersed in V_m, the coordinates y^α of points in V_m are expressible as functions of coordinates x^i in V_n.

Now,

$$\frac{dy^\alpha}{ds} = \frac{\partial y^\alpha}{\partial x^i}\frac{dx^i}{ds} \qquad ...(1)$$

for the curve $x^i = s$, we have

$$\frac{dy^\alpha}{ds} = \frac{\partial y^\alpha}{\partial x^i} \qquad ...(2)$$

Since $\frac{dy^\alpha}{ds}$ is a vector tangential to the curve in V_m. Then from equation (2) it follows that $\frac{\partial y^\alpha}{\partial x^i}$ in V_m is tangential to the coordinate curve of transmeter x^i in V_n. Let the Unit vectors N^α in V_m be normal to each of the above vector fields of V_n then

$$a_{\alpha\beta}\frac{\partial y^\alpha}{\partial x^i}N^\beta = 0$$

$$\Rightarrow \quad (a_{\alpha\beta}N^\beta)\frac{\partial y^\alpha}{\partial x^i} = 0 \quad \Rightarrow \quad a_{\alpha\beta}N^\beta y^\alpha_{,i} = 0 \qquad ...(3)$$

$$\Rightarrow \quad N_\alpha \frac{\partial y^\alpha}{\partial x^i} = 0, \quad i = 1, 2, ..., n \;\&\; \alpha = 1, 2, ..., m$$

The equation (3) are n equations in m unknowns N^α $(m > n)$. The coefficient matrix $\left(\frac{\partial y^\alpha}{\partial x^i}\right)$ is of order $m \times n$ and the rank of this matrix is n.

It means that there will be only $m - n$ linearly independent solution of N_α. This shows that there are $m - n$ linearly independent normals to V_n to V_m.

EXAMPLE 2

Show that

$$[ij, k] = [\alpha\beta, \gamma]\frac{\partial y^\alpha}{\partial x^i}\frac{\partial y^\beta}{\partial x^j}\frac{\partial y^\gamma}{\partial x^k} + a_{\alpha\beta}\frac{\partial^2 y^\alpha}{\partial x^i \partial x^j}\frac{\partial x^\beta}{\partial x^k}$$

where $[\alpha\beta, \gamma]$ and $[i\,j, k]$ are the Christoffel's symbols of first kind relative to metrics $a_{\alpha\beta}dy^a dy^\beta$ and $g_{ij}dx^i dx^j$.

Solution

Since relation between $a_{\alpha\beta}$ and g_{ij} is given by

$$g_{ij} = a_{\alpha\beta}\frac{\partial y^\alpha}{\partial x^i}\frac{\partial y^\beta}{\partial x^j}$$

Differentiating it with respect to x^k, we get

$$\frac{\partial g_{ij}}{\partial x^k} = \frac{\partial a_{\alpha\beta}}{\partial x^\gamma}\frac{\partial y^\alpha}{\partial x^i}\frac{\partial y^\beta}{\partial x^j}\frac{\partial y^\gamma}{\partial x^k} + a_{\alpha\beta}\frac{\partial^2 y^\alpha}{\partial x^i \partial x^k}\frac{\partial y^\gamma}{\partial x^j} + a_{\alpha\beta}\frac{\partial y^\alpha}{\partial x^i}\frac{\partial^2 y^\beta}{\partial x^j \partial x^k} \qquad ...(1)$$

Similarly

$$\frac{\partial g_{jk}}{\partial x^i} = \frac{\partial a_{\beta\gamma}}{\partial y^\alpha}\frac{\partial y^\alpha}{\partial x^i}\frac{\partial y^\beta}{\partial x^j}\frac{\partial y^\gamma}{\partial x^k} + a_{\alpha\beta}\frac{\partial^2 y^\alpha}{\partial x^j \partial x^i}\frac{\partial y^\beta}{\partial x^k} + a_{\alpha\beta}\frac{\partial y^\alpha}{\partial x^j}\frac{\partial^2 y^\beta}{\partial x^k \partial x^i} \qquad \ldots (2)$$

and

$$\frac{\partial g_{ki}}{\partial x^j} = \frac{\partial a_{\gamma\alpha}}{\partial x^\beta}\frac{\partial y^\alpha}{\partial x^i}\frac{\partial y^\beta}{\partial x^j}\frac{\partial y^\gamma}{\partial x^k} + a_{\alpha\beta}\frac{\partial^2 y^\alpha}{\partial x^k \partial x^j}\frac{\partial y^\beta}{\partial x^i} + a_{\alpha\beta}\frac{\partial y^\alpha}{\partial x^k}\frac{\partial^2 y^\beta}{\partial x^i \partial x^j} \qquad \ldots (3)$$

But we know that

$$[ij,k] = \frac{1}{2}\left(\frac{\partial g_{jk}}{\partial x^i} + \frac{\partial g_{ki}}{\partial x^j} - \frac{\partial g_{ij}}{\partial x^k}\right) \qquad \ldots (4)$$

and,

$$[\alpha\beta,\gamma] = \frac{1}{2}\left(\frac{\partial g_{\beta\gamma}}{\partial y^\alpha} + \frac{\partial g_{\gamma\alpha}}{\partial y^\beta} - \frac{\partial g_{\alpha\beta}}{\partial y^\gamma}\right) \qquad \ldots (5)$$

Substituting the value of (1), (2), (3) in equation (4) and using (5) we get,

$$[ij,k] = [\alpha\beta,\gamma]\frac{\partial y^\alpha}{\partial x^i}\frac{\partial y^\beta}{\partial x^j}\frac{\partial y^\gamma}{\partial x^k} + a_{\alpha\beta}\frac{\partial^2 y^\alpha}{\partial x^i \partial x^j}\frac{\partial y^\beta}{\partial x^k}.$$

10.4 PARALLELISM IN A SUBSPACE

THEOREM 10.6 *Let T^α and t^i be the components of the same vector t relative to V_m and V_n respectively. Let the vector t be defined along a curve C in V_n. If p^i and q^α are derived vectors of t along C relative to V_n and V_m respectively. Then*

$$q_\alpha \frac{\partial y^\alpha}{\partial x^i} = p_i$$

Proof: Since from equation (1), theorem, (10.4), Pg. 194 we have

$$T^\alpha = t^i \frac{\partial y^\alpha}{\partial x^i} \qquad \ldots (1)$$

$$\frac{dT^\alpha}{ds} = \frac{dt^i}{ds}\frac{\partial y^\alpha}{\partial x^i} + t^i \frac{\partial^2 y^\alpha}{\partial x^i \partial x^j}\frac{dx^j}{ds} \qquad \ldots (2)$$

Now,

$$p^i = t^i_{,j}\frac{dx^j}{ds} \qquad \ldots (3)$$

$$\begin{aligned} q^\alpha &= T^\alpha_{,\beta}\frac{dy^\beta}{ds} \\ &= \left(\frac{\partial T^\alpha}{\partial x^\beta} + T^\gamma \begin{Bmatrix} \alpha \\ \gamma\beta \end{Bmatrix}\right)\frac{dy^\beta}{ds} \\ &= \frac{\partial T^\alpha}{\partial x^\beta}\frac{dy^\beta}{ds} + T^\gamma \begin{Bmatrix} \alpha \\ \gamma\beta \end{Bmatrix}\frac{dy^\beta}{ds} \end{aligned}$$

$$q^\alpha = \frac{dT^\alpha}{ds} + T^\gamma \begin{Bmatrix} \alpha \\ \gamma\beta \end{Bmatrix}\frac{dy^\beta}{ds} \qquad \ldots (4)$$

Putting the value of $\frac{dT^\alpha}{ds}$ and T^γ from (2) and (1) in equation (4), we get

$$q^\alpha = \frac{dt^i}{ds}\frac{\partial y^\alpha}{\partial x^i} + t^i \frac{\partial^2 y^\alpha}{\partial x^i \partial x^j}\frac{dx^j}{ds} + t^i \frac{\partial y^\gamma}{dx^i}\frac{dy^\beta}{ds}\begin{Bmatrix}\alpha \\ \gamma\beta\end{Bmatrix}$$

$$= \frac{\partial t^i}{\partial x^j}\frac{\partial y^\alpha}{\partial x^i}\frac{dx^j}{ds} + t^i \frac{\partial^2 y^\alpha}{\partial x^i \partial x^j}\frac{dx^j}{ds} + t^i \frac{\partial y^\gamma}{\partial x^i}\frac{dy^\beta}{ds}\begin{Bmatrix}\alpha \\ \gamma\beta\end{Bmatrix}$$

$$= \frac{\partial t^i}{\partial x^j}\frac{\partial y^\alpha}{\partial x^i}\frac{dx^j}{ds} + t^i \frac{\partial^2 y^\alpha}{\partial x^i \partial x^j}\frac{dx^j}{ds} + t^i \frac{\partial y^\gamma}{\partial x^i}\frac{\partial y^\beta}{\partial x^j}\frac{dx^j}{ds}\begin{Bmatrix}\alpha \\ \gamma\beta\end{Bmatrix}$$

$$q^\alpha = \frac{\partial t^i}{\partial x^j}\frac{\partial y^\alpha}{\partial x^i}\frac{dx^j}{ds} + t^i \frac{dx^j}{ds} + \left(\frac{\partial^2 y^\alpha}{\partial x^i \partial x^j} + \frac{\partial y^\beta}{\partial x^j}\frac{\partial y^\gamma}{\partial x^i}\begin{Bmatrix}\alpha \\ \gamma\beta\end{Bmatrix}\right) \quad \dots (5)$$

But we know that

$$[ij,k] = [\alpha\beta,\gamma]\frac{\partial y^\alpha}{\partial x^i}\frac{\partial y^\beta}{\partial x^j}\frac{\partial y^\gamma}{\partial x^k} + a_{\alpha\gamma}\frac{\partial^2 y^\alpha}{\partial x^i \partial x^j}\frac{\partial y^\gamma}{\partial x^k}$$

$$[ji,k] = [\beta\gamma,\delta]\frac{\partial y^\beta}{\partial x^j}\frac{\partial y^\gamma}{\partial x^i}\frac{\partial y^\delta}{\partial x^k} + a_{\alpha\delta}\frac{\partial^2 y^\alpha}{\partial x^i \partial x^j}\frac{\partial y^\delta}{\partial x^k}$$

$$= \begin{Bmatrix}\alpha \\ \beta\gamma\end{Bmatrix} a_{\alpha\delta}\frac{\partial y^\beta}{\partial x^j}\frac{\partial y^\gamma}{\partial x^i}\frac{\partial y^\delta}{\partial x^k} + a_{\alpha\delta}\frac{\partial^2 y^\alpha}{\partial x^i \partial x^j}\frac{\partial y^\delta}{\partial x^k}$$

$$= a_{\alpha\delta}\frac{\partial y^\delta}{\partial x^k}\left(\begin{Bmatrix}\alpha \\ \beta\gamma\end{Bmatrix}\frac{\partial y^\beta}{\partial x^j}\frac{\partial y^\gamma}{\partial x^i} + \frac{\partial^2 y^\alpha}{\partial x^i \partial x^j}\right) \quad \dots (6)$$

Multiplying (5) by $a_{\alpha\delta}\frac{\partial y^\delta}{\partial x^k}$, we get, using (6),

$$a_{\alpha\delta}\frac{\partial y^\delta}{\partial x^k} q^\alpha = \frac{\partial t^i}{\partial x^j}\frac{dx^j}{ds}\frac{\partial y^\alpha}{\partial x^i} a_{\alpha\delta}\frac{\partial y^\delta}{\partial x^k} + t^i \frac{dx^j}{ds}[ij,k]$$

Or $$q_\delta \frac{\partial y^\delta}{\partial x^k} = \frac{dx^j}{ds}\left(\frac{\partial t^i}{\partial x^j} g_{ik} + t^i \begin{Bmatrix}p \\ i\,j\end{Bmatrix} g_{pk}\right), \quad \text{since } g_{ij} = a_{\alpha\beta}\frac{\partial y^\alpha}{\partial x^i}\frac{\partial y^\beta}{\partial x^j}$$

$$= \frac{dx^j}{ds}\left(\frac{\partial t^i}{\partial x^j} g_{ik} + t^a \begin{Bmatrix}i \\ a\,j\end{Bmatrix} g_{ik}\right)$$

$$= g_{ik}\frac{dx^j}{ds} t^i_{,j}$$

$$= (g_{ik}t^i)_{,j}\frac{dx^j}{ds}$$

$$= t_{k,j}\frac{dx^j}{ds}$$

$$q_\alpha \frac{\partial y^\alpha}{\partial x^k} = t_{k,j}\frac{dx^j}{ds}$$

or $$q_\alpha \frac{\partial y^\alpha}{\partial x^k} = p_k \qquad \text{...(7)}$$

Proved.

Properties of V_m

(*i*) If a curve C lies in a subspace V_n of V_m and a vector field in V_n is parallel along the curve C with regard to V_m, then to show that it is also a parallel with regard to V_n.

Proof: If a vector field t is a parallel along C with respect to V_m. then its derived vector q^α vanishes *i.e.,*

$q^\alpha = 0, \quad \alpha = 1, 2,... m.$

Hence equation (7) becomes $p_k = 0$, $k = 1, 2, ..., n$. This shows that the vector field t is also parallel along C with respect to V_n.

(*ii*) To show that if a curve C is a geodesic in a V_n it is a geodesic in any subspace V_n of V_m.

Proff : Science, $q^\alpha \dfrac{\partial y_\alpha}{\partial x^i} = p_i$

Let t be unit tangent vector to the curve C them T^α is unit tangent vector to C relative to V_m and t_i is unit tangent vector to C relative to V_n.

Now, $p_i = 0 \quad \Rightarrow \quad$ curve C is a geodesic in V_n

$q^\alpha = 0 \quad \Rightarrow \quad$ curve C is a geodesic in V_m

Also, $q_\alpha = 0, \forall\alpha \quad \Rightarrow \quad p_i = 0, \forall i$

This shows that if a curve C in V_n is a geodesic relative to V_m then the same curve is also a geodesic relative to V_n.

(*iii*) A necessary and sufficient condition that a vector of constant magnitude be parallel with respect to V_n along C in that subspace, is that its derived vector relative to V_m for the direction of the curve be normal to V_n.

Proof: Let t be a vector of constant magnitude. This vector t is parallel along C relative to Vn iff $p_i = 0, \forall i$

$$\text{iff } q_\alpha \frac{\partial y^\alpha}{\partial x^i} = 0$$

But $q_\alpha \dfrac{\partial y^\alpha}{\partial x^i} = 0$ implies that q_α is normal to V_n.

For $\dfrac{\partial y^\alpha}{\partial x^i} = y^\alpha_{,i}$ lying in V_m is tangential to a coordinate curve of parameter x^i in V_n.

These statements prove that a necessary and sufficient condition that a vector of constant magnitude be parallel along C relative to V_n is that its derived vector *i.e.*, q_α along C relative to V_m be normal to V_n.

(*iv*) A necessary and sufficient condition that a curve be a geodesic in V_n is that its principal subnormal relative to V_m the enveloping space be normal to V_n at all points of the curve.

Proof: In particular let the vector t be unit tangent vector to the curve C. In this case q_α is called principal normal the curve C. Also $p_i = 0, \forall i$ implies that the curve C is a geodesic relative to V_n.

Using result (*iii*), we get at once the result (*iv*).

(*v*) To prove that the tendency of a vector is the same whether it is calculated with respect to V_m.

Proof: Since, we have

$$q_\alpha \frac{\partial y^\alpha}{\partial x^i} = p_i$$

$$q_\alpha \frac{\partial y^\alpha}{\partial x^i}\frac{dx^i}{ds} = p_i \frac{dx^i}{ds}$$

$$q_\alpha \frac{dy^\alpha}{ds} = p_i \frac{dx^i}{ds}$$

$$T_{\alpha,\beta} \frac{dy^\beta}{ds}\frac{dy^\alpha}{ds} = t_{i,j} \frac{dx^j}{ds}\frac{dx^i}{ds}$$

i.e., tendency of T_α along C = tendency of t_i along C.

i.e., tendency of t along C relative to V_m = tendency of t along C relative to V_n.

10.5 THE FUNDAMENTAL THEOREM OF RIEMANNIAN GEOMETRY STATEMENT

With a given Riemannian metric (or fundamental tensor) of a Riemannian manifold there is associated a symmetric affine connection with the property that parallel displacement (or transport) preserves scalar product.

Proof: Let C be a curve in V_n. Let p^i *and* q^i be two unit vectors defined along C. Suppose that the unit vectors p^i and q^i suffer parallel displacement along the curve C in V_n, then we have

$$p^i_{,j} \frac{dx^j}{ds} = 0 \qquad \dots (1)$$

and

$$q^i_{,j} \frac{dx^j}{ds} = 0 \qquad \dots (2)$$

Let g_{ij} be the given fundamental tensor of a Riemannian manifold. Hence, the scalar product of vectors p^i and q^i is $g_{ij}\, p^i\, q^j$.

Now,

$$(g_{ij} p^i q^j)_{,k} \frac{dx^k}{ds} = 0$$

$$g_{ij}\left(p^i_{,k}\frac{dx^k}{ds}\right)q^j + g_{ij}p^i\left(q^j_{,k}\frac{dx^k}{ds}\right) + \left(g_{ij,k}\frac{dx^k}{ds}\right)P^iq^j = 0 \qquad \ldots (3)$$

Using equation (1) and (2), the equation (3) becomes,

$$\left(g_{ij,k}\frac{dx^k}{ds}\right)p^iq^j = 0$$

$$g_{ij,\,k} = 0$$

(Since p^i and q^i are unit vectors and $\frac{dx^k}{ds} \neq 0$)

$$\frac{\partial g_{ij}}{\partial x^k} - g_{mj}\begin{Bmatrix} m \\ i\,k \end{Bmatrix} - g_{im}\begin{Bmatrix} m \\ j\,k \end{Bmatrix} = 0$$

$$\frac{\partial g_{ij}}{\partial x^k} - [ik, j] - [jk, i] = 0$$

$$\frac{\partial g_{ij}}{\partial x^k} = [ik, j] + [jk, i] \qquad \ldots (4)$$

Now, using equation (4), we have

$$\frac{\partial g_{jk}}{\partial x^i} + \frac{\partial g_{ki}}{\partial x^j} - \frac{\partial g_{ij}}{\partial x^k} = [ji,k] + [ki,j] + [kj,i] + [ij,k] - ([ik,i] + [jk,i])$$

since $[ij,k] = [ji,k]$.

So,

$$\frac{\partial g_{jk}}{\partial x^i} + \frac{\partial g_{ki}}{\partial x^j} - \frac{\partial g_{ij}}{\partial x^k} = 2[ij,k]$$

$$[ij,k] = \frac{1}{2}\left(\frac{\partial g_{jk}}{\partial x^i} + \frac{\partial g_{ki}}{\partial x^j} - \frac{\partial g_{ij}}{\partial x^k}\right) \qquad \ldots (5)$$

But we know that $\begin{Bmatrix} k \\ i\,j \end{Bmatrix} = g^{lk}[ij,l]$

from (5), we have

$$\begin{Bmatrix} k \\ i\,j \end{Bmatrix} = \frac{1}{2}g^{lk}\left(\frac{\partial g_{jl}}{\partial x^i} + \frac{\partial g_{li}}{\partial x^j} - \frac{\partial g_{ij}}{\partial x^l}\right)$$

Proved.

EXAMPLE 3

If t^i and T^α are contravariant components in *x's* and *y's* respectively, of *n* vector field in V_n immersed in V_m. Show that

$$T^\alpha_{,j} = y^\alpha_{,ij}t^i + y^\alpha_{,i}t^i_{,j}$$

Solution

Since we know that

$$T^\alpha = t^i\frac{\partial y^\alpha}{\partial x^i}$$

$$T^{\alpha} = t^i y^{\alpha}_{,j}.$$

Taking covariant differentiation of both sides, we get

$$T^{\alpha}_{,j} = (t^i y^{\alpha}_{,i})_{,j}$$

$$= t^i_{,j} y^{\alpha}_{,i} + t^i (y^{\alpha}_{,i})_{,j}$$

$$T^{\alpha}_{,j} = t^i y^{\alpha}_{,ij} + t^i_{,j} y^{\alpha}_{,i}$$

EXAMPLE 4

Show that $g_{ij}\frac{dx^i}{ds}\frac{dx^j}{ds}$ remains constant along a geodesic.

Solution

Let $t^i = \frac{dx^i}{ds}$. Then

$$g_{ij}\frac{dx^i}{ds}\frac{dx^j}{ds} = g_{ij}t^i t^j = t^2$$

Since we know that geodesics are autoparallel curves. Then

$$t^i_{,j}\frac{dx^j}{ds} = 0$$

or

$$t^i_{,j} t^j = 0 \qquad \dots (1)$$

Now,

$$\frac{dt^2}{ds} = \frac{d}{ds}(g_{ij}t^i t^j) = \frac{d}{ds}(t_i t^i)$$

$$= (t_i t^i)_{,j}\frac{dx^j}{ds} = (t_i t^i)_{,j} t^j$$

$$\frac{dt^2}{ds} = (t_{i,j}t^i)t^i + (t^i_{,j}t^i)t_i = 0, \quad \text{from (1)}$$

Integrating it we get

$$t^2 = \text{constant}.$$

So, $g_{ij}\frac{dx^i}{ds}\frac{dx^j}{ds}$ remains constant along a geodesic.

EXAMPLE 5

If t^i are the contravariant components of the unit tangent vector to a congruence of geodesics. Show that

$$t^i(t_{i,j} + t_{j,i}) = 0$$

and also show that $|t_{i,j} + t_{j,i}| = 0$.

Solution

Let t^i denote unit tangent vector to a congruence of geodesic so that

$$t^i_{,j}t^j = 0 \quad \text{... (1)}$$

Since geodesics are auto-parallel curves. Then to prove that

(i) $t^i(t_{i,j} + t_{j,i}) = 0$

(ii) $|t_{i,j} + t_{j,i}| = 0$

Since

$$t^i t_i = t^2 = 1$$

$$t^i t_i = 1$$

Taking covariant derivative of both sides, we get

$$(t^i t_i)_{,j} = 0$$

or $$t^i_{,j}t_i + t^i t_{i,j} = 0$$

Since t is a free index. Then we have

$$t^i_{,j}t^i + t_{i,j}t^i = 0$$

$$2t^i_{,j}t^i = 0$$

$\Rightarrow$ $$t_{i,j}t^i = 0 \quad \text{... (2)}$$

from (1)

$$g_{ik}t^i_{,j}t^j = 0$$

or $$t_{k,j}t^j = 0$$

$\Rightarrow$ $$t_{k,i}t^i = 0 \Rightarrow t_{j,i}t^i = 0 \Rightarrow t^i t_{j,i} = 0$$

Thus $$t^i t_{j,i} = 0 \quad \text{... (3)}$$

Adding (2) and (3), we get

$$t^i(t_{i,j} + t_{j,i}) = 0$$

Also, since $t^i \neq 0, \forall i$.

Taking determinants of both sides we get

$$|t^i(t_{i,j} + t_{j,i})| = 0$$

$\Rightarrow$ $$|t_{i,j} + t_{j,i}| = 0$$

Solved.

EXAMPLE 6

When the coordinates of a V_2 are chosen so that the fundamental forms is $(d\dot{x}') + 2g_{12}dx^1dx^2 + (dx^2)^2$, prove that the tangent to either family of coordinate curves suffers a parallel displacement along a curve of the other family.

Solution

The metric is given by

$$ds^2 = g_{ij}dx^i dx^j \quad i,j = 1,2.$$

Comparing it with

$$ds^2 = (dx^1)^2 + 2g_{12}dx^1 dx^2 + (dx^2)^2 \tag{1}$$

We have,

$$g_{11} = 1, \quad g_{22} = 1.$$

In this case, we have coordinate curves of parameters x^1 and x^2 respectively. The coordinate x^1 curve is defined by

$$x^i = c^i, \quad \forall i, \quad \text{except} \quad i = 1. \tag{2}$$

and the coordinate x^2 curve is defined by

$$x^i = d^i, \quad \forall i, \quad \text{except} \quad i = 2 \tag{3}$$

where c^i and d^i are constants.

Let p^i and q^i be the components of tangents vectors to the curves (2) and (3) respectively. Then we have

$$p^i = dx^i = 0, \quad \forall i, \quad \text{except} \quad i = 1$$

and

$$q^i = dx^i = 0, \ \forall i, \text{ except } i = 2$$

So,

$$p^i = (dx^i, 0) \quad \text{and} \quad q^i = (0, dx^i)$$

Let t be the unit tangent vector to the curve (2).

Hence

$$\frac{dx^i}{ds} = t^i = \frac{p^i}{p} = (1,0) \quad \text{where} \quad p = dx^i$$

So, we have

$$p^i_{,j}\frac{dx^j}{ds} = 0$$

Hence the tangent to the family of coordinates curves (2) suffers a parallel displacement along a curve of the family of curves (3).

EXERCISES

1. Explain Levi-Civita's concept of parallelism of vectors and prove that any vector which undergoes a parallel displacement along a geodesic is inclined at a constant angle to the curve.
2. Show that the geodesics is a Riemannian space are given by

$$\frac{d^2x^m}{ds^2} + \begin{Bmatrix} m \\ i\,k \end{Bmatrix} \frac{dx^i}{ds}\frac{dx^k}{ds} = 0$$

Hence prove that geodesics are auto-parallel curves.

3. Establish the equivalence of the following definitions of a geodesic.

   (*i*) It is an auto–parallel curve.

   (*ii*) It is a line whose first curvature relative to V_n identically zero.

   (*iii*) It is the path extremum length between two points on it.

4. If u and v are orthogonal vector fields in a V_n, prove that the projection on υ of the desired vector of u in its own direction is equal to minus the tendency of v in the direction of u.
5. If the derived vector of a vector u^i is zero then to show that vector u^i has a constant magnitude along curve.
6. Prove that any vector which undergoes a parallel displacement along a geodesic is inclined at a constant angle to the curve.
7. Prove that there are $m - n$ linearly independent vector fields normal to a surface V_n immersed in a Riemannian V_m and they may be chosen in a multiply infinite number of ways. But there is only one vector field normal to the hyperface.
8. Show that the principal normal vector vanishes identically when the given curve in geodesic.
9. Show that if a curve is a geodesic of a space it is a geodesic of any subspace in which it lies.
10. Define parallelism in a subspace of Riemannian manifold. If a curve C lies in a subspace V_n of V_m and a vector field in V_n is parallel along C with respect to V_m. Then show that it is also parallel with respect to V_n.

CHAPTER – 11

RICCI'S COEFFICIENTS OF ROTATION AND CONGRUENCE

11.1 RICCI'S COEFFICIENTS OF ROTATION

Let $e_{h|}^{i}$ (h=1, 2, ...n) be the unit tangents to the n congruences $e_{h|}$, of an orthogonal ennuple in a Riemannian $V_{n'}$. The desired vector of $e_{l|i}$ in the direction of $e_{k|}^{e}$ has components $e_{l|i,j}e_{k|}^{j}$ and the projection of this vector on $e_{h|}^{i}$ is a scalar invariant, denoted by $\gamma_{lhk,}$ so that

$$\gamma_{lhk} = e_{l|i,j}e_{h|}^{i}e_{k|}^{j} \qquad \text{...(1)}$$

The invariants γ_{lhk} are *Ricci's Coefficients of Rotation.*

Since i being a dummy index, has freedom of movement. Then equation (1) may be written as

$$\gamma_{lhk} = e_{l|i,j}e_{h|i}e_{k|}^{j} \qquad \text{...(2)}$$

The indices l, h and k are not tensor indices. But these indices in γ_{lhk} are arranged in proper way, the first index l indicates the congruence whose unit tangent is considered, the second h indicates the direction of projection and the third k is used for differentiation.

THEOREM 11.1 *To prove thast the Riccie's coefficients of rotation are skew-symmetric in the first two indices i.e.,*

$$\gamma_{lhk} = -\gamma_{lhk}$$

Proof: If $e_{h|}^{i}$ (h = 1, 2, ..., n) be n unit tangents to n congruences $e_{h|}$ of an orthgonal ennuple in a V_n then

$$e_{h|i}e_{l|}^{i} = 0$$

convariant differentiation with respect to x^j, we get

$$e_{h|i}e_{l|}^{i},j = 0$$

$$e_{h|i,j}e_{l|}^{i} + e_{l|,j}^{i}e_{h|i} = 0$$

multiplying by $e^j_{k|}$ and summing for j, we get

$$e_{h|i,j}e^i_{l|}e^i_{k|} + e^i_{i|,j}e_{h|i}e^j_{k|} = 0$$

$$\gamma_{hlk} + \gamma_{lhk} = 0$$

or $$\gamma_{hlk} = -\gamma_{lhk} \qquad \ldots(1)$$

Note: Put $l = h$ in equation (1), we get

$$\gamma_{llk} = -\gamma_{llk}$$

or $$2\gamma_{llk} = 0$$

or $$\gamma_{llk} = 0$$

THEOREM 11.2 *To prove that*

$$\sum_h \gamma_{lhk} e_{h|i} = e_{l|i,j} e^j_{k|}$$

Proof: since we know that

$$\gamma_{lhk} = e_{l|i,j} e^i_{h|} e^j_{k|}$$

Multiplying by $e_{h|m}$ and summing for h.

$$\sum_h \gamma_{lhk} e_{h|m} = \sum_h e_{l|i,j} e^i_{h|} e^j_{k|} e_{h|m}$$

$$= e_{l|i,j} e^j_{k|} \sum_h \left(e^i_{h|} e_{h|m}\right)$$

$$= e_{l|i,j} e^j_{k|} \delta^i_m \quad \text{since } \sum_h e^i_{h|} e_{h|m} = \delta^i_m$$

$$= (e_{l|i,j} \delta^i_m)\, e^j_{k|}$$

$$\sum_h \gamma_{lhk} e_{h|m} = e_{l|m,j} e^j_{k|}$$

Replacing m by i, we get

$$\sum_h \gamma_{lhk} e_{i|m} = e_{l|i,j} e^j_{k|}$$

11.2 REASON FOR THE NAME "COEFFICIENTS OF ROTATION"

Let C_m be a definite curve of the congruence whose unit tangent is $e^i_{m|}$ and P_0 a fidxed point on it. Let u^1 be a unit vector which coincides with the vector $e^i_{l|}$ at P_0 and undergoes a parallel displayment along the curve C_m.

Thus

$$u^i = e^i_{l|} \text{ at } P_0 \qquad \ldots (1)$$

and $$u^i_{,j} e^j_{h|} = 0 \qquad \ldots (2)$$

If θ is the angle between the vectors u^i and $e^i_{h|}$, we have

$$\cos\theta = u^i e_{h|i}$$

Differentiating it with respect to arc length s_m along C_m, we get

$$-\sin\theta\frac{d\theta}{ds_m} = (u^i e_{h|i})_{,j}\, e^j_{m|}$$

$$= (u^i e_{h|i,j} + u^i_{,j} e_{h|i}) e^j_{m|}$$

$$-\sin\theta\frac{d\theta}{ds_m} = u^i e_{h|i,j} e^j_{m|} + u^i_{,j} e^j_{m|} \quad \text{...(3)}$$

at the point $P_0, \theta = \frac{\pi}{2}$, we have

$$-\frac{d\theta}{ds_m} = u^i e_{h|i,j} e^j_{m|}$$

$$= u^i e_{h|i,j} e^j_{m|}$$

$$= e^i_{l|} e_{h|i,j} e^j_{m|}; \quad \text{from (1)}$$

$$= e_{h|i,j} e^i_{l|} e^j_{m|}$$

$$-\frac{d\theta}{ds_m} = \gamma_{hlm}$$

$$\Rightarrow \qquad \frac{d\theta}{ds_m} = -\gamma_{hlm} \quad \text{... (4)}$$

In Eucliden space of three dimensions, $\frac{d\theta}{ds_m}$ is the arc-rate of rotation of the vector $e^i_{h|}$ about the curve C_m. Hence the quantities γ_{hlm} are called coefficients of rotation of the ennuple. Since it was discovered by Ricci and hence it is called *Ricci's coefficient of rotation.*

11.3 CURVATURE OF CONGRUENCE

The first curvature vector $p_{l|}$ of curve of the congruence is the derived vector of $e^i_{h|}$ in its own direction. Where $e^i_{h|}$ $(h = 1, 2, ..., n)$ be unit tangents to n congruence $e_{h|}$ of an orthogonal ennuple in a V_n.

If $p^i_{h|}$ be the contravariant component of first curvature vector $p_{l|}$. Then by definition, we have

$$p^i_{h|} = e^i_{h,j} e^j_{h|}$$

from theorem (11.2), we have

$$p^i_{h|} = \sum_l \gamma_{hlh} e^i_{l|} \quad \text{... (1)}$$

The magnitude of $p^i_{h|}$ is called curvature of the curve of congruence $e_{h|}$ and denoted by $K_{h|}$.

Now,

$$K^2_{h|} = g_{ij} p^i_{h|} p^j_{h|}$$

$$= g_{ij}\left(\sum_l \gamma_{hlh} e^i_{l|}\right)\left(\sum_m \gamma_{hmh} e^i_{m|}\right)$$

$$= \sum_l \sum_m g_{ij} e^i_{l|} e^i_{m|} \gamma_{hlh} \gamma_{hmh}$$

$$= \sum_l \sum_m \delta^l_m \gamma_{hlm}\ \gamma_{hmh}, \text{ Since } g_{ij} e^i_{l|} e^i_{m|} = \delta^l_m$$

$$= \sum_m \gamma_{hmh} \gamma_{hmh}$$

$$K^2_{h|} = \sum_m (\gamma_{hmh})^2$$

This is the required formula for $K_{h|}$.

11.4 GEODESIC CONGRUENCE

If all the curves of a congruence are geodesics then the congruence is called a geodesic congruence.

THEOREM 11.3 *A necessary and sufficient condition that congruence C of an orthogonal ennuple be a geodesic congruence is that the tendencies of all the other congruences of the ennuple in the direction of C vanish indentically.*

Or

To obtain necessary and sufficient conditions that a congruence be a geodesic congruence.

Proof: From equation (1), Pg. 207, we have

$$p^i_{h|} = \sum_l \gamma_{hlh} e^i_{l|}$$

Since $e^i_{l|} \neq 0$, $\forall h$ and hence $p^i_{h|} = 0$ iff $\gamma_{hlh} = 0, \forall h$. But $p^i_{l|} = 0$ iff the congruence C is geodesic congruence.

Hence C is a geodesic congruence iff $\gamma_{hlh} = 0,\ \forall h$.

But

$$\gamma_{hlh} = -\gamma_{lhh}$$

So, C is a geodesic congruence iff $-\gamma_{lhh} = 0,\ -\forall h$.

or C is a geodesic congruence iff $\gamma_{lhh} = 0$.

Hence $\gamma_{lhh} = 0$ are the necessary and sufficient conditions that congruence with unit tangents $e^i_{h|}$ be geodesic congruence. Again γ_{lhh} is the tendency of the vector $e^i_{l|}$ in the direction vector $e^i_{h|}$.

Thus a congruence C of an orthogonal ennuple is a geodesic congruence iff the tendency of all other congruences in the direction of C vanish identically.

11.5 NORMAL CONGRUENCE

A normal congruence is one which intersects orthogonally a family of hypersurfaces.

THEOREM 11.4 *Necessary and sufficient conditions that the congruence $e_{h|}$ of an orthogonal ennuple be normal is that*

$$\gamma_{nqp} = \gamma_{npq}$$

Proof: Consider a congruence C of curves in a V_n. Let $\phi(x^1, x^2, ..., x^n)$ = constant be a family of hypersurfaces. To determines a normal congruence whose tangent vector is grad ϕ or $\nabla\phi$.

Let t_i be the convariant components of unit tangent vector to C. The congruence C is a normal congruence to a family of hypersurface $\phi(x^1, x^2, ..., x^n)$ = constant if

$$\frac{\phi_{,1}}{t_1} = \frac{\phi_{,2}}{t_2} = \ldots \frac{\phi_{,n}}{t_n} = y \text{ (say)} \qquad \ldots (1)$$

In order that $(n-1)$ differential equations given by equation (1) admit a solution which is not constant, these must constitute a complete system.

From (1)

$$\frac{\phi_{,i}}{t_i} = y \quad \text{or} \quad \phi_{,i} = yt_i$$

$$\Rightarrow \qquad yt_i = \frac{\partial \phi}{\partial x^i}$$

Differentiating it with respect to x^j, we get

$$\frac{\partial y}{\partial x^j} t_i + y\frac{\partial t_i}{\partial x^j} = \frac{\partial^2 \phi}{\partial x^i \partial x^j} \qquad \ldots (2)$$

Interchanging indices i and j, we get

$$\frac{\partial y}{\partial x^i} t_j + y\frac{\partial t_j}{\partial x^i} = \frac{\partial^2 \phi}{\partial x^j \partial x^i} \qquad \ldots (3)$$

Subtracting (2) and (3) we get

$$\left(\frac{\partial y}{dx^j} t_i + y\frac{\partial t_i}{\partial x^j}\right) - \left(\frac{\partial y}{\partial x^i} t_j + y\frac{\partial t_i}{dx^j}\right) = 0$$

$$y\left(\frac{\partial t_i}{\partial x^j} - \frac{\partial t_j}{\partial x^i}\right) + \left(t_i\frac{\partial y}{\partial x^j} - t_j\frac{\partial y}{\partial x^i}\right) = 0$$

Multiplying by t_k,

$$yt_k\left(\frac{\partial t_i}{dx^j}-\frac{\partial t_j}{\partial x^i}\right)+t_it_k\frac{\partial y}{\partial x^j}-i_kt_i\frac{\partial y}{dx^i}=0 \quad \ldots(4)$$

By cyclic permutation of i, j, k in (4), we get

$$yt_i\left(\frac{\partial t_j}{\partial x^k}-\frac{\partial t_k}{\partial x^j}\right)+t_jt_i\frac{\partial y}{\partial x^k}-i_kt_i\frac{\partial y}{\partial x^j}=0 \quad \ldots(5)$$

and

$$yt_i\left(\frac{\partial t_k}{\partial x^i}-\frac{\partial t_i}{\partial x^k}\right)+t_kt_j\frac{\partial y}{\partial x^i}-t_it_j\frac{\partial y}{\partial x^k}=0 \quad \ldots(6)$$

On adding (4), (5) and (6) we get

$$y\left[t_k\left(\frac{\partial t_i}{\partial x^j}-\frac{\partial t_j}{\partial x^i}\right)+t_i\left(\frac{\partial t_j}{\partial x^k}-\frac{\partial t_k}{\partial x^j}\right)+t_j\left(\frac{\partial t_k}{\partial x^i}-\frac{\partial t_i}{\partial x^k}\right)\right]=0$$

or

$$t_k(t_{i,j}-t_{j,i})+t_i(t_{j,k}-t_{k,j})+t_j\left(t_{k,i}-t_{i,k}\right)=0 \quad \ldots(7)$$

as $y\neq 0$, where $i, j, k = 1, 2, \ldots, n$.

These are the necessary and sufficient conditions that the given congruence be a normal congruence.

Now suppose that the congruence is one of an orthogonal ennuple in V_n. Let $e_{n|i}$ be the unit tangents of given congruence C so that $t_i = e_{n|i}$

Then equation (7) becomes

$$e_{n|k}(e_{n|i,j}-e_{n|j,i})+e_{n|i}(e_{n|j,k}-e_{n|k,j})+e_{n|j}(e_{n|k,i}-e_{n|i,k})=0 \quad \ldots(8)$$

Now, multiplying equation (8) by $e^i_{p|}e^R_{q|}$, we get

$$e_{n|k}(e_{n|i,j}-e_{n|j,i})e^i_{p|}e^k_{q|}+e_{n|i}(e_{n|j,k}-e_{n|k,j})e^i_{p|}e^k_{q}+e_{n|j}(e_{n|k,i}-e_{n|i,k})e^i_{p|}e^k_{q|}$$

where p and q are two new indices chosen from $1, 2, \ldots, n-1$. *i.e.*, p,q and n are unequal.

But

$$e_{n|k}e^k_{q|} = \delta^q_h = 0,\ q\neq n$$

$$e_{n|i}e^k_{p|} = \delta^p_h = 0,\ p\neq n$$

so, we have

$$e_{n|j}(e_{n|k,i}-e_{n|j,k})e^i_{p|}e^k_{q|}=0$$

$$e_{n|j}(e_{n|k,i}e^i_{q|}e^k_{p|}-e_{n|j,k}e^i_{p|}e^k_{q|})=0$$

$$e_{n|j}(\gamma_{nqp}-\gamma_{npq})=0$$

Since

$$e_{n|k,i}e^i_{p|}e^k_{q|}=\gamma_{npq}$$

$$\Rightarrow \qquad \gamma_{nqp} - \gamma_{npq} = 0, \quad e_{n|j} \neq 0$$

$$\Rightarrow \qquad \gamma_{npq} = \gamma_{nqp}, \qquad \dots (9)$$

Conversely if equation (9) in true then we get equation (8). Which implies that equation (7) are satisfied bty $e_{n|i}$ Hence $e_{n|}$ is a normal congruence.

Thus necessary and sufficient conditions that the congruence $e_{n|}$ of an orthogonal ennuple be a normal congruence are that

$$\gamma_{nqp} = \gamma_{npq} \quad (p, q = 1, 2, \dots, n-1 \text{ such that } p \neq q)$$

THEOREM 11.5 *Necessary and sufficient conditions that all the congruences of an orthogonal ennuple be normal.*

Proof: If all the congruences of a orthogonal ennuple are normal. Then

$$\gamma_{nqp} = \gamma_{npq} \ (p, q = 1, 2, \dots, n-1 \text{ such that } p \neq q)$$

If the indices h, k, l and unequal then

$$\gamma_{hkl} = \gamma_{hlk} \qquad \dots (1)$$

But due to skew-symmetric property *i.e.*, $\gamma_{hlk} = -\gamma_{lhk}$.

So,

$$\gamma_{hkl} = \gamma_{hlk} = -\gamma_{lhk} = -\gamma_{lkh}, \quad \text{from (1)}$$

$$= \gamma_{klh} \quad \text{(skew-symmetric property).}$$

$$= \gamma_{khl}, \quad \text{from (1)}$$

$$\gamma_{hkl} = -\gamma_{hkl}, \text{ (skew-symmetric property).}$$

$$\gamma_{hkl} + \gamma_{hkl} = 0$$

$$\Rightarrow \qquad 2\gamma_{hkl} = 0$$

$$\Rightarrow \qquad \gamma_{hkl} = 0$$

where ($l, h, k = 1, 2, \dots, n$ such that h, k, kl are unequal).

11.6 CURL OF CONGRUENCE

The curl of the unit tangent to a congruence of curves is called the curl of congruence.

If $e_{n|}$ is a given congruence of curves then

$$\text{Curl } e_{n|i} = e_{n|i,j} - e_{n|j,i}$$

If curl $e_{n|i} = 0$ then congruence is irrotational.

THEOREM 11.6 *If a congruence of curves satisfy two of the following conditions it will also satisfy the third*

(a) *that it be a satisfy the third*

(b) *that it be a geodesic congruence*

(c) *that it be irrotational.*

Proof: Consider an orthogonal ennuple and $e^i_{h|}$ ($h = 1, 2, \dots, n$) be n unit tangent to n congruences of this orthogonal ennnuple.

From theorem (11.2), we have

$$\sum_h \gamma_{lhk} e_{h|i} = e_{l|i,j} e^j_{k|}$$

Putting $l = n$ and $j = m$, we have

$$\sum_{h=1}^{n} \gamma_{nhk} e_{h|i} = e_{n|i,m} e_{k|}^{m}$$

Now, multiplying by $e_{k|j}$ and summing with respect to k from 1 to n, we have

$$\sum_{h,k=1}^{n} \gamma_{nhk} e_{h|i} e_{k|j} = e_{n|i,m} e_{k|}^{m} e_{k|j}$$

$$= e_{n|i,m} \delta_{j}^{m} \qquad \text{Since } e_{k|}^{m} e_{k|j} = \delta_{j}^{m}$$

$$\sum_{h,k=1}^{n} \gamma_{nhk} e_{h|i} e_{k|j} = e_{n|i,j} \qquad \dots (2)$$

By definition of curl of congruence, we have

$$\text{curl } e_{n|i} = e_{n|i,j} - e_{n|j,i}$$

$$= \sum_{h,k=1}^{n} \gamma_{nhk} e_{h|i} e_{k|j} - \sum_{n,k=1}^{n} \gamma_{nhk} e_{h|j} e_{k|i}; \quad \text{from (2)}$$

$$\text{curl } e_{n|i} = \sum_{h,k=1}^{n} \gamma_{nhk} e_{h|i} e_{k|j} - \sum_{h,k=1}^{n} \gamma_{nkh} e_{k|j} e_{h|i}$$

$$\text{curl } e_{n|i} = \sum_{h,k=1}^{n} (\gamma_{nhk} - \gamma_{nkh}) e_{h|i} e_{k|j} \qquad \dots (3)$$

This double sum may be separated into two sums as follows.

(*i*) Let h and k take the values 1, 2, ..., $n - 1$.

(*ii*) Either $h = n$ or $k = n$ or $h = k = n$.

Now, the equation (3) becomes

$$\text{curl } e_{n|i} = \sum_{h,k=1}^{n-1} (\gamma_{nhk} - \gamma_{nkh}) e_{h|i} e_{k|j} + \sum_{h=1}^{n-1} (\gamma_{nhn} - \gamma_{nnh}) e_{h|i} e_{n|j}$$

$$+ \sum_{k=1}^{n} (\gamma_{nnk} - \gamma_{nkn}) e_{n|i} e_{n|j} + (\gamma_{nnn} - \gamma_{nnn}) e_{n|i} e_{n|j}$$

Since we know that $\gamma_{nnk} = \gamma_{nnh} = \gamma_{nnn} = 0$.

So,

$$\text{curl } e_{n|i} = \sum_{h,k=1}^{n-1} (\gamma_{nhk} - \gamma_{nkh}) e_{h|i} e_{k|j} + \sum_{h=1}^{n-1} \gamma_{nhn} e_{h|i} e_{n|j} - \sum_{k=1}^{n-1} \gamma_{nkn} e_{n|i} e_{k|j}$$

$$\text{curl } e_{n|i} = \sum_{h,k}^{n-1} (\gamma_{nhk} - \gamma_{nkh}) e_{h|i} e_{k|j} + \sum_{h=1}^{n-1} \gamma_{nhn} (e_{h|i} e_{n|j} - e_{n|i} e_{k|j}) \qquad \dots (4)$$

The first term on R.H.S. of equation (4) vanishes

if $$\gamma_{nhk} - \gamma_{nkh} = 0$$

i.e.,

if $$\gamma_{nhk} = \gamma_{nkh}.$$

i.e., if the congruence $e_{n|}$ is normal.

Again the second term of R.H.S of equation (4) vanishes

if $$\gamma_{nhn} = 0 \quad i.e., \text{ if } \gamma_{hnn} = 0$$

i.e., if the congruence $e_{n|}$ is a geodesic congruence. Further, if first and second term on right hand side of equation (4) both vanishes then

$$\text{curl } e_{n|i} = 0$$

Hence we have proved that if the congruence $e_{n|}$ satisfies any two of the following conditions then it will also satisfy the third.

(*a*) $e_{n|}$ is a normal congruence

(*b*) $e_{n|}$ is irrotational

(*c*) $e_{n|}$ is a geodesic congruence

11.7 CANONICAL CONGRUENCE

It has been shown that given a congruence of curves, it is possible to choose, in a multiply infinite number of ways, $n - 1$ other congruences forming with the given congruence an orthogonal ennuple. Consider the system of $n - 1$ congruence discovered by Ricci, and known as the system canonical with respect to the given congruence.

THEOREM 11.7 *Necessary and sufficient conditions that the $n - 1$ congruences $e_{h|}$ of an orthogonal ennuple be canonical with respect to $e_{n|}$ are*

$$\gamma_{nhk} + \gamma_{nkh} = 0; \quad (h, k = 1, 2, \ldots, n - 1, h \neq k).$$

Proof: Let the given congruence $e_{n|}$ be regarded as n[th] of the required ennuple. Let $e_{n|i}$ be unit tangent to given congruence.

$$X_{ij} = \tfrac{1}{2}\,(e_{n|\,i,j} + e_{n|\,j,i}) \qquad \ldots (1)$$

Let us find a quantity ρ and n quantities e^i satisfying the $n + 1$ equations

$$\left.\begin{array}{r} e_{n|i}e^i = 0 \\ (X_{ij} - \omega g_{ij}e^i + \rho E_{n|j} = 0 \end{array}\right\} i, j = 1, 2, \ldots, n \qquad \ldots (2)$$

where ω is a scalar invariant.

Writing equation (2) in expansion form, we have

$$e_{n|1}e^1 + e_{n|2}e^2 + \ldots\ldots + e_{n|n}e^n = 0 \qquad \ldots (3)$$

and

$$(X_{1j} - \omega g_{1j})e^i + (X_{2j} - \omega g_{2j})e^2 + \ldots (X_{nj} - \omega g_{nj})e^n + \rho e_{n|j} = 0$$

for $j = 1, 2, \ldots, n$.

$$(X_{11}-\omega g_{11})\, e^1 + (X_{21}-\omega g_{21})\, e^2 + \ldots + (X_{n1}-\omega g_{n1})\, e^n + \rho e_{n|1} = 0$$

$$(X_{12}-\omega g_{12})\, e^1 + (X_{22}-\omega g_{22})\, e^2 + \ldots + (X_{n2}-\omega g_{n2}) e^n + \rho e_{n|2} = 0$$

$$\vdots$$

$$(X_{1n}-\omega g_{1n})\, e^1 + (X_{2n}-\omega g_{2n})\, e^2 + \ldots + (X_{nn}-\omega g_{nn})\, e^n + \rho e_{n|n} = 0$$

from (3)

$$e_{n|1}e^1 + e_{n|2}e^2 + \ldots e_{n|n}e^n + \rho.0 = 0$$

eliminating ρ and the quantities $e^1, e^2, \ldots e^n$, we have the equation

$$\begin{vmatrix} (X_{11}-\omega g_{11}) & (X_{21}-\omega g_{21}) & \ldots & (X_{n1}-\omega g_{n1}) & e_{n|1} \\ (X_{12}-\omega g_{12}) & (X_{22}-\omega g_{22}) & \ldots & (X_{n2}-\omega g_{n2}) & e_{n|2} \\ \vdots & & & & \\ (X_{1n}-\omega g_{1n}) & (X_{2n}-\omega g_{2n}) & \ldots & (X_{nn}-\omega g_{nn}) & e_{n|n} \\ e_{n|1} & e_{n|2} & \ldots & e_{n|n} & 0 \end{vmatrix}$$

which is of degree $n-1$ in ω. Hence there will be $n-1$ roots of ω and these roots be $\omega_1, \omega_2, \ldots, \omega_{n-1}$. All roots are real. Let ω_h be one of these roots and let the corresponding values of ρ and e^i be denoted by ρh and $e^i_{h|}$ respectively. Then ρ_h and $e^i_{h|}$ will satisfy the equation (2), we have

$$e_{n|i}e^i_{h|} = 0 \qquad \ldots (4)$$

and $$(X_{ij}-\omega g_{ij})\, e^i_{h|} + \rho_h e_{n|j} = 0 \qquad \ldots (5)$$

Similarly ω_k be another root of these roots, we have

$$e_{n|i}e^i_{k|} = 0 \qquad \ldots (6)$$

$$(X_{ij}-\omega g_{ij})\, e^i_{k|} + \rho_h e_{n|j} = 0 \qquad \ldots (7)$$

Multiplying (5) by $e^j_{k|}$ and (7) $e^j_{h|}$ and using (4) and (6), we get

$$(X_{ij}-\omega_h g_{ij})\, e^i_{h|}e^j_{k|} = 0 \qquad \ldots (8)$$

$$(X_{ij}-\omega_k g_{ij})\, e^i_{k|}e^j_{h|} = 0 \qquad \ldots (9)$$

Since X_{ij} and g_{ij} are symmetric tensor in i and j. Now, interchanging i and j in equation (9), we get

$$(X_{ij}-\omega_k g_{ij})\, e^i_{h|}e^j_{k|} = 0 \qquad \ldots (10)$$

Subtracting (10) and (8), we get

$$(\omega_h-\omega_k)\, g_{ij}e^i_{h|}e^j_{k|} = 0 \qquad \ldots (11)$$

Since $\omega_k - \omega_k \neq 0$ as $\omega_h \neq \omega_h$.

$$\Rightarrow \qquad g_{ij}e^i_{h|}e^j_{k|} = 0 \qquad \ldots (12)$$

This shows that $e^i_{h|}$ and $e^j_{k|}$ unit vectors are orthogonal to each other and hence the congruence $e_{h|}$ and $e_{k|}$ $(h \neq k)$ are orthogonal to each other. Hence the $n-1$ congruence $e_{h|}$ $(h = 1, 2, \ldots, n)$ thus determined form an orthogonal ennuple with $e_{n|}$.

Using equation (12), equation (10) becomes

$$X_{ij}\, e^i_{h|}\, e^j_{k|} = 0$$

Since from (1), $X_{ij} = \frac{1}{2}\left(e_{n|i,j} + e_{n|j,i}\right)$. Then we have,

$$\frac{1}{2}\left(e_{n|i,j} + e_{n|j,i} e^i_{h|} e^j_{k|}\right) = 0$$

$$e_{n|i,j} e^i_{h|} e^j_{k|} + e_{n|j,i} e^i_{k} e^j_{h} = 0$$

$$\gamma_{nhk} + \gamma_{nkh} = 0; \quad (h,\ k = 1, 2, \ldots, n-1 \text{ such that } h \neq k) \qquad \ldots (13)$$

Conversely if equation (13) is true then $(n-1)$ congruences $e_{h|}$ of the orthogonal ennuple and canonical with respect to $e_{n|}$. Hence necessary and sufficient condition that the $n-1$ congruences $e_{h|}$ of an orthogonal ennuple be canonical with respect to $e_{n|}$ are

$$\gamma_{nhk} + \gamma_{nkh} = 0; \quad (h,\ k = 1, 2, \ldots, n \text{ such that } h \neq k)$$

THEOREM 11.8 *Necessary and sufficient conditions that $n-1$ mutually orthogonal congruences $e_{h|}$ orthogonal to a normal congruence $e_{n|}$, be canconical with respect to the later are $\gamma_{nhk} = 0$ where $k, h = 1, 2, \ldots, n-1$ such that $h \neq k$.*

Proof: By theorem (11.7), Necessary and sufficient conditions that $(n-1)$ congruences $e_{h|}$ of an orthogonal ennuple be canonical with respect to $e_{n|}$ are

$$\gamma_{nhk} + \gamma_{nkh} = 0 \quad (h,\ k = 1, 2, \ldots, n-1 \text{ such that } h \neq k).$$

If the congruence $e_{n|}$ is normal. Then

$$\gamma_{nhk} = \gamma_{nkh}$$

The given condition $\gamma_{nhk} + \gamma_{nkh} = 0$ becomes

$$\gamma_{nhk} + \gamma_{nhk} = 0$$

$$2\gamma_{nhk} = 0$$

$$\gamma_{nhk} = 0$$

Proved.

EXAMPLE 1

If $e_{n|}$ are the congruences canonical with respect to $e_{n|}$ prove that (i) $\omega_h = \gamma_{nhk}$ (ii) $\rho_h = \frac{1}{2}\gamma_{hnn}$, (iii) If $e_{n|}$ is a geodesic congruence, the congruences canonical with respect to it are given by

$$(X_{ij} - \omega g_{ij})e^i = 0 = 0$$

Solution

Suppose $(n-1)$ congruences $e_{h|}$ of an orthogonal **ennuple** in a V_n are canonical with respect to the cougruence $e_{h|}$ then

$$e_{n|i} e^i_{h|} = 0 \qquad \ldots (1)$$

and $(X_{ij} - \omega_h g_{ij})\, e^i_{h|} + \rho_h e_{n|j} = 0$ ... (2)

where

$$X_{ij} = \frac{1}{2}\,(e_{n|\,i,j} + e_{n|\,j,i}) \qquad \dots (3)$$

Since $e^i_{n|}$ unit tangents then

$$g_{ij} e^i_{h|} e^j_{h|} = 0 \qquad \dots (4)$$

(*i*) Multiplying equation (2) by $e^i_{h|}$, we get

$$(X_{ij} - \omega_h g_{ij})\, e^i_{h|}\, e^j_{h|} + \rho_h\, e_{n|j}\, e^j_{h|} = 0$$

$$X_{ij} e^i_{h|} e^j_{h|} - \omega_h g_{ij} e^i_{h|} e^j_{h|} = 0 \quad \text{since } e_{n|j} e^j_{h|} = 0 \quad \text{(from (1))}$$

$$X_{ij} e^i_{h|} e^j_{h|} - \omega_h = 0, \quad \text{since } g_{ij} e^i_{h|} e^j_{h|} = 1$$

$$\frac{1}{2}(e_{n|i,j} + e_{n|j,i})\, e^i_{h|} e^j_{h|} - \omega_h = 0; \quad \text{from (3)}$$

$$\frac{1}{2}(e_{n|i,j} e^i_{h|} e^j_{h|} + e_{n|j,i} e^i_{h|} e^j_{h|}) - \omega_h = 0$$

$$\frac{1}{2}\,(\gamma_{nhh} + \gamma_{nhh}) - \omega_h = 0$$

$$\omega_h = \gamma_{nhh}$$

(*ii*) Multiplying (2) by $e^j_{n|}$, we get

$$(X_{ij} - \omega_h g_{ij})\, e^i_{h|}\, e^j_{n|} + \rho_h\, e_{n|j}\, e^j_{n|} = 0$$

$$X_{ij} e^i_{h|} e^j_{n|} - \omega_h g_{ij} e^i_{h|} e^j_{n|} + \rho_h \times 1 = 0 \quad \text{from (1)}$$

$$\frac{1}{2}(e_{n|i,j} + e_{n|j,i})\, e^i_{h|} e^j_{h|} + \rho_h = 0, \quad \text{since } e^i_{h|}\, e^j_{h|} = 0$$

$$\frac{1}{2}(e_{n|i,j}\, e^i_{h|}\, e^j_{n|} + e_{n|j,i}\, e^i_{h|}\, e^j_{h|}) + \rho_h = 0$$

$$\frac{1}{2}\,(\gamma_{nhn} + \gamma_{nnh}) + \rho_h = 0$$

$$\rho_h = -\frac{1}{2}\gamma_{nhn}, \text{ since } \gamma_{nnh} = 0$$

or

$$\rho_h = \frac{1}{2}\gamma_{hnn}$$

(*iii*) If $e_{n|}$ is a geodesic congruence, then $\gamma_{hnn} = 0$ from the result (*ii*), we have

$$\rho_h = \frac{1}{2} \times 0$$

$$\rho_h = 0$$

from equation (2), we get

$$(X_{ij} - \omega_h g_{ij})\, e^i_{h|} + 0 e_{n|j} = 0 \text{ or } \quad (X_{ij} - \omega_h g_{ij})\, e^i_{h|} = 0$$

this gives $(X_{ij} - \omega g_{ij})\, e^i = 0$.

EXAMPLE 2

Prove that when a manifold admits an orthogonal systyem of n normal congruences then any of these in canonical with respect to each other congruence of the system.

solution

Let $e^i_{h|}$ $(h = 1, 2, \ldots, n)$ be unit tangents to n normal congruences of an orthogonal ennuple in a V_n. So that

$$\gamma_{lhk} = 0, \text{ where } l, h, k = 1, 2, \ldots n \text{ such that } l, h, k \text{ being unequal.}$$

It is required to show that a congruence $e_{h|}$ is canonical with respect to the congruence $e_{k|}$ $(h, k = 1, 2, \ldots, h \neq k)$. We know that the $n - 1$ congruence $e_{n|}$ of an orthogonal ennuple be canonical to $e_{n|}$ iff

$$\gamma_{nhk} + \gamma_{nkh} = 0$$

This condition is satisfied by virtue of equation (1). Hence $(n - 1)$ congruences $e_{h|}$ of an orthogonal ennuple are canonical to $e_{n|}$.

Similarly we can show that any $n - 1$ congruences are canonical to the remaining congruence.

It follows from the above results that any one congruence is canonical with respect to each other congruence of the system.

EXERCISE

1. If ϕ is a scalar invariant

$$\sum_{h=1}^{n} \phi_{,ij}\, e^i_{h|} e^j_{h|} = \Delta^2 \phi$$

2. The coefficient of ρ^{n-1} in the expansion of the determinant $|\phi_{,ij} - \rho g_{ij}|$ is equal to $\nabla^2 \phi$.
3. If $e_{h|}$ are the unit tangents to n mutually orthogonal normal congruences and $e_{1|} + b e_{2|}$ is also a normal congruence then $a e_{1|} - b e_{2|}$ is a normal congruence.
4. Show that

$$\frac{\partial}{\partial s_h}\frac{\partial \phi}{\partial s_k} - \frac{\partial}{\partial s_k}\frac{\partial Q}{\partial s_h} = \sum_l (\gamma_{lhk} - \gamma_{lkh}) \frac{\partial Q}{\partial s_l}$$

where S_h denotes the arc length of a curve through a point P of an ennuple and Q is a scalar invariant.

5. If the congruence $e_{h|}$ $(h = 1, 2, \ldots n - 1)$ of an orthogonal ennuple are normal, prove that they are canonical with respect to other congruence $e_{h|}$.

CHAPTER – 12

HYPERSURFACES

12.1 INTRODUCTION

We have already studied (Art 10.3 chapter 10) that if $m > n$ then we call V_n to be a subspace of V_m and consequently V_m is called enveloping space of V_n. We also know that there are $m - n$ linearly independent normals N^α to V_n (Art 10.3 Theorem 10.5, chapter 10). If we take $m = n + 1$ then V_n is said to be hypersurface of the enveloping space V_{n+1}

Let V_n be Riemannian space of n dimensions referred to cordinates x^i and having the metric $ds^2 = g_{ij}\, dx^i\, dx^j$. Let V_n be Riemannian space of m dimensions referred to coordinates y^α and having the metric $ds^2 = a_{\alpha\beta}\, dy^\alpha\, dy^\beta$. Where $m > n$. Let Greek letters α, β, γ ... take the values 1, 2, ..., m and latin indices i, j, k, ... take the values 1, 2, ...n. Then we have, the relation between $a_{\alpha\beta}$ and g_{ij}

$$g_{ij} = a_{\alpha\beta} \frac{\partial y^\alpha}{\partial x^i} \frac{\partial y^\beta}{\partial x^j} \qquad \dots (1)$$

Since the function y^α are invariants for transformations of the coordinates x^i in V_n, their first covariant derivatives with respect to the metric of V_n are the same as their ordinary derivatives with respect to the variables x^i.

i.e., $$y^\alpha_{,i} = \frac{\partial y^\alpha}{\partial x^i}$$

Then equation (1) can be written as

$$g_{ij} = a_{\alpha\beta}\, y^\alpha_{,i}\, y^\beta_{,j} \qquad \dots (2)$$

The vector of V_{n+i} whose contravariant components are $y^\alpha_{,i}$ is tangential to the curve of parameter x^i in V_n. Consequently if N^α are the contravariant components of the unit vector normal to V_n. Then we have

$$a_{\alpha\beta} N^\beta\, y^\alpha_{,i} = 0, \quad (i = 1, 2, \dots, n) \qquad \dots (3)$$

and $$a_{\alpha\beta} N^\alpha N^\beta = 1 \qquad \dots (4)$$

12.2 GENERALISED COVARIANT DIFFERENTIATION

Let C be any curve in V_n and s its arc length. The along this curve the x's and the y's may be expressed as function of s only. Let u_α and v^β be the components in the y's of two unit vector fields which are parallel along C with respect to V_m. Similarly w^i the components in x's of a unit vector field which is parallel along C with respect to V_n. Now, u_α is parallel along C relative to V_m then we have,

$$u_{\alpha,\beta}\frac{dy^\beta}{ds} = 0$$

$$\left[\frac{\partial u_\alpha}{\partial y^\beta} - u_\gamma \begin{Bmatrix} \gamma \\ \alpha\beta \end{Bmatrix}\right]\frac{dy^\beta}{ds} = 0$$

$$\frac{\partial u^\alpha}{\partial y^\beta}\frac{dy^\beta}{ds} - u_\gamma \begin{Bmatrix} \gamma \\ \alpha\beta \end{Bmatrix}\frac{dy^\beta}{ds} = 0$$

$$\frac{du_\alpha}{ds} - u_\gamma \begin{Bmatrix} \gamma \\ \alpha\beta \end{Bmatrix}\frac{dy^\beta}{ds} = 0$$

$$\frac{du_\alpha}{ds} = u_\gamma \begin{Bmatrix} \alpha \\ \beta\gamma \end{Bmatrix}\frac{dy^\beta}{ds} \quad \ldots (1)$$

similarly v^β is parallel along C relative to V_m then

$$\frac{dv^\beta}{ds} = -v^\gamma \begin{Bmatrix} \gamma \\ \alpha\beta \end{Bmatrix}\frac{dy^\alpha}{ds} \quad \ldots (2)$$

and w^i is parallel along C relative to V_n then

$$\frac{dw^i}{ds} = -w^j \begin{Bmatrix} i \\ jk \end{Bmatrix}\frac{dx^k}{ds} \quad \ldots (3)$$

The Christoffel symbol with Greek indices being formed with respect to the $a_{\alpha\beta}$ and the y's and christoffel symbol with Latin indices with respect to the g_{ij} and the x's.

Let $A^\alpha_{\beta i}$ be a tensor field, defined along C, which is mixed tenser of the second order in the y's and a covariant vector in the x's. Then the product $u_\alpha\, v^\beta\, w^i\, A^\alpha_{\beta i}$ is scalar invariant and it is a function of s along c. Its derivative with respect to s is also a scalar invariant.

Differentiating $u_\alpha\, v^\beta\, w^i\, A^\alpha_{\beta i}$ with respect to s, we have

$$\frac{d}{dt}(u_\alpha v^\beta w^i A^\alpha_{\beta i}) = u_\alpha v^\beta w^i \frac{dA^\alpha_{\beta i}}{ds} + \frac{du_\alpha}{ds} v^\beta w^i A^\alpha_{\beta i} + \frac{dv^\beta}{ds} u_\alpha w^i A^\alpha_{\beta i} + \frac{dw^i}{ds} u_\alpha v^\beta A^\alpha_{\beta i}$$

$$= u_\alpha v^\beta w^i \frac{dA^\alpha_{\beta i}}{ds} + \frac{du_\delta}{ds} v^\beta w^i A^\delta_{\beta i} + \frac{dv^\delta}{ds} u_\alpha w^i A^\alpha_{\delta i} + \frac{dw^j}{ds} u_\alpha v^\beta A^\alpha_{\beta j}$$

$$= u_\alpha v^\beta w^i \frac{dA^\alpha_{\beta i}}{ds} + A^\delta_{\beta i}.u_\alpha v^\beta w^i \begin{Bmatrix} \alpha \\ \delta\gamma \end{Bmatrix} \frac{dy^\gamma}{ds} - A^\alpha_{\delta i}.u_\alpha v^\beta w^i \begin{Bmatrix} \delta \\ \beta\gamma \end{Bmatrix} \frac{dy^\gamma}{ds}$$

$$- A^\alpha_{\beta j}\, u_\alpha\, v^\beta\, w^i \begin{Bmatrix} j \\ ik \end{Bmatrix} \frac{dx^k}{ds}, \text{ by equation (1), (2) \& (3).}$$

$$= u_\alpha v^\beta w^i \left[\frac{dA^\alpha_{\beta i}}{ds} + A^\delta_{\beta i}. \begin{Bmatrix} \alpha \\ \delta\gamma \end{Bmatrix} \frac{dy^\gamma}{ds} - A^\alpha_{\delta i} \begin{Bmatrix} \delta \\ \beta\gamma \end{Bmatrix} \frac{dy^\gamma}{ds} - A^\alpha_{\beta j}. \begin{Bmatrix} j \\ ik \end{Bmatrix} \frac{dx^k}{ds} \right]$$

$$= \text{Scalar, along } C$$

Since the outer product $u_\alpha v^\beta w^i$ is a tensor and hence from quotient law that the expression within the bracket is tenser of the type $A^\alpha_{\beta i}$ and this tensor is called *intrinsic derivative of* $A^\alpha_{\beta i}$ with respect to s.

The expression within the bracket can also be expressed as

$$\frac{dx^k}{ds} \left[\frac{\partial A^\alpha_{\beta i}}{\partial x^k} + A^\delta_{\beta i} \begin{Bmatrix} \alpha \\ \delta\gamma \end{Bmatrix} y^\gamma_{,k} - A^\alpha_{\delta i} \begin{Bmatrix} \delta \\ \beta\gamma \end{Bmatrix} y^\gamma_{,k} - A^\alpha_{\beta j} \begin{Bmatrix} j \\ ik \end{Bmatrix} \right]$$

Since $\frac{dx^k}{ds}$ is arbitrary.

So, by Quotient law the expression within the bracket is a tensor and is called tensor derivative of $A^\alpha_{\beta i}$ with respect to x^k. It is denoted by $A^\alpha_{\beta i;k}$. Then we have

$$A^\alpha_{\beta i;k} = \frac{\partial A^\alpha_{\beta i}}{dx^k} + A^\delta_{\beta i} \begin{Bmatrix} \alpha \\ \delta\gamma \end{Bmatrix} y^\gamma_{,k} - A^\alpha_{\delta i} \begin{Bmatrix} \delta \\ \beta\gamma \end{Bmatrix} y^\gamma_{,k} - A^\alpha_{\beta j} \begin{Bmatrix} j \\ ik \end{Bmatrix}.$$

$A^\alpha_{\beta i;k}$ is also defined as generalised covariant derivative $A^\alpha_{\beta i}$ with respect to x^k.

Note:– Semi-colon(;) is used to denote tensor differentiation.

12.3 LAWS OF TENSOR DIFFERENTIATION

THEOREM 12.1 *Tensor differentiation of sums and products obeys the ordinary rules of differentiation.*

Proof: Suppose $A^\alpha_\beta, B^\alpha_\beta$ and B_γ are tensors in V_m.

(*i*) To prove that

$$(A^\alpha_\beta + B^\alpha_\beta)_{;k} = A^\alpha_{\beta;k} + B^\alpha_{\beta;k}$$

Let the sum $A^\alpha_\beta + B^\alpha_\beta$ be denoted by the tensor C^α_β.

Now,

$$C^\alpha_{\beta;k} = \frac{\partial C^\alpha_\beta}{\partial x^k} + C^a_\beta \begin{Bmatrix} \alpha \\ a\ c \end{Bmatrix} y^c_{,k} - C^\alpha_a \begin{Bmatrix} \alpha \\ \beta c \end{Bmatrix} y^c_{,k}$$

$$\therefore \quad (A^\alpha_\beta + B^\alpha_\beta)_{;k} = \frac{\partial(A^\alpha_\beta + B^\alpha_\beta)}{dx^k} + (A^a_\beta + A^\alpha_\beta)\begin{Bmatrix}\alpha\\ a\ c\end{Bmatrix} y^c_{,k} - (A^\alpha_a + B^\alpha_a)\begin{Bmatrix}a\\ \beta\ c\end{Bmatrix} y^c_{,k}$$

$$= \left[\frac{\partial A^\alpha_\beta}{\partial x^k} + A^a_\beta\begin{Bmatrix}\alpha\\ ac\end{Bmatrix} y^c_{,k} - A^\alpha_a\begin{Bmatrix}a\\ \beta c\end{Bmatrix} y^c_{,k}\right] +$$

$$\left[\frac{\partial B^\alpha_\beta}{\partial x^k} + B^a_\beta\begin{Bmatrix}\alpha\\ ac\end{Bmatrix} y^c_{,k} - B^\alpha_a\begin{Bmatrix}a\\ \beta c\end{Bmatrix} y^c_{,k}\right]$$

$(A^\alpha_\beta + B^\alpha_\beta)_{;k} = A^a_{\beta;\alpha} + B^a_{\beta;k}$ Hence the result (*i*)

(*ii*) Prove that

$$(A^\alpha_\beta + B_\gamma);\, k = A^\alpha_{\beta;k} B_\gamma + A^\alpha_\beta B_{\gamma;k}$$

Let $\quad A^\alpha_\beta B_\gamma = D^\alpha_{\beta\gamma} \quad$ Then $D^\alpha_{\beta\gamma}$ is a tensor

we have

$$D^\alpha_{\beta\gamma;k} = \frac{\partial D^\alpha_{\beta\gamma}}{\partial x^k} + D^a_{;k}\begin{Bmatrix}\alpha\\ ac\end{Bmatrix} y^c - D^\alpha_{a\gamma}\begin{Bmatrix}a\\ \beta c\end{Bmatrix} y^c_{,k} - D^\alpha_{\beta a}\begin{Bmatrix}a\\ \gamma c\end{Bmatrix} y^c_{,k}$$

$$\therefore \quad (A^\alpha_\beta B_\gamma)_{;k} = \frac{\partial(A^\alpha_\beta B_\gamma)}{dx^k} + A^a_{\beta\gamma} B_\gamma\begin{Bmatrix}\alpha\\ ac\end{Bmatrix} y^c_{,k} - A^\alpha_a B_\gamma\begin{Bmatrix}a\\ \beta c\end{Bmatrix} y^c_{,k} - A^\alpha_\beta B_a\begin{Bmatrix}a\\ \beta c\end{Bmatrix} y^c_{,k}$$

$$= \left[\frac{\partial A^\alpha_\beta}{dx^k} + A^a_\beta\begin{Bmatrix}\alpha\\ ac\end{Bmatrix} y^c_{,k} - A^\alpha_a\begin{Bmatrix}a\\ \beta c\end{Bmatrix} y^c_{,k}\right] B_\gamma + A^\alpha_\beta\left[\frac{\partial B_\gamma}{dx^k} - B_a\begin{Bmatrix}a\\ \gamma c\end{Bmatrix} y^c_{,k}\right]$$

$$(A^\alpha_\beta B_\gamma)_{;k} = A^\alpha_{\beta;k} B_\gamma + A^\alpha_a B_{\gamma\,;\,k}$$

Hence the result (*ii*).

Note:– α, β, γ, a, c take values from 1 to m while k take values from 1 to n.

THEOREM 12.2 *To show that* $a_{\alpha\beta;j} = 0$

or

To prove that the metric tensor of the enveloping space is generalised covariant constant with respect to the Christoffel symbol of the subspace.

Proof: We have (see pg. 220)

$$a_{\alpha\beta;i} = \frac{\partial a_{\alpha\beta}}{dx^i} - a_{\gamma\beta}\begin{Bmatrix}\gamma\\ \alpha\beta\end{Bmatrix} y^\delta_{,i} - a_{\alpha\gamma}\begin{Bmatrix}\gamma\\ \beta\delta\end{Bmatrix} y^\delta_{,i}$$

$$= \frac{\partial a_{\alpha\beta}}{dx^i} - \left([\alpha\delta, \beta] + [\beta\delta, \alpha]\right) y^\delta_{,i}$$

or $$a_{\alpha\beta;i} = \frac{\partial a_{\alpha\beta}}{dx^i} - \frac{\partial a_{\alpha\beta}}{\partial y^\delta} y^\delta_{,i}$$

or $$a_{\alpha\beta;i} = \frac{\partial a_{\alpha\beta}}{dx^i} - \frac{\partial a_{\alpha\beta}}{\partial y^\delta} \cdot \frac{\partial y_\delta}{dx^i} = \frac{\partial a_{\alpha\beta}}{\partial x^i} - \frac{\partial a_{\alpha\beta}}{\partial x^i} = 0$$ **Proved.**

12.4 GAUSS'S FORMULA

At a point of a hypersurface V_n of a Riemannian space V_{n+1}, the formula of Gauss are given by

$$y^\alpha_{;ij} = \Omega_{ij} N^\alpha$$

Proof: Since y^α is an invariant for transformation of the $x's$ and its tensor derivative is the same as its covariant derivative with respect to the x's, so that

$$y^\alpha_{;i} = y^\alpha_{,i} = \frac{dy^\alpha}{dx^i} \qquad \dots (1)$$

Again tenser derivative of equation (1) with respect to x's is

$$y^\alpha_{;ij} = (y^\alpha_{;i})_{;j} = (y^\alpha_{,i})$$

$$= \frac{\partial}{dx^j}(y^\alpha_{,i}) - y^\alpha_{,l}\begin{Bmatrix} l \\ ij \end{Bmatrix} + y^\beta_{,i} y^\gamma_{,j}\begin{Bmatrix} \alpha \\ \beta\gamma \end{Bmatrix}$$

$$= \frac{\partial}{\partial x^j}\left(\frac{\partial y^\alpha}{\partial x^i}\right) - y^\alpha_{,l}\begin{Bmatrix} l \\ ij \end{Bmatrix} + y^\beta_{,i} y^\gamma_{,j}\begin{Bmatrix} \alpha \\ \beta\gamma \end{Bmatrix}$$

$$y^\alpha_{;ij} = \frac{\partial^2 y^\alpha}{\partial x^i \partial x^j} - y^\alpha_{,l}\begin{Bmatrix} l \\ ij \end{Bmatrix} + y^\beta_{,i} y^\gamma_{,j}\begin{Bmatrix} \alpha \\ \beta\gamma \end{Bmatrix} \qquad \dots (2)$$

Interchanging j and i in (2), we have

$$y^\alpha_{;ji} = \frac{\partial^2 y^\alpha}{\partial x^j \partial x^i} - y^\alpha_{,l}\begin{Bmatrix} l \\ ji \end{Bmatrix} + y^\beta_{,j} y^\gamma_{,i}\begin{Bmatrix} \alpha \\ \beta\gamma \end{Bmatrix}$$

$$y^\alpha_{;ji} = \frac{\partial^2 y^\alpha}{\partial x^i \partial x^j} - y^\alpha_{,l}\begin{Bmatrix} l \\ ij \end{Bmatrix} + y^\beta_{,l} y^\gamma_{,j}\begin{Bmatrix} \alpha \\ \beta\gamma \end{Bmatrix} \qquad \dots (3)$$

[On interchanging β and γ in third term of R.H.S. of (3) and using $\begin{Bmatrix} \alpha \\ \beta\gamma \end{Bmatrix} = \begin{Bmatrix} \alpha \\ \gamma\beta \end{Bmatrix}$].

On comparing equation (2) and (3), we have

$$y^\alpha_{;ij} = y^\alpha_{;ji}$$

So, $y^\alpha_{;ij}$ is symmetrical with respect to indices i & j.

Let $g_{ij}dx^i dx^j$ and $a_{\alpha\beta}dy^\alpha dy^\beta$ be fundamental forms corresponding to V_n and V_{n+1} respectively. Then

$$g_{ij} = a_{\alpha\beta}\frac{\partial y^\alpha}{\partial x^i}\frac{\partial y^\beta}{\partial x^j}$$

or $$g_{ij} = a_{\alpha\beta}\, y^\alpha_{;i} y^\beta_{;j}$$

Taking tensor derivative of both sides with respect to x^k

$$g_{ij;k} = a_{\alpha\beta;k} y^\alpha_{;i} y^\beta_{;j} + a_{\alpha\beta} y^\alpha_{;i} y^\beta_{;jk} + a_{\alpha\beta} y^\alpha_{;ik} y^\beta_{;j}$$

But $$g_{ij;k} = 0 \quad \text{and} \quad a_{\alpha\beta;k} = 0.$$

we have,

$$0 = a_{\alpha\beta}\, y^\alpha_{;i} y^\beta_{;jk} + a_{\alpha\beta} y^\alpha_{;ik} y^\beta_{;j}$$

or $$a_{\alpha\beta}\, y^\alpha_{;ik} y^\beta_{,j} + a_{\alpha\beta} y^\alpha_{,i} y^\beta_{;jk} = 0, \text{ using (1), Art. 12.4} \qquad \dots (4)$$

By cyclic permutation on i, j, k in (4), we have

$$a_{\alpha\beta}\, y^\alpha_{;ji} y^\beta_{,k} + a_{\alpha\beta} y^\alpha_{,j} y^\beta_{;ki} = 0 \qquad \dots (5)$$

and $$a_{\alpha\beta}\, y^\alpha_{;kj} y^\beta_{,i} + a_{\alpha\beta} y^\alpha_{,k} y^\beta_{;ij} = 0 \qquad \dots (6)$$

subtracting equation (4) from the sum of (5) and (6), we get.

$$2a_{\alpha\beta}\, y^\alpha_{;ij} y^\beta_{,k} = 0$$

or $$a_{\alpha\beta}\, y^\alpha_{;ij} y^\beta_{,k} = 0 \qquad \dots (7)$$

This shows that $y^\alpha_{;ij}$ is normal (orthogonal) to $y^\beta_{,k}$. Since $y^\beta_{,k}$ is tangential to V_n and hence $y^\alpha_{;ij}$ is normal to V_n. Then we can write

$$y^\alpha_{;ij} = N^\alpha \Omega_{ij} \qquad (8)$$

where N^α is unit vector normal to V_n and Ω_{ij} is a symmetric covariant tensor of rank two. Since $y^\alpha_{;ij}$ is a function of $x's$ the tenser Ω_{ij} is also a function of $x's$.

The equation (8) are called Gauss's formula.

From equation (8)

$$y^\alpha_{;ij} = \Omega_{ij} N^\alpha$$

$$y^\alpha_{;ij}\, a_{\alpha\beta} N^\beta = \Omega_{ij}\, a_{\alpha\beta}\, N^\beta\, N^\alpha$$

$$= \Omega_{ij}\, N^2$$

$$y^{\alpha}_{;ij}\, a_{\alpha\beta}\, N^{\beta} = \Omega_{ij}, \quad N = 1$$

or $$\Omega_{ij} = y^{\alpha}_{;ij} a_{\alpha\beta} N^{\beta}.$$

The quadratic differential form

$$\Omega_{ij} dx^i dx^j$$

is called the *second fundamental form* for the hypersurface V_n of V_{n+1}. The components of tenser Ω_{ij} are said to be *coefficient of second fundamental* form.

Note: The quadratic differential form $g_{ij}dx^i dx^j$ is called *first fundamental form.*

12.5 CURVATURE OF A CURVE IN A HYPERSURFACE AND NORMAL CURVATURE

If U^{α} and u^i be the contravariant components of the vector u relative to V_n and V_{n+1} respectively then we have (from chapter 10, Theorem 10.4)

$$U^{\alpha} = \frac{\partial y^{\alpha}}{\partial x^i} u^i = y^{\alpha}_{,i} u^i \qquad \text{... (1)}$$

Let the derived vector to vector u along C with respect to metric of V_n *and* V_{n+1} are denoted by p and q respectively. Then

$$p^i = u^i_{,j} \frac{dx^j}{ds}$$

and $$q^{\alpha} = U^{\alpha}_{,\beta} \frac{dy^{\beta}}{ds}$$

$$= U^{\alpha}_{;j} \frac{dx^j}{ds}, \text{ (from equation 1, Art 12.4)} \qquad \text{... (2)}$$

Taking the tensor derivative of each side of equation (1) with respect to $x's$, we have

$$U^{\alpha}_{;j} = y^{\alpha}_{;ij}\, u^i + y^{\alpha}_{,i}\, u^i_{,j}$$

By Gauss's formula, we have $y^{\alpha}_{;ij} = \Omega_{ij} N^{\alpha}$

Then $$U^{\alpha}_{;j} = \Omega_{ij}\, N^{\alpha}\, u^i + y^{\alpha}_{,i} u^i_{,j}$$

Putting the value of $u^{\alpha}_{;j}$ in equation (2), we have

$$q^{\alpha} = (\Omega_{ij} N^{\alpha} u^i + y^{\alpha}_{,i} u^i_{,j}) \frac{dx^j}{ds}$$

or $$q^{\alpha} = \left(\Omega_{ij}\, u^i \frac{dx^j}{ds}\right) N^{\alpha} + y^{\alpha}_{,i}\, p^i \qquad \text{... (3)}$$

where $$p^i = u^i_{,j} \frac{dx^j}{ds}$$

Now suppose that vector u is a the unit tangent t to the curve C. Then the derived vectors q and p are the curvature vectors of a C relatively to V_{n+1} and V_n respectively. Then equation (3) becomes.

$$q^\alpha = \left(\Omega_{ij} \frac{dx^i}{ds} \frac{dx^j}{ds}\right) N^\alpha + y^\alpha_{,i}\, p^i \quad \text{... (4)}$$

Taking $$K_n = \Omega_{ij} \frac{dx^i}{ds} \frac{dx^j}{ds}, \quad \text{we get}$$

$$q^\alpha = K_n N^\alpha + y^\alpha_{,i}\, p^i \quad \text{...(5)}$$

K_n is called Normal curvature of V_n at any point P of the curve C and $K_n N^\alpha$ is called normal curvature vector of V_{n+1} in the direction of C.

Meaunier's Theorem

If K_a and K_n are the first curvature of C relative to V_{n+1} and normal curvature of V_n respectively and w is the angle between $\vec{N}$ and $\vec{C}$ (C being the unit vector of V_{n+1} then the relation between K_a, K_n and $\overline{w}$ is given by

$$K_n = K_a \cos\overline{\omega}$$

Proof: We know that

$$q^\alpha = K_n N^\alpha + p^i y^\alpha_{,i} \quad \text{...(1)}$$

where $$K_n = \Omega_{ij} \frac{dx^i}{ds} \frac{dx^j}{ds}$$

Let K_a and K_g be the first curvatures of C with respect V_{n+1} and V_n respectively then

$$K_a = \sqrt{a_{\alpha\beta} q^\alpha q^\beta}, \; K_g = \sqrt{g_{ij} p^i p^j}$$

Let $\overline{\omega}$ be the angle between $\vec{N}$ and $\vec{C}$.

Then $$\vec{N}\cdot\vec{C} = |N|\cdot|C|\cos\overline{\omega}$$

$$= \cos\overline{\omega} \quad \text{as } |N| = |C| = 1$$

$$\vec{N}.\vec{C} = \cos\overline{\omega} \quad \text{...(2)}$$

If $\vec{b}$ is a unit vector of V_{n+1} in the direction principal normal of $\vec{C}$ with respect to V_n then equation (1) becomes

$$K_a\vec{C} = K_g\vec{b} + K_n\vec{N} \quad \text{... (3)}$$

Taking scalar product of equation (3) with $\vec{N}$, we have

$$K_a \vec{N}.\vec{C} = K_g \vec{N}.\vec{b} + K_n \vec{N}.\vec{N}$$

$$K_a \cos\overline{\omega} = K_g \cdot 0 + K_n \cdot 1; \quad \text{from (2)}$$

$$K_n = K_a \cos\overline{\omega} \quad \text{... (4)}$$

Proved.

EXAMPLE 1

Show that the normal curvature is the difference of squares of geodesic curvatures.

Solution

We know that (from Meurier's Theorem, equation 3)

$$K_a\vec{C} = K_g\vec{b} + K_n\vec{N}$$

Taking modulus of both sides, we get

$$K_a^2 = K_g^2 + K_n^2$$

or $$K_n^2 = K_a^2 - K_g^2.$$

Theorem 12.3 *To show that the first curvature in V_{n+1} of a geodesic of the hypersurface V_n is the normal curvature of the hypersurface in the direction of the geodesic.*

Proof: From, example 1, we have

$$K_a^2 = K_g^2 + K_n^2$$

If C is a geodesic of V_n then $p^i = 0 \Rightarrow K_g = 0$

Then we have

Proved.

Dupin's Theorem

The sum of normal curvatures of a hypersurface V_n for n mutually orthogonal directions is an invariant and equal to $\Omega_{ij}g^{ij}$.

Proof: Let $e^i_{h|}$ $(h = 1, 2, \ldots, n)$ be unit tangents to n congruences of an orthogonal ennuple in a V_n. Let K_{nh} be normal curvature of the hypersurface V_n in the direction of the congruence $e_{h|}$. Then

$$K_{nh} = \Omega_{ij}\, e^i_{h|}\, e^j_{h|}$$

The sum of normal curvatures for n mutually orthogonal directions of an orthogonal ennuple is a V_n is

$$\sum_{h=1}^{n} K_{nh} = \sum_{h=1}^{n} \Omega_{ij}\, e^i_{h|}\, e^j_{h|}$$

$$= \Omega_{ij} \sum_{h=1}^{n} e^i_{h|}\, e^j_{h|}$$

$$= \Omega_{ij}\, g^{ij}$$

= Scalar invariant **Proved.**

12.6 DEFINITIONS

(a) *First curvature (or mean curvature) of the hypersurface V_n at point P.*

It is defined as the sum of normal curvatures of a hypersurface V_n form mutually orthogonal directions at P and is denoted by M. Then

$$M = \Omega_{ij}\, g^{ij}$$

(b) *Minimal Hypersurface*

The hypersurface V_n is said to be minimal if $M = 0$

i.e., $$\Omega_{ij}\, g^{ij} = 0$$

(c) *Principle normal curvatures*

The maximum and minimum values of K_n are said to be the principle normal curvatures of V_n at P. Since these maximum and minimum values of K_n correspond to the principal directions of the symmetric tensor Ω_{ij}.

(d) *Principal directions of the hypersurface at a point P.*

The principal directions determined by the symmetric tensor Ω_{ij} at P are said to be principal directions of the hypersurface at P.

(e) Line of curvature in V_n

A line of curvature in a hypersurface V_n is a curve such that its direction at any point is a principal direction.

Hence we have n congruences of lines of curvature of a V_n.

THEOREM 12.4 *To show that the mean curvature of a hypersurface is equal to the negative of the divergence of the unit normal.*

or

To show that the first curvature of a hypersurface is equal to the negative of the divergence of the unit normal.

or

To show that the normal curvature of a hypersurface for any direction is the negative of the tendency of the unit normal in that direction.

Proof: Let N be the unit normal vector to the hypersurface V_n in y's and let N_α be its covariant components. Let t be the unit tangent vector to N congruences $e_{h|}$ $(h = 1, 2, ..., n)$ of an orthogonal ennuple in V_n and let $T^\alpha_{h|}$ be the contravariant components in V_{n+1} of t. Since t is orthogonal to N, therefore

$$T^\alpha_{h|}\, N_\alpha = 0 \qquad \ldots (1)$$

Taking covariant derivative of equation (1) with regard to y^β provides

$$T^\alpha_{h|,\beta}\, N_\alpha + T^\alpha_{h|}\, N_{\alpha,\beta} = 0 \qquad \ldots (2)$$

Multiplying equation (2) by T^β_h, we get

$$T^\alpha_{h|,\beta}\, T^\beta_{h|}\, N_\alpha + T^\alpha_{h|}\, T^\beta_{h|}\, N_{\alpha,\beta} = 0$$

$$(T^{\alpha}_{h|,\beta} T^{\beta}_{h|}) N_{\alpha} = - N_{\alpha,\beta}\, T^{\beta}_{h|}\, T^{\alpha}_{h|} \quad \ldots (3)$$

Now $T^{\alpha}_{h|,\beta} T^{\beta}_{h|}$ is the first curvature of the curve $e_{h|}$ and $N_{\alpha,\beta} \,.\, T^{\beta}_{h|} \,.\, T^{\alpha}_{h|}$ is the tendency of N_{α} is the direction of $e_{h|}$. Hence equation (3) implies that the normal component of the first curvature of the $e_{h|}$ relative to V_{n+1} or the normal curvature of V_n in the direction of the curve $e_{h|}$

$$= - \text{tendency of } N \text{ in the direction of the curve } e_{h|} \quad \ldots (4)$$

Taking summation of both of (3) and (4) for $h = 1, 2, \ldots, n$

we have

i.e., mean curvature or first curvature of a hypersurface = – divergence of the unit normal.

Corollary: To prove that

$$M = - \text{div}_{n+1} N$$

Proof: Since N is a vector of unit magnitude *i.e.*, constant magnitude, its tendency is zero. also by definition, the $\text{div}_{n+1} N$ and $\text{div}_n N$ differ only by the tendency of the vector N in its direction. But the tendency of N is zero hence it follows that

$$\text{div}_n N = - \text{div}_{n+1} N$$

Hence $$M = - \text{div}_n N = - \text{div}_{n+1} N.$$

12.7 EULER'S THEOREM

Statement

The normal curvature K_n of V_n for any direction of $\vec{a}$ in V_n is given by

$$K_n = \sum_{h=1}^{n} K_h \cos^2 \alpha_h$$

where K_h are the principal curvature and α_h are the angles between direction of $\vec{a}$ and the congruence $e_{h|}$.

Proof: The principal directions in V_n determined by the symmetric covariant tensor teser Ω_{ij} are given by

$$(\Omega_{ij} - K_h\, g_{ij})\, p^i_{h|} = 0 \quad \ldots (1)$$

where K_h are the roots of the equation

$$\left| \Omega_{ij} - K g_{ij} \right| = 0 \quad \ldots (2)$$

and $p^i_{h|}$ are the unit tangents to n congruences of lines of curvature.

The roots of the equation (2) are the maximum and minimum values of the quality K_n defined by

$$K_n = \frac{\Omega_{ij}\, p^i p^j}{g_{ij}\, p^i p^i} \quad \ldots (3)$$

Multiplying equation (1) by $p^i_{k|}$, $(K \neq h)$, we get

$$(\Omega_{ij} - K_n g_{ij})\, p^i_{h|}\, p^i_{k|} = 0$$

The principal directions satisfy the equation

$$\Omega_{ij}\, p^i_{h|}\, p^j_{k|} = 0 \quad (h \neq k) \qquad \ldots (4)$$

Let $e_{h|}$ be the unit tangents to the n congruences of lines of curvature. Then the principal curvatures are given by

$$K_h = \Omega_{ij}\, e^i_{h|}\, e^i_{h|} \qquad (h = 1, 2, \ldots, n) \qquad \ldots (5)$$

Any other unit vector $\vec{a}$ in V_n is expressible in the form

$$\vec{a} = \sum_h e_{h|} \cos\alpha_h$$

or

$$a^i = \sum_{h=1}^{n} e^i_{h|} \cos\alpha_h \qquad \ldots (6)$$

where

$$\cos\alpha_{h|} = \vec{a} \cdot \vec{e}_{h|} = a^i e_{h|i}$$

a^i being the contravariant components of $\vec{a}$ and α_h being the inclination of vector a to $e_{h|}$. The normal curvature of V_n for the direction of $\vec{a}$ is given by

$$K_n = \Omega_{ij} a^i a^j$$

from (6), we get

$$K_n = \Omega_{ij} \left(\sum_h e^i_{h|} \cos\alpha_h\right) \cdot \left(\sum_h e^j_{k|} \cos\alpha_h\right)$$

$$= \sum_{h,k=1}^{n} (\Omega_{ij} e^i_{h|} e^j_{k|}) \cos\alpha_h \cos\alpha_k$$

$$= \sum_{h=1}^{n} (\Omega_{ij} e^i_{h|} e^j_{h|}) \cos^2\alpha_h$$

$$K_n = \sum_{h=1}^{n} K_h \cos^2\alpha_h \qquad \ldots (7)$$

This is a *generalisation* of *Euler's Theorem.*

12.8 CONJUGATE DIRECTIONS AND ASYMPTOTIC DIRECTIONS IN A HYPERSURFACE

The directions of two vector, $\vec{a}$ and $\vec{b}$, at a point in V_n are said to be conjugate if

$$\Omega_{ij} a^i b^j = 0 \qquad \ldots (1)$$

and two congruences of curves in the hypersurface are said to be conjugate if the directions of the two curves through any point are conjugate.

A direction in V_n which is self-conjugate is said to be asymptotic and the curves whose direction are along asymptotic directions are called asymptotic lines.

Therefore the direction the vector $\vec{a}$ at a point of V_n be asymptotic if

$$\Omega_{ij} a^i a^j = 0 \qquad \dots (2)$$

The asymptotic lines at a point of a hypersurface satisfies the differential equation

$$\Omega_{ij} dx^i dx^j = 0 \qquad \dots (3)$$

THEOREM 12.5 *If a curve C in a hypersurface V_n has any two of the following properties it has the third*

(*i*) *it is a geodesic in the hypersurface V_n*

(*ii*) *it is a geodesic in the enveloping space V_{n+1}*

(*iii*) *it is an asymptotic line in the hypersurface V_n.*

Proof: Let C be a curve in the hypersurface V_n. The normal curvature K_n of the hypersurface V_n in the direction of C is given by

$$K_a^2 = K_g^2 + K_n^2 \qquad \dots (1)$$

where K_a and K_g are the first curvatures of C relative to enveloping space V_{n+1} and hypersurface V_n respectively.

Suppose C is a geodesic in the hypersurface V_n [*i.e.*, (*i*) holds] then $K_g = 0$.

If C is also a geodesic in the enveloping space V_{n+1} [*i.e.*, (*ii*) holds] then $K_a = 0$.

Now, using these values in equation (1), we have

$$K_n^2 = 0 \quad \Rightarrow K_n = 0$$

Implies that C is an asymptotic line is the hypersurface V_n *i.e.*, (*iii*) holds.

Hence we have proved that (*i*) and (*ii*) $\Rightarrow$ (*iii*)

Similarly we have proved that

(*ii*) and (*iii*) $\Rightarrow$ (*i*)

and (*i*) and (*iii*) $\Rightarrow$ (*ii*)

12.9 TENSOR DERIVATIVE OF THE UNIT NORMAL

The function y^α are invariants for transformations of the coordinates x^i in V_n their first covariant derivatives with respect to the metric of V_n are the same as their ordinary derivatives with respect to the variables x^i.

i.e.,
$$y^\alpha_{;i} = y^\alpha_{,i} = \frac{\partial y^\alpha}{\partial x^i} \qquad \dots (1)$$

The unit Normal N^α be the contravariant vector in the y's whose tensor derivative with respect to the x's is

$$N^\alpha_{;i} = \frac{\partial N^\alpha}{\partial x^i} + \begin{Bmatrix} \alpha \\ \beta\delta \end{Bmatrix} N^\beta y^\delta_{,i} \qquad \dots (2)$$

Since $$a_{\alpha\beta}N^{\beta}N^{\alpha} = 1 \qquad \ldots (3)$$

Tensor derivative of this equation with respect to x^i gives

$$a_{\alpha\beta}N^{\alpha}N^{\beta}_{;i} + a_{\alpha\beta}N^{\alpha}_{;i}N^{\beta} = 0$$

Interchanging α and β in Ist term, we get

or $$a_{\beta\alpha}N^{\beta}N^{\alpha}_{;i} + a_{\alpha\beta}N^{\alpha}_{;i}N^{\beta} = 0$$

$$a_{\alpha\beta}N^{\beta}N^{\alpha}_{;i} + a_{\alpha\beta}N^{\alpha}_{;i}N^{\beta} = 0$$

$$2a_{\alpha\beta}N^{\beta}N^{\alpha}_{;i} = 0$$

$$a_{\alpha\beta}N^{\beta}N^{\alpha}_{;i} = 0 \qquad \ldots (4)$$

which shows that $N^{\alpha}_{;i}$ is orthogonal to the normal and therefore tangential to the hypersurface. Thus $N^{\alpha}_{;i}$ can be expressed in terms to tangential vectors $y^{\alpha}_{,k}$ to V_n so that

$$N^{\alpha}_{;i} = A^{\alpha}_{i}y^{\alpha}_{,k} \qquad \ldots (5)$$

where A^{k}_{i} is a mixed tensor of second order in V_n to be determined.

Since unit normal N^{α} is orthogonal to tangential vector $y^{\alpha}_{,i}$ in V_n. Then

$$a_{\alpha\beta}N^{\alpha}y^{\beta}_{,i} = 0$$

Taking tensor derivative with respect to x^j, we get

$$a_{\alpha\beta}N^{\alpha}_{;j}y^{\beta}_{,i} + a_{\alpha\beta}N^{\alpha}y^{\beta}_{,ij} = 0 \quad \text{since } y^{\beta}_{,ij} = \Omega_{ij}N^{\beta}$$

or $$a_{\alpha\beta}N^{\alpha}_{;j}y^{\beta}_{,i} + a_{\alpha\beta}N^{\alpha}\Omega_{ij}N^{\beta} = 0$$

or $$a_{\alpha\beta}N^{\alpha}_{;j}y^{\beta}_{,i} + \Omega_{ij}(a_{\alpha\beta}N^{\alpha}N^{\beta}) = 0$$

from equation (3), $$a_{\alpha\beta}N^{\alpha}N^{\beta} = 1$$

or $$a_{\alpha\beta}N^{\alpha}_{;j}y^{\beta}_{,i} + \Omega_{ij} = 0$$

or $$g_{ki}A^{k}_{j} + \Omega_{ij} = 0$$

$$\left[\text{since } a_{\alpha\beta}\,y^{\alpha}_{,k}\,y^{\alpha}_{,i} = a_{\alpha\beta}\frac{\partial y^{\alpha}}{\partial x^{k}}\frac{\partial y^{\beta}}{\partial x^{i}} = g_{ki}\right]$$

Multiplying this equation by g^{im}, we get

$$g_{ki}\,g^{im}A^{k}_{j} + \Omega_{ij}\,g^{im} = 0$$

$$\delta_k^m A_j^k + \Omega_{ij} g^{im} = 0$$

$$A_i^m + \Omega_{ij} g^{im} = 0$$

$$A_i^m = -\Omega_{ij} g^{im}$$

Substituting this value in equation (5), we get

$$N_{;i}^{\alpha} = \Omega_{ij}\, g^{ik}\, y_{,k}^{\alpha} = -\Omega_{ji}\, g^{jk}\, y_{,k}^{\alpha}$$

$$N_{;i}^{\alpha} = -\Omega_{ij}\, g^{jk}\, y_{,k}^{\alpha} \qquad \ldots (6)$$

This is the required expression for the tensor derivative of N^{α}.

Theorem 12.6 *The derived vector of the unit normal with respect to the enveloping space, along a curve provided it be a line of curvature of the hypersurface.*

Proof: Since the tensor derivative of N^{α} is

$$N_{;i}^{\alpha} = -\Omega_{ij}\, g^{jk}\, y_{,k}^{\alpha} \qquad \ldots (1)$$

Consider a unit vector e^i tangential to the curve C.

Then
$$N_{;i}^{\alpha} e^i = -\Omega_{ij}\, g^{jk}\, y_{,k}^{\alpha} e^i \qquad \ldots (2)$$

The direction of $N_{;i}^{\alpha} e^i$ is identical with that of e^i

Then
$$N_{;i}^{\alpha} e^i = -\lambda y_{,i}^{\alpha} e^i, \quad (\lambda \text{ is scalar constant})$$

from (2), we have

$$\Omega_{ij} g^{jk} y_{,k}^{\alpha} e^i = \lambda y_{,i}^{\alpha} e^i$$

Multiplying both sides of this equation by $a_{\alpha\beta} y_{,l}^{\beta}$

$$\Omega_{ij}\, g^{jk} e^i (a_{\alpha\beta}\, y_{,k}^{\alpha}\, y_{,l}^{\beta}) = \lambda e^i (a_{\alpha\beta} y_{,i}^{\alpha} y_{,l}^{\beta})$$

$$\Omega_{ij} g^{jk} e^i g_{kl} = \lambda e^i g_{il} \quad \text{since } g_{il} = a_{\alpha\beta}\, y_{,i}^{\alpha} y_{,l}^{\beta}$$

$$\Omega_{ij} \delta_l^j e^i = \lambda e^i g_{il}$$

$$\Omega_{il} e^i - \lambda e^i g_{il} = 0$$

$$(\Omega_{il} - \lambda g_{il}) e^i = 0 \quad (l = 1, 2, \ldots, n)$$

This equation implies that the direction of e^i is a principal direction for the symmetric tensor Ω_{il} *i.e.*, e^i is a principal direction for the hypersurface V_n. Hence by definition the curve C is a line of curvature V_n.

12.10 THE EQUATION OF GAUSS AND CODAZZI

since we know that (pg. 86, equation 5)

$$A_{i,jk} - A_{i,kj} = A_p R^p_{ijk} \quad \dots (1)$$

where A_i is a covariant tenser of rank one and difference of two tensers $A_{i,jk} - A_{i,kj}$ is a covariant tenser of rank three.

It $y^\alpha_{,i}$ are components of a covariant tensor of rank one in $x's$. Then replacing A_i by $y^\alpha_{,i}$ in equation (1), we get

$$y^\alpha_{,ijk} - y^\alpha_{,ikj} = y^\alpha_{,p} R^p_{ijk} = y^\alpha_{,p} g^{ph} R_{hijk}, \quad \text{Since } R^p_{ijk} = g^{ph} R_{hijk} \quad \dots (2)$$

where R_{hijk} are Riemann symbols for the tensor g_{ij}

We know that

$$y^\alpha_{,ij} = \Omega_{ij} N^\alpha \quad \dots (3)$$

and

$$N^\alpha_{,i} = -\Omega_{ij} g^{ik} y^\alpha_{,k} \quad \dots (4)$$

Let $\overline{R}^\alpha_{\gamma\delta\varepsilon}$ are Riemann symbols for the tensor $a_{\alpha\beta}$ and evaluated at points of the hypersurface using equation (3) and (4), equation (2) becomes

$$y^\alpha_{,p} g^{ph}[R_{hijk} - (\Omega_{hj}\Omega_{ik} - \Omega_{hk}\Omega_{ij})] - N^\alpha(\Omega_{ij,k} - \Omega_{ik,j}) - \overline{R}^\alpha_{\gamma\delta\varepsilon} y^\gamma_{,i} y^\delta_{,j} y^\varepsilon_{,k} \quad \dots (5)$$

Multiplying equation (5) by $a_{\alpha\beta} y^\beta_{,l}$ and summed with respect to α. Using the relations

$$a_{\alpha\beta} y^\alpha_{,i} y^\beta_{,l} = 0$$

and

$$g_{\alpha\beta} = N^\beta y^\alpha_{,l} = 0,$$

we get

$$R_{lijk} = (\Omega_{lj}\Omega_{ik} - \Omega_{lk}\Omega_{ij}) + \overline{R}_{\beta\gamma\delta\varepsilon} y^\beta_{,l} y^\gamma_{,i} y^\delta_{,j} y^\varepsilon_{,k} \quad \dots (6)$$

Multiplying (6) by $a_{\alpha\beta} N^\beta$ and summing with respect to α. Using relations

$$a_{\alpha\beta} N^\beta y^\alpha_{,l} = 0$$

and

$$a_{\alpha\beta} N^\beta B^\beta = 0$$

we get $\Omega_{ij,k} - \Omega_{ik,j} + \overline{R}_{\beta\gamma\delta\varepsilon} N^\beta y^\gamma_{,i} y^\delta_{,j} y^\varepsilon_{,k} = 0$ ... (7)

Hence, The equation (6) are *generalisation of the Gauss Characteristic equation and equation* (7) *of the Mainardi-Codazzi equations.*

12.11 HYPERSURFACES WITH INDETERMINATE LINES OF CURVATURE

A point of a hypersurface at which the lines of curvature are indeterminate is called an *Umbilical Point*.

The lines of *curvature may* be *indeterminate* at every point of the hypersurface iff

$$\Omega_{ij} = \omega g_{ij} \quad \ldots (8)$$

where ω is an invariant

The mean *curvature M of such a hypersurface* is given by

$$M = \Omega_{ij} g^{ij} = \omega g_{ij} g^{ij} = \omega n$$

$$\Rightarrow \quad \omega = \frac{M}{n}.$$

So that the conditions for indeterminate lines of curvature are expressible as

$$\Omega_{ij} = \frac{M}{n} g_{ij}, \quad \text{from } (i) \quad \ldots (9)$$

If all the geodesics of a hypersurface V_n are also geodesics of an enveloping V_{n+1}. They hypersurface V_n is called a *totally geodesic hypersurface of the hypersurface* V_{n+1}.

THEOREM 12.7 *A totally geodesic hypersurface is a minimal hypersurface and its-lines of curvature are indeterminate.*

Proof: We know that

$$K_a^2 = K_n^2 + K_g^2 \quad \ldots (1)$$

and a hypersurface is said to be minimal if

$$M = 0 \quad \ldots (2)$$

and the lines of curvature are indeterminate if

$$\Omega_{ij} = \frac{M}{n} g_{ij} \quad \ldots (3)$$

If a hypersurface V_n is totally geodesic then geodesics of V_n are also geodesics of V_{n+1}.

i.e., $\quad K_a = 0 = k_g$

Now, from (1), we have

$$K_n = 0$$

But normal curvature K_n is zero for an asymptotic direction. Hence a hypersurface V_n is totally geodesic hypersurface iff the normal curvature K_n zero for all directions in V_n and hence

$$\Omega_{ij} = 0$$

$$M = \Omega_{ij}\, g^{ij} = 0$$

i.e., equation (2) is satisfied.

Hence, the totally geodesic hypersurface is minimal hypersurface.

In this case equation (3) are satisfied hence the lines of curvature are indeterminate.

12.12 CENTRAL QUADRATIC HYPERSURFACE

Let x^i be the cartesian in Euclidean space S_n, so that the components g_{ij} of the fundamental tensor are constants. Let y^i be the Riemmannian coordinates. If a fixed point O is taken as a pole and s the distance of any point P then Riemannian coordinates y^i of P with pole O are given by

$$y^i = s\xi^i \qquad \ldots (1)$$

ξ^i is unit tangent in the direction of OP.

Let a_{ij} be the components in the $x's$ of a symmetric tensor of the rank two and evaluated at the pole O. Then the equation

$$y^i a_{ij} y^j = 1 \qquad \ldots (2)$$

represents a central quadratic hypersurface

Substituting the value of equation (1) in equation (2), we get

$$s\xi^i a_{ij} s\xi^i = 1$$

$$\xi^i a_{ij} \xi^i = \frac{1}{s^2} \qquad \ldots (3)$$

The equation (3) showing that the two values of s are equal in magnitude but opposite in sign.

The positive value of s given by equation (3) is the length of the radius of the quadric (2) for the direction ξ^i.

THEOREM 12.8 *The sum of the inverse squares of the radii of the quadric for n mutually orthogonal directions at O is an invariant equal to* $a_{ij}g^{ij}$.

Proof: If $e^i_{h|}$, $(h = 1, 2, \ldots n)$ are the contravariant components of the unit tangents at O to the curves of an orthogonal ennuple in S_n. The radius S_h relative to the direction $e^i_{h|}$ is given by

$$a_{ij}\, e^i_{h|}\, e^j_{h|} = \frac{1}{s_h^2}$$

or

$$\sum_{h=1}^{n} (s_h)^{-2} = \sum_{h=1}^{h} a_{ij}\, e^i_{h|}\, e^j_{h|}$$

$$= a_{ij} \sum_{h=1}^{n} e^i_{h|}\, e^j_{h|}$$

$$\sum_{h=1}^{n} (s_h)^{-2} = a_{ij} g^{ij}$$

Proved.

THEOREM 12.9 *The equation of hyperplane of contact of the tangent hypercone with vertex at the point* $Q(\bar{y})$.

Proof: Given (from equation 2, pg. 235)

$$a_{ij} y^i y^j = 1$$

Differentiating it

$$a_{ij}\, dy^i y^j + a_{ij} y^i dy^j = 0$$

or

$$a_{ij} dy^i y^j + a_{ij} y^j dy^i = 0$$

$$2a_{ij}\, dy^i y^j = 0$$

$$a_{ij}\, dy^i y^j = 0$$

This shows that dy^i is tangential to the quadric. Hence y^j is normal to the quadric.

The tangent hyperplane at the point $P\,(y^j)$ is given by

$$(Y^i - y^i) a_{ij}\, y^j = 0$$

$$a_{ij} Y^i y^j = a_{ij} y^i y^j$$

$$a_{ij} Y^i y^j = 1 \quad \text{since } a_{ij} y^i y^j = 1 \qquad \dots (4)$$

This equation represents the equation of tangent hyperplane at $P\,(y^i)$.

If the tangent hyperplane $P\,(y^j)$ passes through the point $Q\,(y^{-i})$. Then we have

$$\bar{y}^i\, a_{ij}\, y^j = 1 \qquad \dots (5)$$

Thus all points of the hyperquadric, the tangent hyperplanes at which pass through Q lie on the hyperplane (5) on which y^i is the current point. This is the hyperplane of contact of the tangent hypercone whose vertex is $Q(\bar{y}^i)$.

12.13 POLAR HYPERPLANE

The polar hyperplane of the point $R(\tilde{y}^i)$ with respect to quadric (2) is the locus of the vertices of the hypercones which touch the hyperquadric along its intersections with hyperplanes through R. If $Q(\bar{y}^i)$ is the vertex of such tangent hypercone, then R lies on the hyperplane of contact of Q so that

$$\bar{y}^i a_{ij}\, \tilde{y}^j = 1$$

Consequently for all positions of the hyperplane through R, Q lies on the hyperplane

$$y^i a_{ij}\, \tilde{y}^j = 1 \qquad \dots (6)$$

This is required equations of the polar hyperplane of R and R is the pole of this hyperplane.

12.14 EVOLUTE OF A HYPERSURFACE IN EUCLIDEAN SPACE

Consider a hypersurface V_n of Euclidean space S_{n+1} and let x^i ($i = 1, 2, \dots n$) be coordinates of an arbitrary point P of V_n whose components relative to S_{n+1} are

$$y^\alpha (\alpha = 1, 2, \cdots n+1)$$

Let N^α be a unit normal vector at P relative to S_{n+1} so that tensor derivate N^α becomes covariant derivative.

So,
$$N^\alpha_{;i} = N^\alpha_{,i} = -\Omega_{ij} g^{jk} y^\alpha_{,k} \quad \dots (1)$$

$$y^\alpha_{;ij} = y^\alpha_{,ij} = \Omega_{ij} N^\alpha \quad \dots (2)$$

and
$$g_{ij} = \sum_{\alpha=1}^{n+1} y^\alpha_{,i} y^\alpha_{,j} \quad \dots (3)$$

Let $\overline{P}(\overline{y}^\alpha)$ be a point on the unit normal N^α such that distance of $\overline{P}$ from P is ρ in the direction of N^α such that

$$\overline{y}^\alpha = y^\alpha + \rho N^\alpha \quad \dots (4)$$

Suppose P undergoes a displacement dx^i in V_n then the corresponding displacement $d\overline{y}^\alpha$ of $\overline{P}$ is given by

$$d\overline{y}^\alpha = (y^\alpha_{,i} + \rho N^\alpha_{,i}) dx^i + N^\alpha d\rho \quad \dots (5)$$

The vector $(y^\alpha_{,i} + \rho N^\alpha_{,i}) dx^i$ is tangential to V_n whereas $N^\alpha d\rho$ is a normal vector. Therefore if the displacement of $\overline{P}(\overline{y}^\alpha)$ be along the normal to the hypersurface then we have

$$(y^\alpha_{,i} - \rho N^\alpha_{,i} 0) dx^i = 0 \quad \dots (6)$$

Using equation (1) in equation (6), we get

$$(y^\alpha_{,i} - \rho g^{jk} y^\alpha_{,k}) dx^i = 0$$

Multiplying it by $y^\alpha_{,i}$ and summing with respect to α, we get, using equation (3), as

$$(g_{il} - \rho \Omega_{ij} g^{jk} g_{lk}) dx^i = 0$$

$$(g_{il} - \rho \Omega_{ij} \delta^j_l) dx^i = 0$$

$$(g_{il} - \rho \Omega_{il}) dx^i = 0$$

$$(g_{ij} - \rho \Omega_{ij}) dx^i = 0 \quad \dots (7)$$

This shows that the directions dx^i given by equation (7) are principal directions of the hypersurface where the roots ρ of the equation $|g_{ij} - \rho\Omega_{ij}| = 0$ are called *principal radii of normal curvature*. The locus of $\overline{P}(\overline{y}^\alpha)$ satisfying the condition (4) is called evolute of the hypersurface V_n of S_{n+1} where ρ is a root of (7). *The evolute is also a hypersurface of* S_{n+1}.

12.15 HYPERSPHERE

The locus of a point in S_n which moves in such way that it is always at a fixed distance R from a fixed point $C(b^\alpha)$ is called a hypersphere of radius R and centre C. Therefore the equation of such a hypersphere is given by

$$\sum_{\alpha=1}^{n}(y^\alpha - b^\alpha)^2 = R^2 \qquad \ldots (1)$$

THEOREM 12.10 *The Riemannian curvatrure of a hypersphere of radius R is constant and equal to* $\frac{1}{R^2}$.

Proof: Let the hypersurface be a V_n of S_{n+1} and let its centre be taken as origin of Euclidean coordinates in S_{n+1}. Then the hypersphere is given by

$$\sum_{\alpha}(y^\alpha)^2 = R^2, \quad (\alpha = 1, 2, \ldots, n+1) \qquad \ldots (2)$$

For the point in V_n the y's are functions of the coordinates x^i on the hypersphere. Differentiating equation (2) with respect to x^i, we get

$$\sum_{\alpha} y^\alpha y^\alpha_{,i} = 0 \qquad \ldots (3)$$

and again differentiating it with respect to x^j, we get.

$$\sum_{\alpha} y^\alpha_{,j} y^\alpha_{,i} + \sum_{\alpha} y^\alpha y^\alpha_{,ij} = 0 \qquad \ldots (4)$$

By Gauss formula,

$$y^\alpha_{,ij} = \Omega_{ij} N^\alpha \qquad \ldots (5)$$

Using (5), equation (4) becomes

$$g_{ij} + \sum_{\alpha} y^\alpha \Omega_{ij} N^\alpha = 0 \qquad \ldots (6)$$

From equation (3) it follows that $y^\alpha_{,i}$ is perpendicular to y^α. But $y^\alpha_{,i}$ is tangential to V_m. Hence y^α is normal to V_n. The equation (2) implies that the components of the unit vector N^α are given by

$$N^\alpha = \frac{y^\alpha}{R} \Rightarrow y^\alpha = RN^\alpha \qquad \ldots (7)$$

Using (7), equation (6) becomes

$$g_{ij} + \sum_{\alpha} RN^\alpha N^\alpha \Omega_{ij} = 0$$

or $$g_{ij} + R\Omega_{ij} \sum_{\alpha}(N^\alpha)^2 = 0$$

or $$g_{ij} + R\Omega_{ij} = 0 \quad \text{since } (N^\alpha)^2 = 1$$

or $$R\Omega_{ij} = -g_{ij} \qquad \ldots (8)$$

Suppose R_{lijk} and $\overline{R}_{\beta\gamma\delta\varepsilon}$ are Reimann's symbols with respect to metrics $g_{ij}\, dx^i\, dx^j$ and $a_{\beta\gamma}\, dy^\beta\, dy^\gamma$ respectively.

we know that $$a_{\beta\gamma} = \delta^\beta_\gamma = \begin{cases} 1, & \beta = \gamma \\ 0, & \beta \neq \gamma \end{cases}$$

This shows that Christoffel brackets vanish

$$i.e. \begin{Bmatrix} \delta \\ \beta\gamma \end{Bmatrix} = 0, [\beta\gamma, \delta] = 0$$

Hence $$\overline{R}_{\beta\gamma\delta\rho} = 0$$

This Gauss characteristic equation,

$$R_{lijk} = (\Omega_{lj}\Omega_{ik} - \Omega_{lk}\Omega_{ij}) + \overline{R}_{\beta\gamma\delta\varepsilon}\, y^\beta_{,l}\, y^\gamma_{,i}\, y^\delta_{,j}\, y^\varepsilon_{,k}$$

becomes

$$R_{lijk} = \Omega_{ij}\, \Omega_{ik} - \Omega_{lk}\, \Omega_{ij}$$

$$= \frac{1}{R}[g_{lj}\, g_{ij} - g_{lk}\, g_{ij}], \quad \text{(using equation 8)}$$

or $$\frac{R_{lijk}}{g_{lj}g_{ik} - g_{lk}g_{ij}} = \frac{1}{R^2} = \text{constant}$$

Now, the formula for K, the Riemannian curvature to V_n at the origin P corresponding to the orientation determined by unit vector p^i and q^i is given by

$$K = \frac{R_{lijk}\, p^l q^i p^j q^k}{(g_{lj}g_{ik} - g_{lk}g_{ij})p^l q^i p^j q^k}$$

$$K = \frac{1}{\rho^2} \frac{p^l q^i p^j q^k}{p^l q^i p^j q^k}$$

$$K = \frac{1}{\rho^2} \qquad \ldots (9)$$

Proved.

Note: 1. Point coordinates of Weierstrass for V_n. The function y^α satisfying the following conditions are called point coordinates of Weierstrass for V_n: We have

$$ds^2 = \sum_{\alpha=1}^{n+1} (dy^\alpha) \quad \text{with} \quad \frac{1}{R^2} = K$$

such that $$\sum_{\alpha=1}^{n+1} (y^\alpha)^2 = R^2$$

2. To prove that for a space V_n with positive constant Riemannian curvature K these exists sets of $n + 1$ real coordinate y^α satisfying the condition

$$\sum_{\alpha=1}^{n+1}(y^\alpha) = \frac{1}{K} \qquad \text{where } R^2 = \frac{1}{K}$$

Proof: Using (9) in equation (1), we get

$$\sum_{\alpha}(y^\alpha)^2 = \frac{1}{K}$$ **Proved.**

EXAMPLE 2

Show that the directions of two lines of curvature at a point of a hypersurface are conjugate.

Solution

The principal direction $e^i_{h|}$ are given by

$$\Omega_{ij}\, e^i_{h|}\, e^j_{h|} = 0 \qquad \text{... (1)}$$

The shows that principal directions at a point of a hypersurface are conjugate.
Thus we say that two congruences of lines of curvature are conjugate.

EXAMPLE 3

Show that the normal curvature of hyper surface V_n in an asymptotic direction vanishes.

Solution

Let us consider a curve C in a V_n. If C is an asymptotic line then it satisfies the differential equation

$$\Omega_{ij}\, dx^i\, dx^j = 0$$

i.e.,

$$\Omega_{ij}\frac{dx^i}{ds}\frac{dx^j}{ds} = 0 \qquad \text{... (1)}$$

Now the normal curvature K_n of the hypersurface V_n in an asymptotic direction in a V_n is given by

$$K_n = \Omega_{ij}\frac{dx^i}{ds}\frac{dx^j}{ds}$$

i.e.,

$$K_n = 0 \quad \text{from (1)}$$

EXAMPLE 4

To prove that if the polar hyperplane of the point R passes through a point P then that of P passes through R.

Solution

Let $P(\overline{y}^i)$ and $R(\overline{\overline{y}}^i)$ be two points. The polar hyperplane of $R(\overline{\overline{y}}^i)$ is

$$Y^i a_{ij} \bar{\bar{y}}^j = 1 \qquad \text{... (1)}$$

If equation (1) passes through $P(\bar{\bar{y}}^i)$ then

$$\bar{y}^i a_{ij} \bar{\bar{y}}^j = 1 \qquad \text{... (2)}$$

Again polar hyperplane of $P(\bar{\bar{y}}^i)$ is

$$Y^i a_{ij} \bar{y}^j = 0 \qquad \text{... (3)}$$

If equation (3) passes through $R(\bar{y}^i)$ then

$$\bar{\bar{y}}^i a_{ij} \bar{y}^j = 1$$

or $$\bar{\bar{y}}^j a_{ji} \bar{y}^i = \text{(on interchanging } i \ \& \ j)$$

or $$\bar{y}^i a_{ij} \bar{\bar{y}}^j = 1 \qquad \text{... (4)}$$

Clearly the relation (2) and (4) are same. **Proved.**

EXERCISES

1. Show that the normal to a total geodesic hypersurface is parallel in the enveloping manifold.
2. Obtain an expression for the derived vector of the unit normal N^α to a hypersurface V_n along a curve C in V_n and prove that it will be tangential to the curve provided that C be a line of curvature of V_n.
3. Deduce that the first curvature in V_{n+1} of geodesic of the hypersurface V_n is the normal curvature of the hypersurface in the direction of the geodesic.
4. Prove that when a geodesic of V_{n+1} lies in a hypersurface V_n it is both geodesic and an asymptotic line in the hypersurface.
5. Prove that a surface C in a subspace V_n is a geodesic in the enveloping space V_m if and only if it is both a geodesic and an asymptotic line in the subspace V_m.
6. Show that any two distinct principal directions relative to the normal N of a hypersurface in the neighbourhood U of a point are conjugate direction.
7. Prove that conjugate directions in a hypersurface are such that the derived vector of the unit normal in either direction is orthogonal to the other direction.
8. Prove that when the line of curvature of a hypersurface of a space of constant curvature are indeterminate, the hypersurface has constant curvature.
9. Show that if, in a hypersurface $g_{ij} = 0$ and $\Omega_{ij} = 0 \ (i \neq j)$ the coordinate curves are lines of curvature.
10. Show that for a hypersurface in Euclidean space the Gauss and Codazzi equations reduce to

$$R_{lijk} = \Omega_{lj}\,\Omega_{ik} - \Omega_{lk}\,\Omega_{ij}$$

and $$\Omega_{ij,k} - \Omega_{ik,j} = 0$$

11. Prove that the necessary and sufficient condition that system of hypersurface with unit normal N be isothermic is that

$$\rho(N \operatorname{div} N - N.\nabla N) = 0.$$

12. Show that the normal curvature of a subspace in an asymptotic direction is zero.
13. If straight line through a point P in S_n meets a hyperquadric in A and B and the polar hyperplane of P in Q prove that P, Q are harmonic conjugates to A, B.
14. What are evolutes of a hypersurface in an Euclidean space. Show that the varieties ρ_1 = constant in evolute are parallel, having the curves of parameter ρ_1 as orthogonal geodesics of V_n.

INDEX

S

T

U

W